Springer Tracts in Modern Physics 100

## Springer Tracts in Modern Physics

73 **Excitons at High Density** Edited by H. Haken, S. Nikitine. With contributions by V. S. Bagaev, J. Biellmann, A. Bivas, J. Goll, M. Grosmann, J. B. Grun, H. Haken, E. Hanamura, R. Levy, H. Mahr, S. Nikitine, B. V. Novikov, E. I. Rashba, T. M. Rice, A. A. Rogachev, A. Schenzle, K. L. Shaklee

74 **Solid-State Physics** With contributions by G. Bauer, G. Borstel, H. J. Falge, A. Otto

75 **Light Scattering by Phonon-Polaritons** By R. Claus, L. Merten, J. Brandmüller

76 **Irreversible Properties of Type II Superconductors** By H. Ullmaier

77 **Surface Physics** With contributions by K. Müller, P. Wißmann

78 **Solid-State Physics** With contributions by R. Dornhaus, G. Nimtz, W. Richter

79 **Elementary Particle Physics** With contributions by E. Paul, H. Rollnick, P. Stichel

80* **Neutron Physics** With contributions by L. Koester, A. Steyerl

81 **Point Defects in Metals I:** Introductions to the Theory (2nd Printing) By G. Leibfried, N. Breuer

82 **Electronic Structure of Noble Metals, and Polariton-Mediated Light Scattering** With contributions by B. Bendow, B. Lengeler

83 **Electroproduction at Low Energy and Hadron Form Factors** By E. Amaldi, S. P. Fubini, G. Furlan

84 **Collective Ion Acceleration** With contributions by C. L. Olson, U. Schumacher

85 **Solid Surface Physics** With contributions by J. Hölzl, F. K. Schulte, H. Wagner

86 **Electron-Positron Interactions** By B. H. Wiik, G. Wolf

87 **Point Defects in Metals II:** Dynamical Properties and Diffusion Controlled Reactions With contributions by P. H. Dederichs, K. Schroeder, R. Zeller

88 **Excitation of Plasmons and Interband Transitions by Electrons** By H. Raether

89 Giant Resonance Phenomena in **Intermediate-Energy Nuclear Reactions** By F. Cannata, H. Überall

90* **Jets of Hadrons** By W. Hofmann

91 **Structural Studies of Surfaces** With contributions by K. Heinz, K. Müller, T. Engel, and K. H. Rieder

92 **Single-Particle Rotations in Molecular Crystals** By W. Press

93 **Coherent Inelastic Neutron Scattering in Lattice Dynamics** By B. Dorner

94 **Exciton Dynamics in Molecular Crystals and Aggregates** With contributions by V. M. Kenkre and P. Reineker

95 **Projection Operator Techniques** in Nonequilibrium Statistical Mechanics By H. Grabert

96 **Hyperfine Structure in 4d- and 5d-Shell Atoms** By S. Büttgenbach

97 **Elements of Flow and Diffusion Processes in Separation Nozzles** By W. Ehrfeld

98 **Narrow-Gap Semiconductors** With contributions by G. Nimtz and B. Schlicht

99 **Property of IV—VI Compounds** With contributions by H. Bilz, A. Bussmann-Holder and W. Jantsch

100* **Quarks and Nuclear Forces** Edited by D.C. Fries and B. Zeitnitz

---

* denotes a volume which contains a Classified Index starting from Volume 36.

# Quarks and Nuclear Forces

Editors: D. C. Fries and B. Zeitnitz

With 69 Figures

Springer-Verlag
Berlin Heidelberg New York 1982

Dr. Dietrich C. Fries
Professor Dr. Bernhard Zeitnitz

Kernforschungszentrum Karlsruhe, Institut für Kernphysik 1
D-7500 Karlsruhe, Fed. Rep. of Germany

---

*Manuscripts for publication should be addressed to:*

Gerhard Höhler
Institut für Theoretische Kernphysik der Universität Karlsruhe
Postfach 6380, D-7500 Karlsruhe 1, Fed. Rep. of Germany

*Proofs and all correspondence concerning papers in the process of publication should be addressed to:*

Ernst A. Niekisch
Haubourdinstrasse 6, D-5170 Jülich 1, Fed. Rep. of Germany

---

ISBN 3-540-11717-2 Springer-Verlag Berlin Heidelberg New York
ISBN 0-387-11717-2 Springer-Verlag New York Heidelberg Berlin



Offset printing and bookbinding: Brühlsche Universitätsdruckerei, Giessen
2153/3130 – 5 4 3 2 1 0

# Foreword

In recent years an impressive amount of experimental data has been accumulated, supporting the theoretical concept, that matter on a subnuclear scale consists essentially of two entities: quarks and leptons and that there is a universal way to describe forces and interactions between them.

Although tested in the domain of high energies, it is the hope, that the new description of hadronic matter, whose theoretical basis is quantum chromodynamics (QCD), can also be applied to the low energy interactions of hadrons and nucleons, and that QCD, which appears to describe the asymptotic region of small quark distances well, will also work on the scale of large, that is nuclear dimensions.

The 6-day program of lectures and seminars of the KfK-Summerschool 81 has been devoted to the introduction to the quark picture with its symmetries and dynamics and to its application in the domain of nuclear forces. This volume contains the main lectures and a summary.

The school at Burg Liebenzell has benefitted greatly from the knowledge, the expedience and dedication of lecturers and seminar-speakers to address the subject as well as the audience. The audience was delightfully alert and interested. We are very grateful to both of them.

Our thanks also go to the secretaries V. Lallemand and G. Grundel and to our colleagues Drs. P. Doll, W. Herringa, H. Klages, and R. Maschuw for their assistance in organizing and conducting the school.

We wish finally to acknowledge the financial support of our sponsors, the Kernforschungszentrum Karlsruhe and the Bundesministerium für Forschung und Technologie.

Karlsruhe, May 1982 *D.C. Fries and B. Zeitnitz*

# Contents

# List of Contributors

**Baym, Gordon**
University of Illinois, Urbana, IL 61801, USA

**Brodsky, Stanley J.**
Stanford Linear Accelerator Center, Stanford University, Stanford, CA 94305, USA

**Close, Frank E.**
Rutherford Appleton Laboratory, Chilton, Didcot, Oxon OX 11 OOX, United Kingdom

**Faessler, Amand**
Institut für Theoretische Physik, Universität Tübingen, D-7400 Tübingen, Fed. Rep. of Germany

**Flügge, Günter**
Kernforschungszentrum und Universität Karlsruhe, D-7500 Karlsruhe, Fed. Rep. of Germany

**Heller, Leon**
Theoretical Division, Los Alamos National Laboratory, University of California, Los Alamos, NM 87545, USA

# 1. Experimental Evidence of Quarks and Gluons

G. Flügge

With 47 Figures

## 1.1 Introduction

When the word 'quark' in elementary particle physics was first coined by Gell-Mann and Zweig[1] in 1964 it was rather a synonym for the amazingly successful symmetry scheme of SU(3) than a name for real physical entities. In the 70's however, mainly due to results from deep inelastic scattering and $e^+e^-$ experiments, the quarks have become (almost) as real as their precursors, the 'elementary' particles. Today we believe we have overwhelming experimental evidence for the existence of quarks as pointlike constituents of hadronic matter.

In 1964 all particles fitted into the SU(3) symmetry scheme of which the three quarks u, d, and s form the fundamental representation (table 1).

All new particles could be attributed to simple higher representations until in 1974 new heavy stable mesons (J/$\psi$) were found[2] which led to an extention to SU(4) with a new charm quark (c). The story repeated itself in 1977 with the discovery

| Name | Symbol | Baryon number | charge | third comp.of isospin | strange-ness | charm | beauty |
|---|---|---|---|---|---|---|---|
| up | u | 1/3 | 2/3 | + 1/2 | 0 | 0 | 0 |
| down | d | 1/3 | -1/3 | - 1/2 | 0 | 0 | 0 |
| strange | s | 1/3 | -1/3 | 0 | -1 | 0 | 0 |
| charm | c | 1/3 | 2/3 | 0 | 0 | 1 | 0 |
| bottom (beauty) | b | 1/3 | -1/3 | 0 | 0 | 0 | 1 |

Table 1. The five known quarks in particle physics and some of their quantum numbers

| Leptons | | | Quarks | | |
|---|---|---|---|---|---|
| $\nu_e$ | $\nu_\mu$ | $(\nu_\tau)$ | u | c | ((t)) |
| $e^-$ | $\mu^-$ | $\tau^-$ | d | s | b |

Table 2. Generations of Leptons and Quarks. The evidence for $\nu_\tau$ is only indirect[43] The quark t has not yet been seen.

of the $\Upsilon$ meson[3] which was taken as evidence for the fifth quark b (bottom, beauty)[38]. Meanwhile, the number of leptons also went up from 4 to 6 (table 2).

The sixth quark t (top, truth) which most physicists would like to exist for theoretical and esthetical reasons has so far evaded experimental observation.

With the advent of quantum chromodynamics (QCD)[4] in the early 70's a solution of the old problem of strong interaction came into sight. QCD is based on the gauge group of colour SU(3), which had as early as 1964 been suggested to solve the problem of Fermi statistics in hadron spectroscopy. In QCD the strong interaction between quarks is mediated by the exchange of 'gluons' which couple to the colour charges of quarks. After indirect experimental evidence for colour and gluons had come from many sources, direct observation of these new quanta was announced in 1979 in $e^+e^-$ annihilation events at PETRA[5].

I will try to describe some of these developments in the following three lectures on

- deep inelastic scattering
- quarks in $e^+e^-$ annihilation
- gluons in $e^+e^-$ annihilation.

## 1.2 Deep Inelastic Scattering

### 1.2.1 Form Factors and Structure Functions

Nonrelativistic elastic scattering on spin 0 target

The simple case of nonrelativistic elastic scattering of an electron on a spin 0 target is illustrated in fig.1.

The cross section for this process is

$$\frac{d\sigma}{d\Omega} = \left(\frac{d\sigma}{d\Omega}\right)_{Mott} \cdot |F(q^2)|^2$$

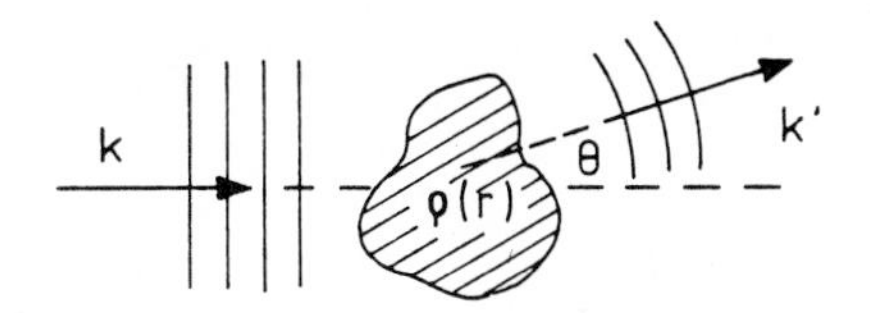

Fig.1. Elastic scattering on an extended target

where $q^2 = (k-k')^2$ is the momentum transfer and the form factor $F(q^2)$ is related to the charge distribution $\rho(r)$ via a Fourier transformation

$$\int d^3r\ \rho(r)\ e^{i/r\ (\vec{q}\cdot\vec{r})} = F(\vec{q}^{\,2})\ .$$

If we want to probe a charge structure inside hadrons we need ($\langle r\rangle$ being the nuclear radius)

$$q^2 \gg \frac{\hbar c}{\langle r\rangle} \simeq 1\ \mathrm{GeV}^2$$

i.e. we cannot use the simple nonrelativistic approach. Still the qualitative rules we learn from the above picture prevail:

- an extended charge distribution in space leads to a decreasing and vanishing $F(q^2)$ as $q^2 \to \infty$
- only a pointlike structure allows for $F(q^2)$ = const. for $q^2 \to \infty$

Electron-nucleon scattering (spin 1/2 target)

Let us consider the more general process of (elastic or inelastic) electron-proton scattering (fig.2).

We need two variables to describe the process, $Q^2$ and $\nu$.

$$Q^2 = -q^2 = (k-k')^2$$

is the momentum transfer described above. Since the photon is spacelike in this process ($q^2$ is negative), the variable $Q^2 = -q^2$ is taken for convenience.

$$\nu = E - E'$$

is the energy transfer, which is of course zero in the elastic case (in the CM

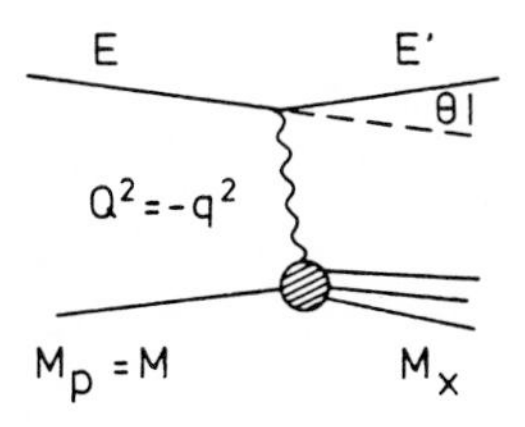

Fig.2. Electron-nucleon scattering and definition of variables

system). The differential cross sections are for the elastic case (Rosenbluth formula):

$$\frac{d\sigma}{d\Omega} = \left(\frac{d\sigma}{d\Omega}\right)_{Mott} \left\{ \frac{G_E^2 + bG_M^2}{1+b} + 2bG_M^2 \tan^2\frac{\Theta}{2} \right\} \qquad b = \frac{Q^2}{4M}$$

where $G_{E/M}(Q^2)$ are the two form factors (electric and magnetic) introduced in order to describe the scattering on an extended spin 1/2 object.

For the inelastic case one finds

$$\frac{d\sigma}{dE' d\Omega} = \frac{4\alpha^2 E'^2}{Q^2} \left\{ \cos^2\frac{\Theta}{2}\frac{F_2}{\nu} + 2\sin^2\frac{\Theta}{2}\frac{F_1}{M} \right\}$$

$\alpha$ = fine structure constant

where $F_{1,2}$ $(Q^2,\nu)$ are called structure functions. Again we need two functions which, however, now depend on two variables, in order to take the inelasticity into account.

### 1.2.2 Deep Inelastic Scattering and Bjorken Scaling

As stated above, one would expect a decrease of $F_i$ $(Q^2,\nu)$ as $Q^2$ increases. This would be a consequence of the spatial extension of the target charge. In fact, the first results showed[6)] that

$$F_{1,2}(M_x,Q^2) \to 0 \quad \text{for} \quad Q^2 \to \infty$$

if the structure function was measured for the case where after the scattering process the target nucleon remained in an excited (resonant) state; i.e. for $M_x$ (mass of the excited nucleon) being constant.

The surprising observation was, however, that

$$\left.\begin{matrix} x\,F_1\,(x,Q^2) \\ \\ F_2\,(x,Q^2) \end{matrix}\right\} \begin{matrix} Q^2 \to \infty \\ \to \quad \text{const.} \\ x \text{ fixed} \end{matrix}$$

if $F_i$ were measured at fixed values of the Bjorken variable

$$x = \frac{Q^2}{2M\nu}$$

without any restriction on the final state of the nucleon.

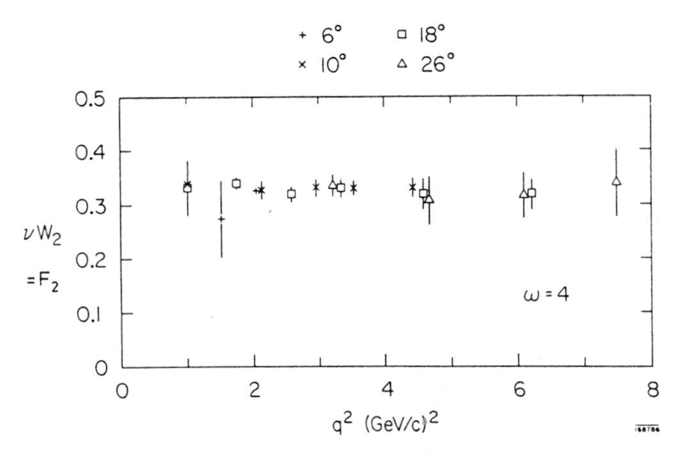


Fig.3. The structure function $F_2(Q^2,x)$ at fixed x for large $Q^2$. Deep inelastic electron scattering results at SLAC (1968)

This behaviour, predicted by Feyman and Bjorken[7] in 1968, is clear evidence for electron scattering from pointlike structures in the nucleon. Since the variable x is dimensionless, there is no "scale" in the theory. This effect known as Bjorken scaling was experimentally discovered in 1968 at SLAC[8] (fig.3). It led to the invention of the parton model.

### 1.2.3 Parton Scattering and Significance of Bjorken x

The parton model[9] assumes that the nucleon consists of subcomponents called partons. These partons are considered to be pointlike; furthermore it is assumed that one can neglect their mutual interaction during and after the hard scattering process (fig.4).

If we define $\xi p$ as the momentum fraction carried by a parton at the instant when it is struck by the photon we can calculate $\xi$ in terms of $Q^2$ and $\nu$:

$$\mu^2 = (\xi p+q)^2$$

from which we can derive

$$\xi = \frac{Q^2+\mu^2}{M(\nu+\sqrt{\nu^2+Q^2+\mu^2})}$$

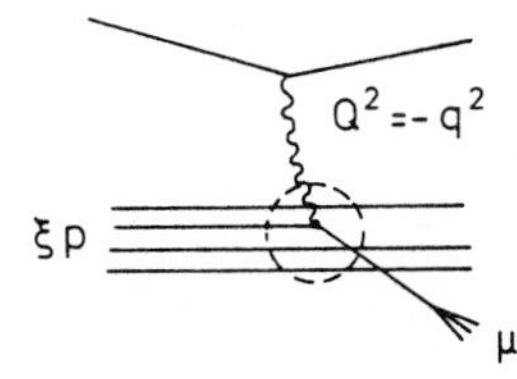


Fig.4. Parton scattering

which in the Bjorken limit $Q^2$, $\nu \to \infty$ goes to $(\nu^2 >> Q^2 >> \mu^2)$

$$\xi \to \frac{Q^2}{2M\nu} = x .$$

This means that for large values of $Q^2$ and $\nu$ the Bjorken variable x is a good approximation to the momentum fraction carried by the struck parton.

The structure function $F_2$ can for the general case of elastic (parton) scattering be decomposed into

$$F_2(x,Q^2) = f(x) \times G(Q^2)$$

where f(x) describes the kinematical target structure and $G(Q^2)$ stands for a potential deviation from a pointlike target.+) In the case of pointlike partons one has: $G(Q^2)$ = const.

$$\Rightarrow F(x,Q^2) \to F(x)$$

whereas in the case of elastic scattering from an extended target nucleon one has f(x) = 1

$$\Rightarrow F(x,Q^2) = G(Q^2)$$

which in this case is the well-known dipole form factor.

Fig.5 illustrates the kinematical effect of the target structure in the case of electron scattering at different momentum transfer. One observes peaks as a function of x, the fractional momentum of the struck object, representing scattering

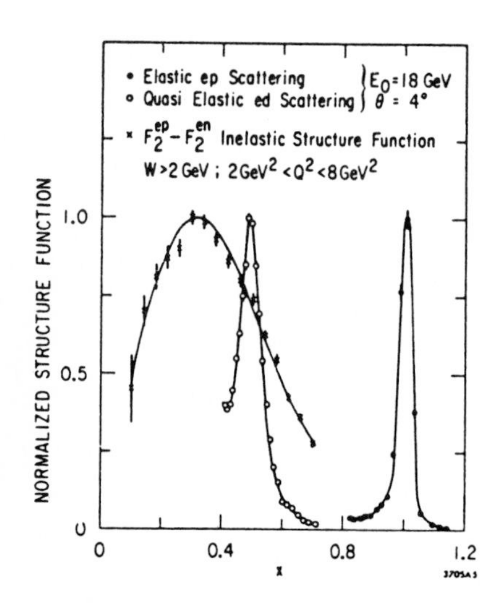


Fig.5. Structure functions as measured in three different (quasi-) eleastic scattering experiments

---

+) Spin-averaged combination of $G_E$ and $G_M$ in the case of spin 1/2 particles.

from the entire proton (x=1) in the case of elastic scattering, from the two nucleons in quasielastic ed scattering (x=1/2), and from the parton substructure in deep inelastic eN scattering (x=1/3). The broadening of the peaks reflects the 'Fermi motion' inside the target. We come back to the details of the parton structure function later.

### 1.2.4 Deep Inelastic Scattering Experiments

The formalism of deep inelastic scattering can easily be extended to the weak interaction of neutrinos. Fig.6 illustrates the general case of lepton scattering.

In addition to the variables already defined a new quantity $y$ is commonly used to describe the angular dependence:

$$Q^2 = -(k-k')^2;\ \nu = E-E',\ x = \frac{Q^2}{2M\nu}$$

$$y = \frac{\nu}{E} = \frac{1}{2}(1-\cos\Theta^*),\ \Theta^* = \text{scattering angle in CMS.}$$

The final choice of variables is then

$$Q^2 = -q^2 : \text{momentum transfer}$$

$$x = \frac{Q^2}{2M\nu} \underset{\text{limit}}{\xrightarrow{\text{Bjorken}}} \frac{p_{part}}{p} : \text{parton momentum fraction}$$

$$y = \frac{\nu}{E} : \text{fractional energy transfer to target}$$

In terms of these variables the electromagnetic cross section for charged leptons (electrons and muons) is then given by

$$\frac{d^2\sigma^1}{dxdy} = \frac{2\pi\alpha^2 s}{Q^4} F_2(x,Q^2)\ \{1+(1-y)^2 - \frac{y^2(F_2-2xF_1)}{F_2}\} \tag{1}$$

$s$ = CM energy squared

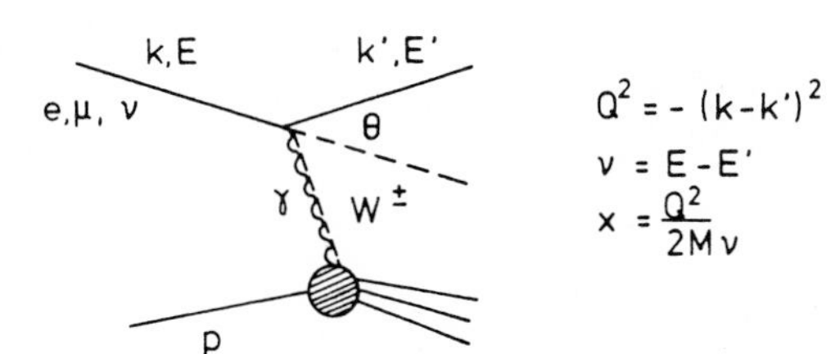


$$Q^2 = -(k-k')^2$$
$$\nu = E-E'$$
$$x = \frac{Q^2}{2M\nu}$$

Fig.6. Definition of variables in deep inelastic lepton scattering

The weak cross sections for neutrino ($\nu$ and $\bar{\nu}$) scattering can be conveniently written

$$\frac{d^2(\sigma^{\nu}+\sigma^{\bar{\nu}})}{dxdy} = \frac{G^2 s}{2\pi} F_2(x,Q^2) \{1+(1-y)^2 - \frac{y^2(F_2-2xF_1)}{F_2}\} \quad (2)$$

$$\frac{d^2(\sigma^{\nu}-\sigma^{\bar{\nu}})}{dxdy} = \frac{G^2 s}{2\pi} x F_3(x,Q^2) \{1-(1-y)^2\} \quad (3)$$

G = weak coupling constant

In (2) and (3) propagator effects of the exchanged $W^{\pm}$ are neglected. The formulae are written for an isoscalar target, i.e. equal number of neutrons and protons, and assuming charge symmetry.

The experimental task to determine $F_i(x,Q^2)$ is - as simple it may look in principle - rather formidable. In principle one only has to measure the energy (x) and angular (y) distribution of the inelastically scattered lepton. Since it will turn out that $F_2 \simeq 2xF_1$ one can immediately see from equations (1) to (3) how the structure functions are determined from simple combinations of the differential cross sections.

The formidable problems one encounters in practice are illustrated in figs. 7, 8 with two modern detectors for charged and neutral lepton scattering experiments. The main problems of high-energy ($\simeq$ 200 GeV) muon scattering (fig.7) are the electromagnetic background and the large instrumentation needed for muon measurements at these energies. Neutrino detectors (fig.8) have to combine large target masses (minute cross section) with high precision on the $\nu$ interaction vertex and the $\mu$ scattering angle and energy.

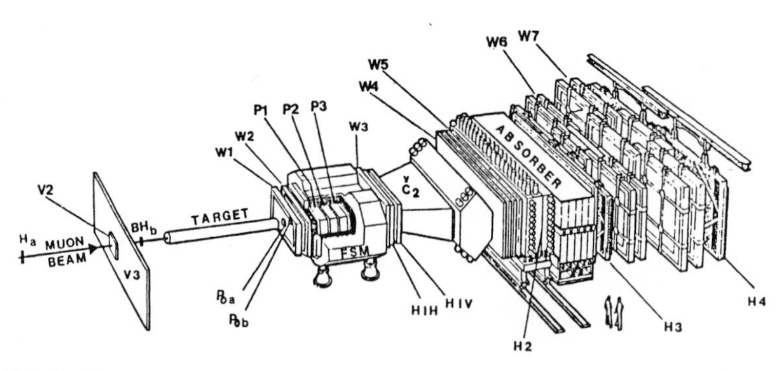


Fig.7. European Muon Collaboration: a detector for deep inelastic muon scattering

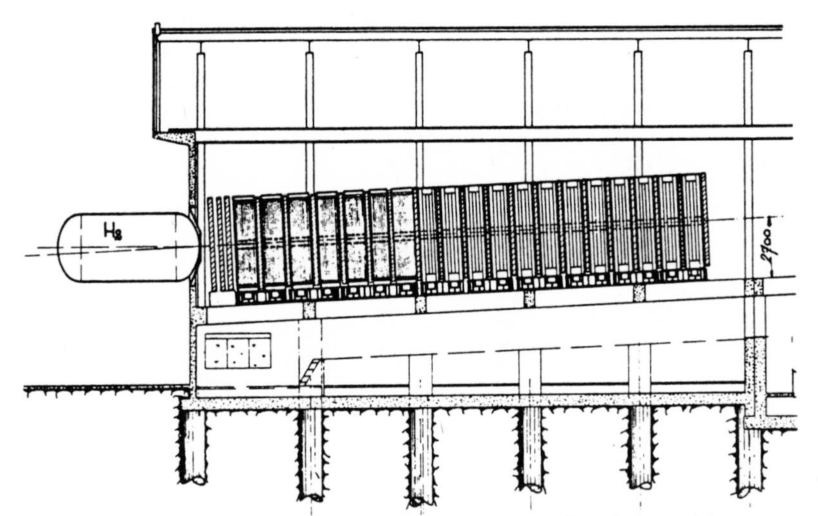


Experiment WA1: High-energy neutrino interactions

Fig.8. CERN-Dortmund-Heidelberg-Saclay Collaboration: a detector for neutrino scattering

## 1.2.5 Results on the Structure Functions

It can be shown that the structure functions $F_1$ and $F_2$ are related to the ratio of the cross section for the two helicity states of the exchanged particle by

$$\frac{\sigma_L}{\sigma_T} = \frac{F_2(1+Q^2/\nu^2) - 2xF_1}{2xF_1} \equiv R(x,Q^2) \tag{4}$$

Experimentally it turns out that $R(x,Q^2)$ - measured through $F_i(x,Q^2)$ - is small

$R = 0.10 \pm 0.07$ CDHS $\nu$-scattering[10)]

$R = 0.21 \pm 0.10$ SLAC-MIT e-scattering[11)].

This has two important consequences:

(i) If $R \to 0$ and in the limit $Q^2$, $\nu \to \infty$, we can deduce from equation (4)

$$F_2(x,Q^2) = 2xF_1(x,Q^2). \tag{5}$$

(Callon-Gross-relation[12)])

This means that deep inelastic lepton scattering is in this limit completely described by $F_2$ and $F_3$ only. Equations (1) to (3) get a very simple form which allows for direct extraction of the structure functions $F_2$ and $xF_3$.

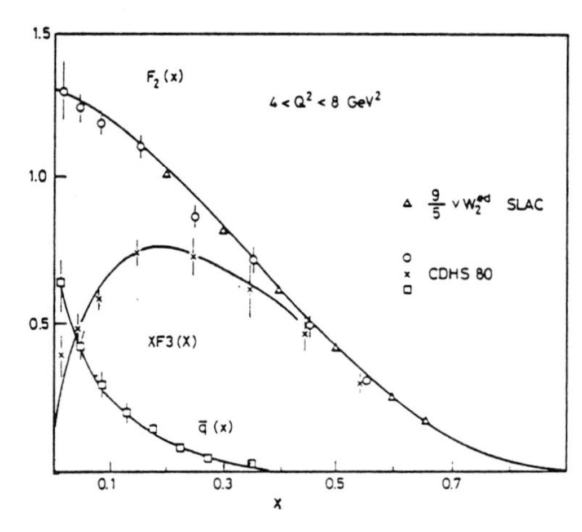


Fig.9. Structure functions $F_2$ and $xF_3$ from electron and neutrino data. The ed data are scaled by a factor 9/5 relative to the $\nu$ data (see sction 6)

(ii) In the parton model the ratio $R = \sigma_L/\sigma_T$ renders a definite statement on the spin of the partons. In the Bjorken limit one has

$$R = {}^{\sigma_L}/_{\sigma_T} \quad \begin{array}{l} \to 0 \text{ for spin } 1/2 \\ \to \infty \text{ for spin } 0 \end{array} \quad .$$

The experimental results clearly rule out spin 0. The small deviation from zero can be accounted for by residual transverse momentum. Detailed QCD calculations show good agreement with the spin 1/2 hypothesis[13].

The measured results on the structure functions $F_2$ and $xF_3$ from different experiments are summarized in fig.9[10,11,13].

Note that the electron data on $F_2$ coincide with the $\nu$-data once they are scaled by a factor of 9/5. The significance of this factor, and the interpretation of $F_2$ and $xF_3$ will be explained in the next section.

## 1.2.6 Interpretation of Structure Functions in the Quark-Parton Model

We have seen in section 3 that the inevitable consequence of Bjorken scaling was elastic scattering on pointlike constituents. The structure function reflects the momentum distribution of the partons inside the nucleon.

It is natural to identify these spin 1/2 pointlike partons with the quarks of Gell-Mann and Zweig. This quark-parton model (QPM) assumes that the deep inelastic scattering data can be understood from elastic scattering on quarks inside the nucleon via electromagnetic or weak V-A interaction. In this simple form, the model assumes that the quarks behave as free particles when they are struck.

Let us call the momentum distribution function of the quarks of flavour i $f_i(x)$. Then simply from summing incoherently over all elastic quark scattering terms one can derive[9]:

$$F_2^{\nu N}(x) = F_2^{\bar{\nu} N}(x) = x \sum_i f_i(x) \qquad \text{(neutrinos)} \tag{6}$$

where charge symmetry and an isoscalar target in $F_2^{\nu}$ (see above) are assumed. In case of charged lepton the quark charges $e_i$ enter:

$$F_2^{eN} = F_2^{\mu N} = F_2^{e}(x) = x \sum_i e_i f_i(x) \qquad \text{(charged leptons)}. \tag{7}$$

In the quark model nucleons are composed of u and d quarks (valence quarks $q_v$, fig.10a). In the further discussion it will turn out that this is not the whole story. Valence quarks interact with each other by the exchange of quanta which we will call gluons (fig.10b). In these interactions quark-anti-quark pairs may be excited from the Dirac sea (sea quarks $q_s$, fig.10c). In the following discussion we shall assume that only the light quarks (u,d,s) play an essential role in this pair production (vacuum polarisation) process.

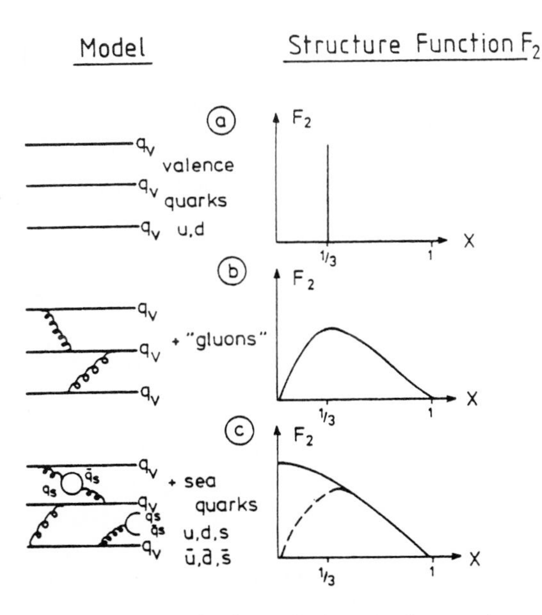


Fig.10a-c. Evolution of the simple quark parton model and consequences for the structure function in three steps. a) 3 valence quarks moving independently would lead to a $\delta$ function in $F_2$ at $x = 1/3$. b) Quark-quark interaction leads to a broadening of $F_2$ (Fermi motion of quarks inside the nucleon). c) $q_s\bar{q}_s$ pairs are created from the Dirac sea. They fill up $F_2$ at small x

We will then use the following nomenclature (assuming charge symmetry):

$u(x)$ = x distribution of a u quark in a proton
= x distribution of a d quark in a neutron

$d(x)$ = x distribution of a d quark in a proton
= x distribution of a u quark in a neutron

$s(x)$ = x distribution of a s quark in proton or neutron

and corresponding definitions for $\bar{u}$, $\bar{d}$ and $\bar{s}$.

It is convenient to further define:

$$q(x) = u(x) + d(x) + s(x),$$
$$\bar{q}(x) = \bar{u}(x) + \bar{d}(x) + \bar{s}(x).$$

*Relation Between the Charged and Neutral Structure Functions* $F_2^e$ *and* $F_2^\nu$

It is now straightforward to write down the structure functions for isoscalar targets $F_2^e$ and $F_2^\nu$ for charged and neutral leptons:

$$F_2^e = x\left\{\frac{1}{9}(d+\bar{d}+s+\bar{s}) + \frac{4}{9}(u+\bar{u})\right\}$$
$$F_2^\nu = x\,\{d+\bar{d}+u+\bar{u}+s+\bar{s}\}\ .$$

A simple calculation yields

$$\frac{F_2^e}{F_2^\nu} = \frac{F_2^{en} + F_2^{ep}}{F_{2\bar{s}}^{\nu N} + F_2^{\bar{\nu}N}} = \frac{5}{18}\left\{\frac{q+\bar{q} - \frac{3}{5}\,(s+\bar{s})}{q+\bar{q}}\right\} \tag{8}$$

i.e. for small $\bar{s}$ contributions the charged and neutral structure functions differ only by a factor 5/18, the mean quark charge. Fig.11 shows that this simple expression is in good agreement with experimental observation[10,13,14]. This is strong support for our simple model and shows that s quark contributions are small beyond $x \gtrsim 0.1$.

*Significance of* $F_2$ *and* $F_3$ *in terms of Quark Distributions*

We have seen above that $F_2^\nu$ measures the sum of all quark distributions

$$F_2^\nu = x(q+\bar{q}) \tag{9}$$

It turns out that in the quark model the neutrino structure function $xF_3$ measures the difference of quark and antiquark distributions:

$$\frac{1}{2}(xF_3^{\nu N} + xF_3^{\bar{\nu}N}) = \overline{xF_3} = x\ (q-\bar{q})\ . \tag{10}$$

If we rewrite (9) and (10) in terms of valence ($q_v$) and sea quarks ($q_s$) we get

$$F_2^{\nu}=x(u+d+s+\bar{u}+\bar{d}+\bar{s}) = x(u_v+d_v+\underbrace{u_s+d_s+s_s+\bar{u}_s+\bar{d}_s+\bar{s}_s}_{=\,2\,q_s\,=\,2\,\bar{q}_s})$$

$$xF_3 = x(u+d+s-\bar{u}-\bar{d}-\bar{s}) = x(u_v+d_v+\underbrace{u_s+d_s+s_s-\bar{u}_s-\bar{d}_s-\bar{s}_s}_{=\,q_s-\bar{q}_s=0})\ .$$

Observing that $q_s = \bar{q}_s$ the two structure functions take the very simple form

$$F_2^{\nu} = x(u_v+d_v+2q_s) = x(q_v+2q_s)$$

$$\overline{xF_3} = x(u_v+d_v) = x\ q_v\ .$$

The physical interpretation of fig.9 is now simple: $\overline{xF_3}$ measures directly the momentum distribution of the valence quarks inside the nucleon. The difference between $F_2$ and $\overline{xF_3}$ turns out to be twice the momentum distribution of the sea quarks, simply because $F_2$ measures all quarks and $q_s$ and $\bar{q}_s$ are equal due to their pair production origin.

The form of the quark distributions is made plausible in fig.10. The mutual interaction of valence quarks leads to a broadening of the momentum distribution of the valence quarks around $x=1/3$. The exact form of this Fermi motion of valence quarks inside the nucleon is measured by $xF_3$ in fig.9. The additional effect from $q_s\bar{q}_s$ pairs is expected to occur predominantly at small x. Quantitatively $xq_s = x\bar{q}_s$ is given by 1/2 $(F_2-xF_3)$ as indicated in fig.9.

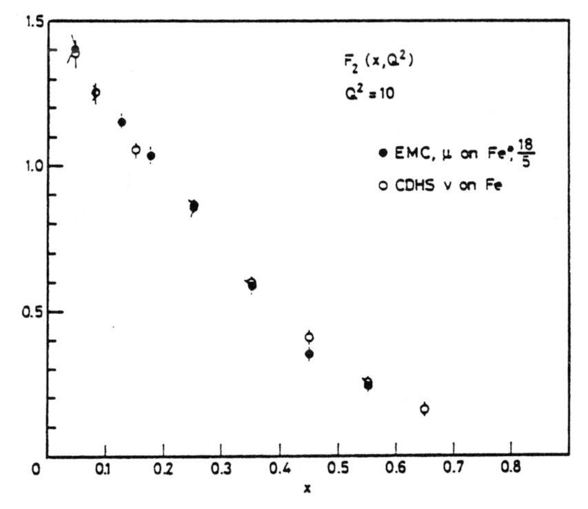


Fig.11. Comparison of the structure function $F_2$ as measured in neutral and charged lepton scattering (weak and e/m interaction of quarks)

Some simple consequences of this picture can be deduced:

*Number of Valence Quarks*

The number of valence quarks inside the nucleon is of course 3 in the quark parton model. This can be tested experimentally:

$$\int_0^1 (q-\bar{q})dx = \int_0^1 \frac{xF_3}{x} dx = 3.2 \pm 0.5$$

CDHS[10,13]

*Fraction of Momentum Carried by Valence Quarks*

This number can again be deduced from $xF_3$:

$$\int_0^1 x\ (q-\bar{q})\ dx = \int_0^1 xF_3\ dx = 0.32 \pm 0.01$$

CDHS[10,13]

*Fraction of Momentum Carried by all Quarks*

This number can be deduced from $F_2$:

$$\int_0^1 x\ (q+\bar{q})\ dx = \int_0^1 F_2(x)\ dx = 0.45 \pm 0.01$$

CDHS[10,13]

*Fraction of Momentum Carried by Sea Quarks*

$$\int_0^1 x\ (2\bar{q})\ dx = \int_0^1 (F_2 - xF_3)\ dx = 0.13 \pm 0.01$$

CDHS[10,13]

All observations lead to a consistent picture of the nucleon made up of three valence quarks and additional sea quarks. The amazing figures are, however, the momentum fractions carried by quarks. Valence quarks carry only about 1/3 of the total momentum and even if one adds the sea quarks more than 1/2 of the momentum is missing! The inevitable consequence is that there are additional quanta inside the nucleon which do not participate in the weak or electromagnetic interaction. The best candidates are the quanta of quark-quark interaction, the 'gluons', which according to the above results carry 55% of the total momentum of the nucleon.

### 1.2.7 Scaling Violation and QCD[15]

It has early been realized that scaling is only an approximate property of deep inelastic scattering[16]. Such scaling violations are in fact expected if the model developed in the previous section is correct[17].

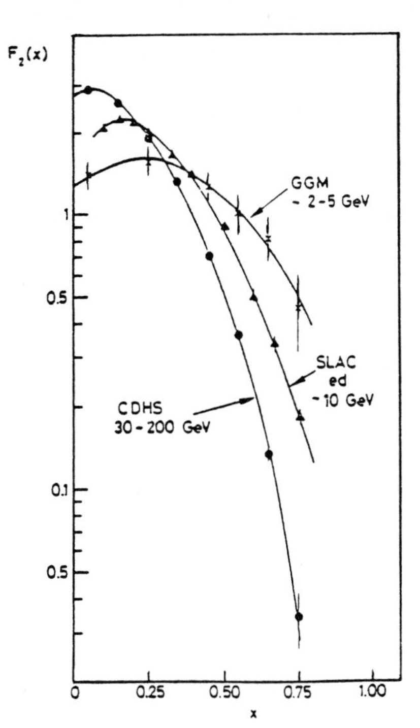


Fig.12: The structure function $F_2(x)$ at different values of $Q^2$

With rising $Q^2$ the resolving power improves in deep inelastic scattering. If we assume that the bare partons are 'dressed' with a cloud of gluons and $q_s\bar{q}_s$ pairs (see fig.10c), then we expect a $Q^2$ dependence of the structure function as shown in fig.12.

Whereas at moderate $Q^2$ the dressed parton appears as an integral object, the bare parton is resolved from its cloud with increasing $Q^2$. Since the quarks and gluons in the cloud carry part of the momentum of the dressed quarks, the x distribution gets shifted towards lower values. Or in an alternative view, with rising $Q^2$ the parton behaves like an extended object with a corresponding depression of the high $Q^2/2M\nu$ values in the structure function.

One may try to understand these scaling violation effects in perturbative QCD[18]. The graphs contributing in leading order are shown in fig.13b-e. These QCD

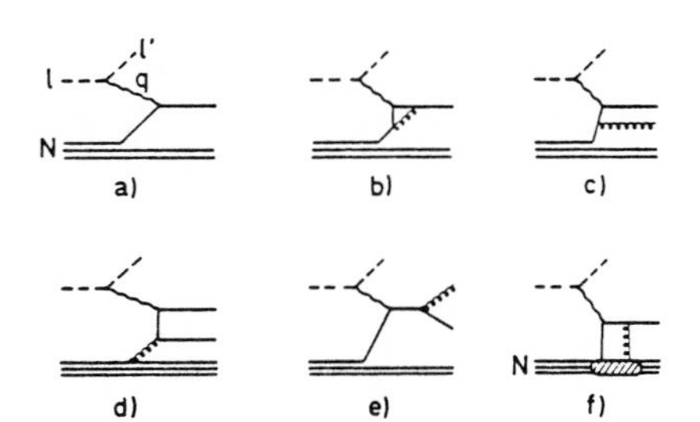


Fig.13. High-order QCD corrections

corrections have been calculated. They lead to a ln $Q^2$ dependence of the structure functions.

There are two other sources of scale breaking which obscure the perturbative QCD effects.

- charm production:
  With increasing $Q^2$ charm quark pairs will be created. They occur at low x.

- higher-order twist:
  An important contribution at large x comes from graphs of the form 13f, where final-state interaction occurs in the quark system. There contributions behave like powers of $1/Q^2$.

Unfortunately these effects contribute mainly to the large and small x region, which is most sensitive to perturbative QCD effects.

Figs. 14, 15 show fits of leading-order QCD to the deep inelastic neutrino and muon data. The resulting values for the QCD scale parameter $\Lambda$ (see other lectures and chapter IV) are in leading-order QCD

$\Lambda$ = (100 $\pm$ 100) MeV EMC[14)]

$\Lambda$ = (300 $\pm$ 100(stat.) $\pm$ 100(syst.)) MeV CDHS[10)] .

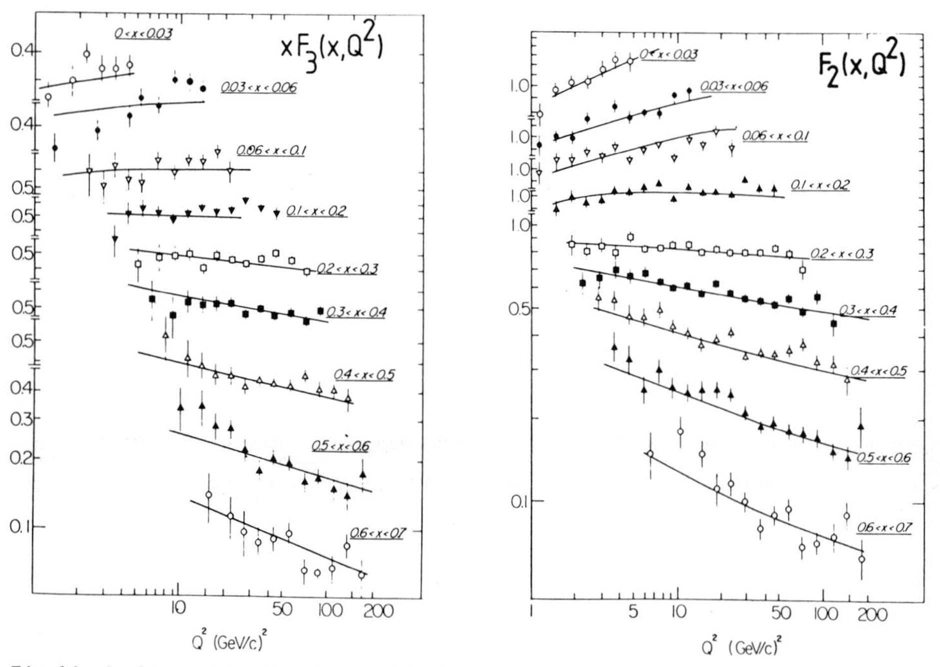


Fig.14. Scaling violation in $\nu$ scattering compared to QCD

We will come back to some details of QCD and to the relation between $\Lambda$ and the strong coupling constant $\alpha_s$ in chapter IV.

## 1.3 Quarks in $e^+e^-$ Annihilation

### 1.3.1 $e^+e^-$ Annihilation into Hadrons in the Quark-Parton Model

In QPM the process of $e^+e^-$ annihilation[19] is viewed as the incoherent sum over quark pair production diagrams as shown in fig.16a. The quarks fragment into hadrons (with probability 1) and the total hadronic cross section is then

$$\sigma_{had} = \sum_q \sigma_{q\bar{q}} \tag{1}$$

where the sum runs over all species of quarks for which pair production is energetically allowed. The cross section can be simply calculated from QED (see fig.16):

$$\sigma_{had} = \sigma_{QED} \cdot \sum_q e_q^2$$

$$\text{or} \qquad R = \frac{\sigma_{had}}{\sigma_{QED}} = \sum_q e_q^2 \; . \tag{2}$$

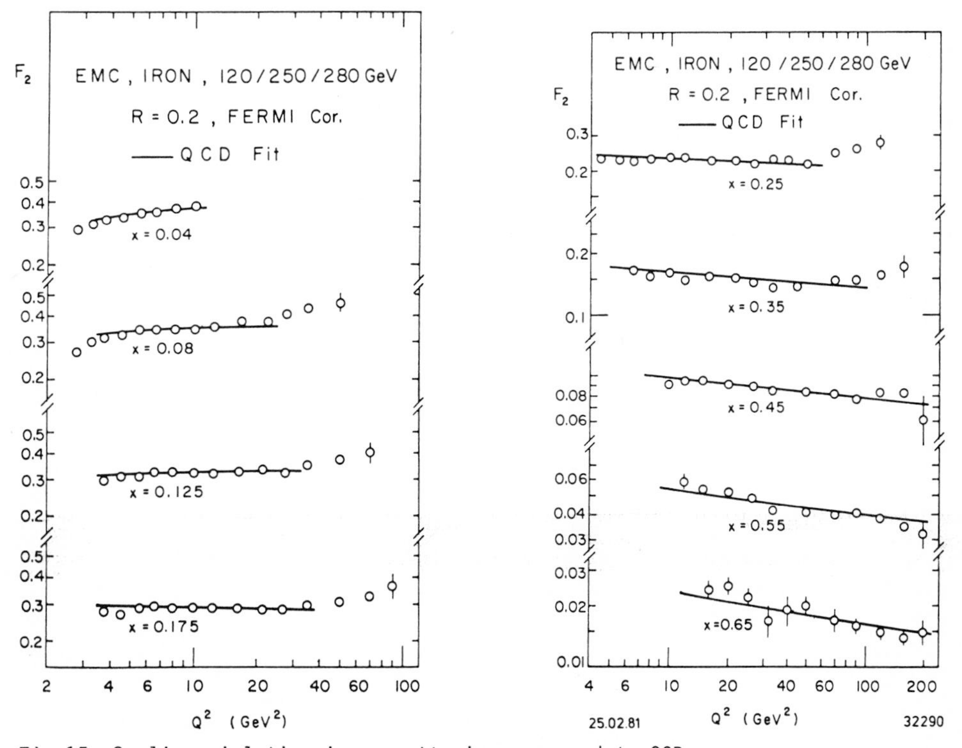


Fig.15. Scaling violation in $\mu$ scattering compared to QCD

a) $e^+e^-$ annihilation

$$\sigma_{had} = \sum_q \left| \; \right|^2$$

$$\frac{\sigma_{had}}{\sigma_{QED}} = R = \sum_q e_q^2$$

b) deep inelastic scattering

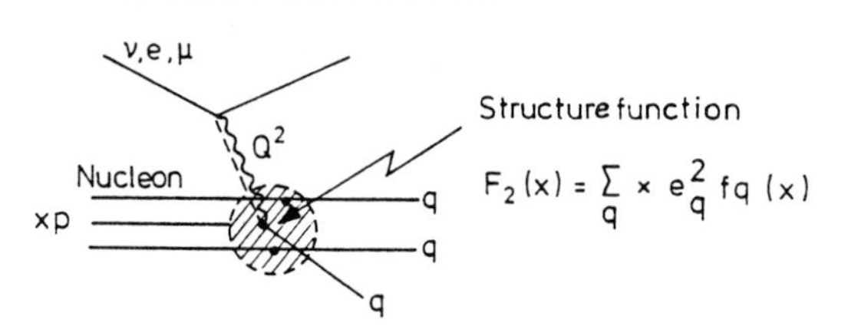


Fig.16. $e^+e^-$ annihilation into hadrons compared to deep inelastic scattering

The total hadronic cross section is usually measured in terms of R in which the trivial kinematical terms from QED are taken out.

Let us compare again to the deep inelastic process of the previous chapter (fig.16b). The spacelike photon ($q^2 < 0$) is now replaced by a timelike one ($q^2 > 0$). Since we deal now only with pair production at the origin of the CMS, the 'structure function' is now simply the sum over all quark charges, whereas in deep inelastic scattering the momentum distribution of quarks inside the nucleon had to be taken into account.

The two processes - although similar in the above sense - are quite complementary in their physics aims: whereas deep inelastic scattering is confined to testing the properties of quarks inside nucleons, $e^+e^-$ annihilation can excite all existing quark-antiquark pairs from the Dirac sea provided there is enough energy in the $e^+e^-$ system. In this sense deep inelastic scattering tests the nucleon structure whereas $e^+e^-$ annihilation tests the vacuum structure.

### 1.3.2 $e^+e^-$ Storage Rings and Detectors

$e^+e^-$ storage rings provide the appropiate tool to measure $e^+e^-$ annihilation processes. Fig.17 shows as an example the PETRA storage ring at DESY in Hamburg, which yields a CMS energy of up to 36.6 GeV, the highest energy nowadays available in $e^+e^-$ collisions. Similar devices are - in order of decreasing energy - PEP at Stanford, CESR at Cornell, DORIS at Hamburg and SPEAR at Stanford. New larger machines are planned: TRISTAN at Tokio (2x30 GeV), HERA at Hamburg (2x35 GeV), SLC at Stanford (2x50 GeV) and LEP at CERN (up to 2x130 GeV).

Experimental equipment can be mounted in 4 intersection areas at PETRA. 5 experiments have taken data: CELLO, JADE, MARK-J, PLUTO, TASSO. As an example, the CELLO detector[20] is shown in fig.18. Its inner part consists of cylindrical proportional and drift chambers which measure charged particle tracks in a large sole-

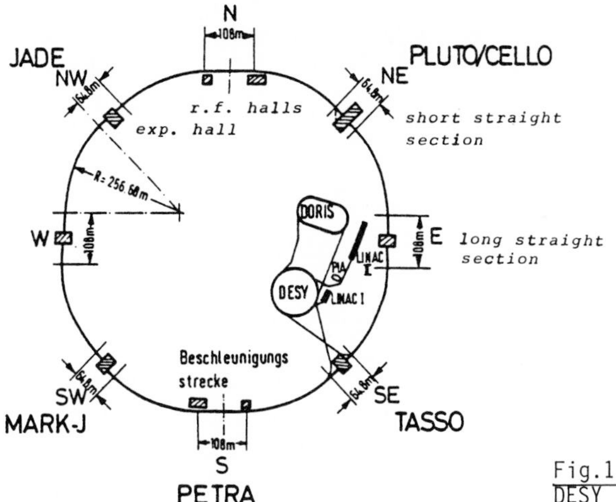


Fig.17. $e^+e^-$ storage ring PETRA at DESY in Hamburg

noidal field of a superconducting magnet. Outside the magnet a lead-liquid argon calorimeter provides photon and electron detection through electromagnetic showers. The iron return yoke acts as a filter for muon/hadron separation. Muons penetrating the iron are detected in large area drift chambers outside the yoke.

Fig.19 shows two typical events taken at CELLO at highest PETRA energy. A cross section perpendicular to the beam is displayed. Charged tracks are visualised by their bits in the cylindrical chamber and their computer reconstruction. The dashes in the liquid argon system correspond to electromagnetic energy deposits.

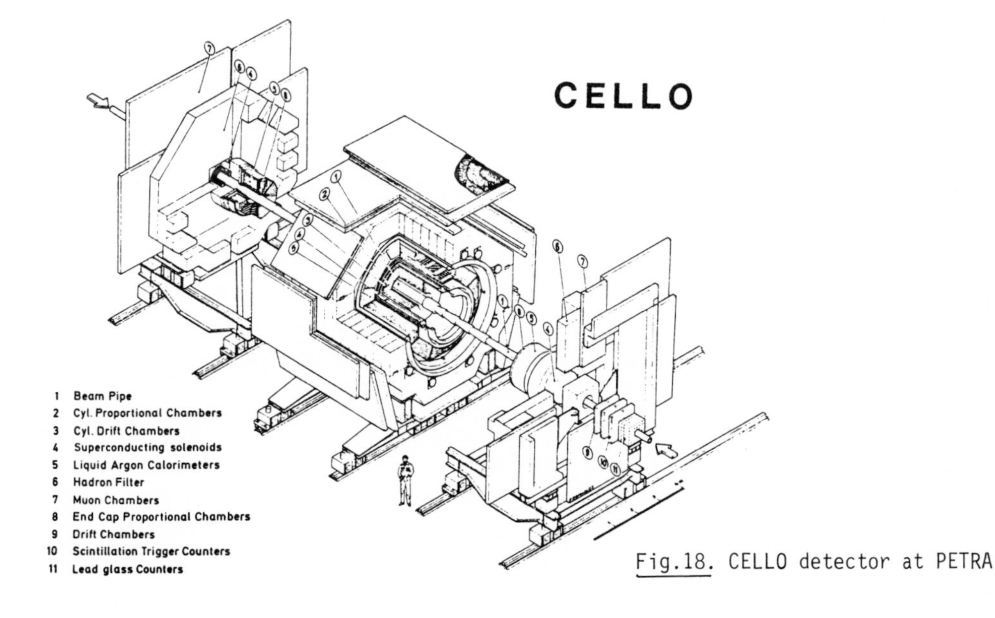


Fig.18. CELLO detector at PETRA

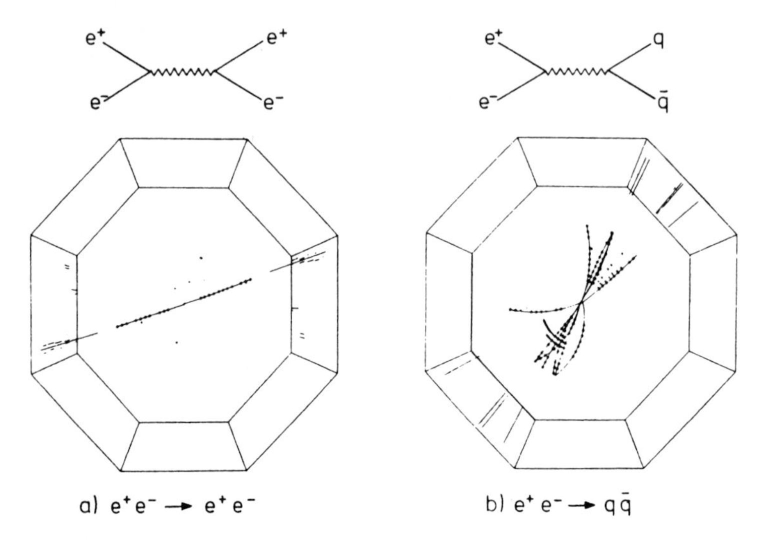


Fig.19a,b. Typical events in $e^+e^-$ reactions. a) Eleastic scattering, b) hadronic annihilation

### 1.3.3 Total Hadronic Cross Section

The total hadronic cross section in terms of $R = \sigma_{had}/\sigma_{\mu\mu}$ as a function of CMS energy is summarized +) in fig.20 ($\sigma_{\mu\mu} = \sigma_{QED}$ for charge 1).

The three most striking features are

- R is flat between 10 and 36 GeV
- R exhibits a clear step and spikes structure around 4 GeV
- another series of spikes lies around 9.5 GeV. The corresponding step in R is not so obvious++).

The first observation can be readily interpreted in the QPM. It shows that R has no $Q^2$ dependence, hence that the quarks are pointlike. We can quantify this statement by introducing form factors $G_i(Q^2)$ into equation (2).

$$R = \Sigma\, e_i^2\, G_i(Q^2)$$

where $G_i(Q^2)$ are again the spin-averaged form factors. If we parametrize $G_i(Q^2)$ by a dipole form we get cutoff parameters corresponding to $\langle r\rangle \leq 2\cdot 10^{-16}$ cm for the 'quark radius'.

---

+) P.Söding, DESY report 81-070 (1981)

++) Recent data from CESR indicate a clear step in the relative cross section measured below and above the resonance region.

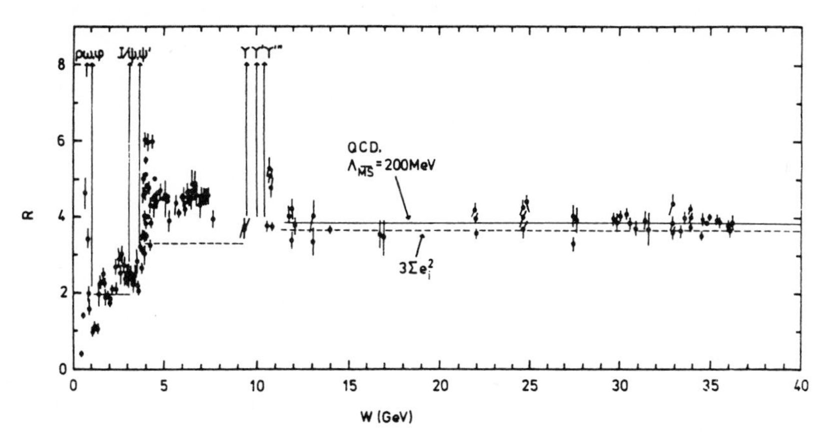


Fig.20. Total hadronic cross section in terms of $R = \sigma_{had}/\sigma_{\mu\mu}$ as a function of $e^+e^-$

Let us next try to understand the steps and the absolute values of R. We know that charm is produced around 4 GeV (see below). Hence we expect from (2)

$$R = \Sigma\, e_i^2 = \underbrace{\underbrace{e_u^2 + e_d^2 + e_s^2}_{2/3} \; (+e_c^2)}_{10/9}$$

2/3 below charm threshold

10/9 above charm threshold

This is in clear disagreement with the data. Only if we introduce an additional factor of 3 for the three different colours of quarks do we get reasonable agreement with the data

$$R = 3\,\Sigma\, e_i^2 = 2 \text{ below, } 3\,1/3 \text{ above charm threshold.}$$

This is one of the most important experimental confirmations of colour. We will come back to this point later, in particular to QCD corrections which further improve the agreement with data.

### 1.3.4 Jets in $e^+e^-$ Annihilation

We have seen that the asymptotic behaviour of R is in good agreement with the simple description of the quark-parton model. Let us, therefore, assume that quark-pair production really governs the process

$$e^+e^- \rightarrow \text{quark-antiquark} \rightarrow \text{hadrons.}$$

In this picture the two quarks should fragment to form two back-to-back jets of particles (fig.21).

What are these jets like ? In the quark-parton model jets are described in a phenomenological way by a fragmentation of quarks with limited transverse momentum with respect to the original quark axis.

In 1975 first evidence for a two jet structure was reported at SLAC[21]. At the upgraded DORIS the phenomenon was confirmed by the PLUTO group[22].

In fig.18b we saw a very clean jetlike event from the CELLO detector. Two distinct back-to-back bunches of particles are clearly visible. Also the neutral energy of the two jets is clustered and follows the charged energy.

*Jet Measures*

Several quantities have been proposed to measure jets. I will only use two of them here, namely sphericity[23]

$$S = \frac{3}{2} \min \frac{\Sigma\, p_{Ti}^2}{\Sigma_i\, p_i^2}$$

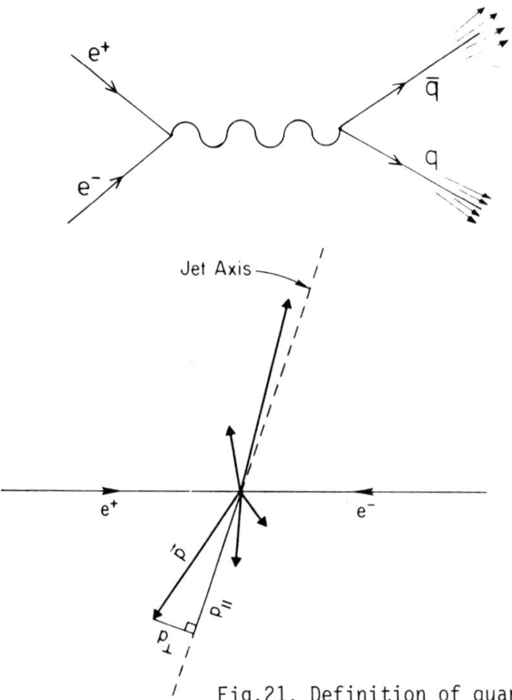


Fig.21. Definition of quantities used in the jet analysis, $p_\perp \equiv p_T$, $p_\parallel \equiv p_L$

and thrust[24])

$$T = \max \frac{\sum_i |p_{Li}|}{\sum |p_i|} .$$

(This definition is slightly different from the original one, where the sum for $p_L$ runs over one hemisphere only.)

Both quantities simultaneously define the jet axis and give a measure for the topological structure of the event. The axis is found in a variational method by either minimizing the sum of the transverse momentum squared ($p_T^2$) or maximizing the sum of the absolute longitudinal momentum component ($|p_L|$) with respect to a given axis (fig.21). Extreme values of the two quantities for isotropic or ideally jetlike events are summarized in table 3.

| Events | S | T |
|---|---|---|
| isotropic | $\rightarrow 1$ | $\rightarrow 1/2$ |
| ideally jetlike | $\rightarrow 0$ | $\rightarrow 1$ |

Table 3. Values of S and T in extreme topologies

If we assume that $p_T$ is about constant the quantities $\langle S \rangle$ or $\langle 1-T \rangle$ will both fall with increasing energy (assuming that the multiplicity is only slowly varying).

*Measurements*

Fig.22 shows the mean longitudinal and transverse momentum of charged particles with respect to the thrust axis as measured in the PLUTO and TASSO group[25,26]. The tendency at lower energies prevails. $p_L$ increases whereas $p_T$ stays small, i.e., the jet structure becomes more and more pronounced.

The energy dependence of the mean sphericity (fig.23) quantifies this observation. The figure includes data from PLUTO and TASSO.

The angular dependence of the thrust axis[28]) is shown in fig.38a (LENA data). Data are in fair agreement with $a = 1 + \cos^2\Theta$ behaviour (full curve) expected for

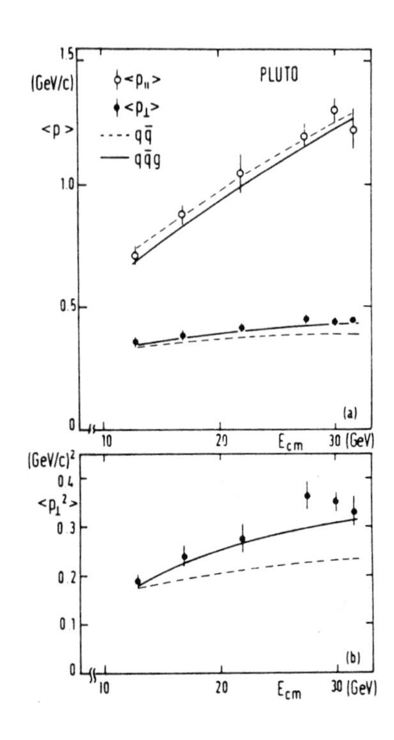


Fig.22. Energy dependence of the mean momentum components $\langle p_T \rangle$, $\langle p_L \rangle$ and $\langle p_T^2 \rangle$

pair production of spin 1/2 objects. Polarisation measurements in the MARK I detector yield $\alpha = 0.97 \pm 0.14$[21].

Thus all measurements are in quantitative agreement with the expectations of the quark-parton model, where $e^+e^- \rightarrow$ hadrons proceeds through the production of a spin 1/2 quark pair.

### 1.3.5 Resonances in $e^+e^-$ Annihilation

Spikes in $e^+e^-$ annihilation occur near to quark-antiquark thresholds. The effect can be understood as resonance production in a potential model. In fig.24

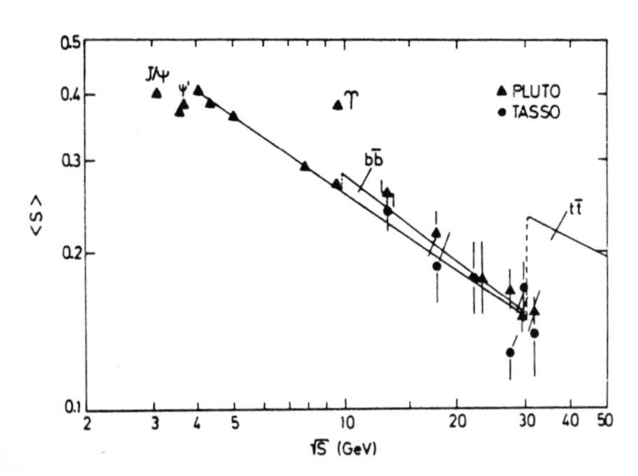


Fig.23. Mean sphericity as a function of energy

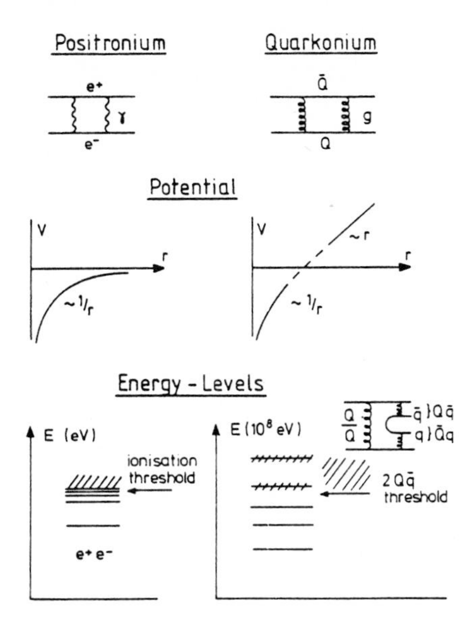


Fig.24. Potential picture of quarkonium

the evolving quarkonium system is sketched in analogy to positronium. There are, besides the nontrivial factor of $10^8$ in energy scale, some important differences: The potential in the $Q\bar{Q}$ system which may be Coulomb-like due to 1 gluon exchange at small distances has to take confinement into account at large distances. The simplest guess supported by string models is a potential linear in r. Consequently the energy levels in the $Q\bar{Q}$ system extend to infinity. The threshold which is just the $e^+e^-$ ionisation energy in the positronium system gets another meaning now. The $Q\bar{Q}$ system cannot disintegrate directly due to the Zweig rule[29] (the reason for this will become clear in section IV).

Only if sufficient energy is available to generate another light $q\bar{q}$ pair can the system decay into $\bar{Q}q$ + $Q\bar{q}$. Above this threshold, which is hence at twice the $Q\bar{q}$ mass (charm: twice the D meson mass), the resonances get broader and finally smear out to the flat continuum observed at large energies. The transition region with broad resonances is clearly seen above charm threshold (fig.20).

### 1.3.6 Charm and Carmonium

For several reasons, the heavy quarkonium systems are particularly important:

- Since the quarks are relatively slow (e.g. $\beta$ = 0.2 for $c\bar{c}$), relativistic corrections are small,
- heavy quarks probe deep into the potential, where one gluon exchange (Coulomb potential) is supposed to dominate, and

- in the heavy systems (charmonium, bottonium) several narrow resonances lie below threshold. This implies a rich spectroscopy of different spin and angular momentum states.

Consequently in good approximation heavy quarkonia systems can be treated as nonrelativistic two-body systems using simple quantum mechanics (Schrödinger equation).

The expected level spectrum for charmonium is shown in fig.25a.

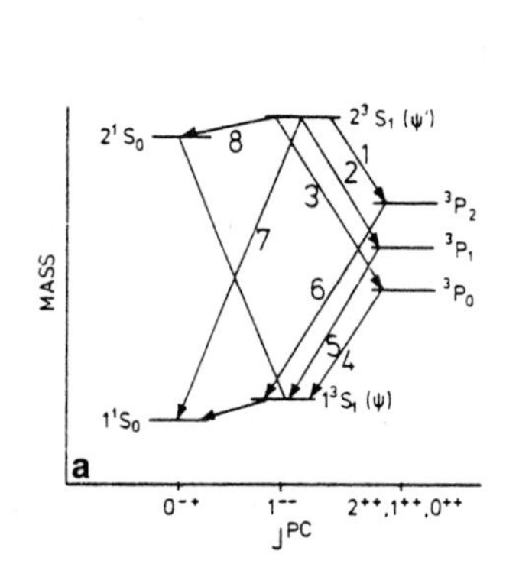


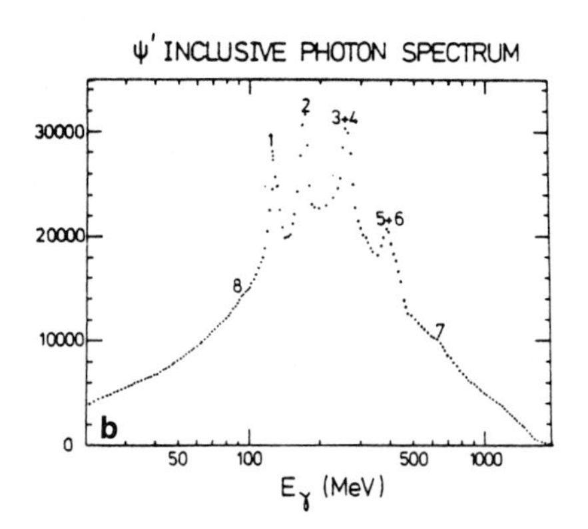


Fig.25. Lowest energy levels in charmonium a) and their radiative transitions b)

Analogous to atomic physics the states are labeled as

$$n^{2S+1}L_J$$

where n = radial quantum number

S = spin of diquark system, 0 or 1

L = relative angular momentum

J = total angular momentum.

Only the two $^3S_1$ states are directly accessible in $e^+e^-$ annihilation$^{+)}$. Radiative transitions between the states are indicated in the figure.

The spectrum of these lowest charmonium states below threshold was fully accessed experimentally by the beautiful results of the Crystal Ball detector at SPEAR[30]. Fig.26 shows the detector, which consists of NaJ crystals covering nearly the full solid angle around an $e^+e^-$ interaction area. The radiative spectrum obtained when the $e^+e^-$ energy was tuned to the second $^3S_1$ state $\psi'$ is shown in fig.25b.

---

$^{+)}$The $^3D_1$ state (not shown in fig.25a) can also be excited in $e^+e^-$ annihilation, since it has $J^{PC} = 1^{--}$ and mixes with the nearby $2\ ^3S_1$ state.

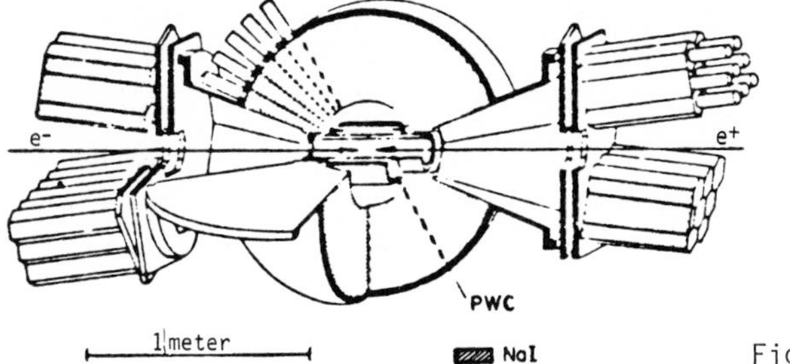


Fig.26. The Crystal Ball detector at SPEAR

The lines in this spectrum can be identified as radiative transitions as indicated by the numbers. The least and the most energetic photon lines, 7 and 8, are of particular importance, since they are most probably related to the two $^1S_0$ states $\eta_c$ and $\eta_c'$. These two states had fooled experimentalists for a long time.

Since the level spacing to the $^3S_1$ states is very small, the radiative transitions are correspondingly rare. In addition the detection of low-energy photons poses a problem in itself. After some erroneous measurements for both states (which caused a lot of trouble to the theoreticians) the problem seems finally to be solved by the Crystal Ball measurements. A candidate state for $\eta_c$ has clearly been established in inclusive[30)]

$$J/\psi \rightarrow \eta_c \, \gamma$$

$$\psi' \rightarrow \eta_c \, \gamma \qquad \text{(see fig.25)}$$

and in exclusive channels[31,32)]

$$\psi' \rightarrow \gamma\eta_c$$
$$\hookrightarrow \pi^{\pm} K^{+} K_s, \; \pi^{+}\pi^{-}\eta \, .$$

The mass for this state was found to be 2983 $\pm$ 5 MeV. The $\eta_c'$ candidate (decay 8 in fig.25) was only discovered recently[33)]. Fig.27 shows this part of the spectrum enlarged and with background subtraction. The mass of the $\eta_c'$ candidate is 3592 $\pm$ 5 MeV. Both states fit well into the expected charmonium level scheme. However, for a final identification of these states with $\eta_c$ and $\eta_c'$ a spin-parity determination is still missing.

The situation of charm and charmonium is summarized in fig.28. The charmonium system seems to be in good shape, although spin-parity assignments for the $^1S$ states are still missing. Above charm threshold, a complicated structure of broad resonances is observed in the annihilation cross section, of which only the $\psi''$ (3.77) is clearly resolved. In this region, open charm production sets in:

pseudoscalar (S=0) mesons D ($c\bar{u}$, $c\bar{d}$, $c\bar{u}$, $c\bar{d}$)[34] and F ($c\bar{s}$, $\bar{c}s$)[35] and their vector (S=1) companions $D^{*}$[34] and $F^{*}$[35]. Only D and $D^{*}$ are well established[34].

In addition, evidence for charm baryons ($\Lambda_c$)[36] and measurements of charm lifetimes[37] came from neutrino and hadron experiments.

1.3.7 Bottonium

At about 9.5 GeV cms energy a new series of narrow states occurs in $e^+e^-$ annihilation[38]. These are the quarkonium states of a new heavy quark[3] b (bottom or beauty). Three narrow and one broad resonance have been clearly identified at the $e^+e^-$ storage ring CESR[39] (fig.29).

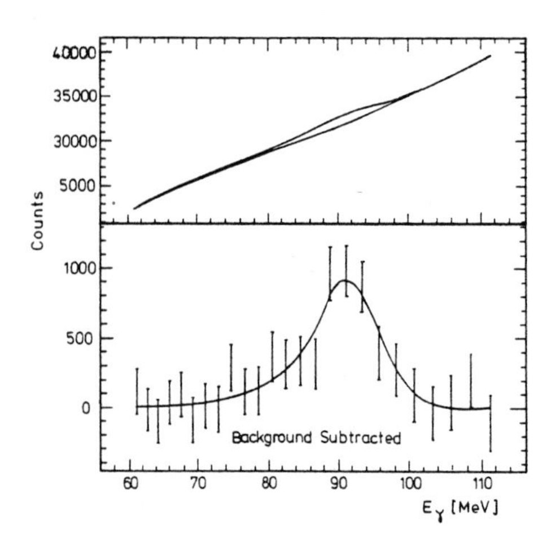


Fig.27. Experimental evidence for the radiative decay of ψ' into an $\eta_c'$ candidate state at a mass of 3592 ± 5 MeV

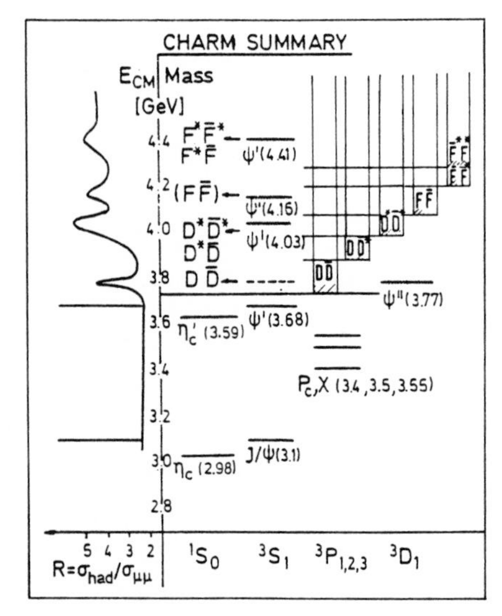


Fig.28. Summary of the experimental knowledge on charm

The experimental situation is again summarized in fig.30. Only the $^3S$ states which are directly accessible in $e^+e^-$ annihilation have been seen. Pion decays between these states have been measured[40)] with the expected branching ratios indicating that the resonances are indeed related. However, no radiative transitions to potential $^1S$ and $^3P$ states have so far been observed. A spectroscopy much richer than charm still waits for discovery.

## 1.3.8 Toponium

As indicated in table 2 a sixth quark with charge + 2/3 would make everybody happy. Apart from filling a hole in a nicely symmetric scheme it would also

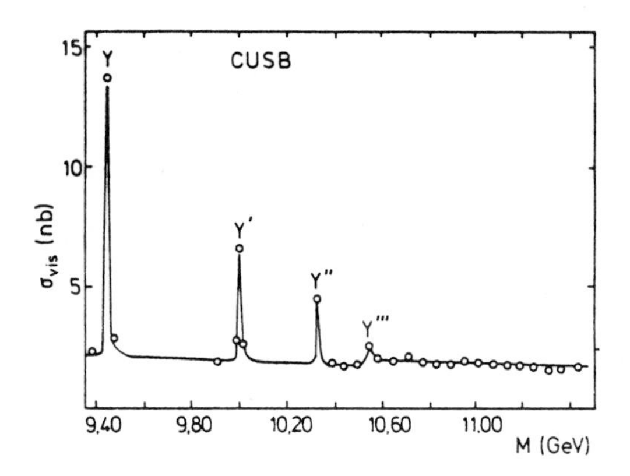


Fig.29. Visible cross section of the bottonium resonances at CESR. The fourth resonance is broad, indicating that it lies above $B\bar{B}$ threshold

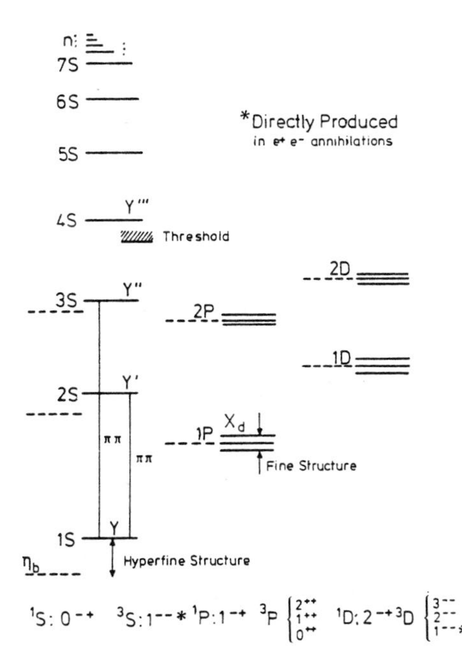


Fig.30. Expected level scheme (schematically) for bottonium. Only the $\pi\pi$ transitions from $\Upsilon''$ and $\Upsilon'$ to the ground state have been observed experimentally

reveal an even richer spectroscopy (see figs. 31, 32) and probe deeper into the Coulomb-like region of the quarkonium potential. Also it would probably yield much better separated gluon jets than now available in bottonium decay (see chapter IV). Furthermore Higgs particles[41] would have a good chance to be seen in toponium decays (if they exist in the right mass region).

Experimental measurements have, however, so far only shown the paucity of toponium or top thresholds up to 36.7 GeV (fig.33). Also a new charge 1/3 quark is very unlikely up to these energies[42].

## 1.4 Gluons in $e^+e^-$ Annihilation

### 1.4.1 Introduction

In analogy to QED two main sources for gluon production in $e^+e^-$ annihilation have been predicted in Quantum Chromo Dynamics (QCD)[4].

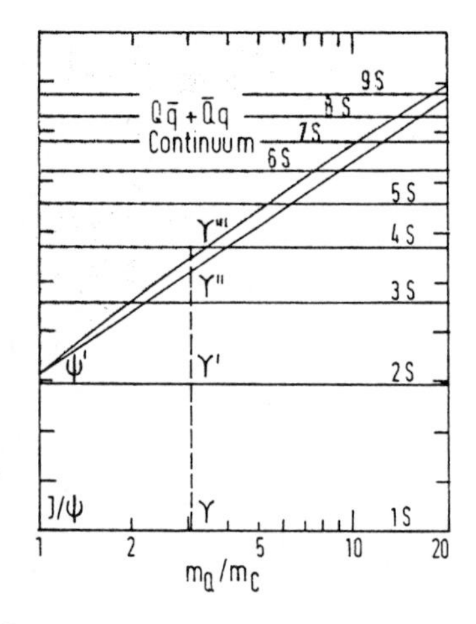


Fig.31. Number of narrow resonances below threshold as a function of quark mass

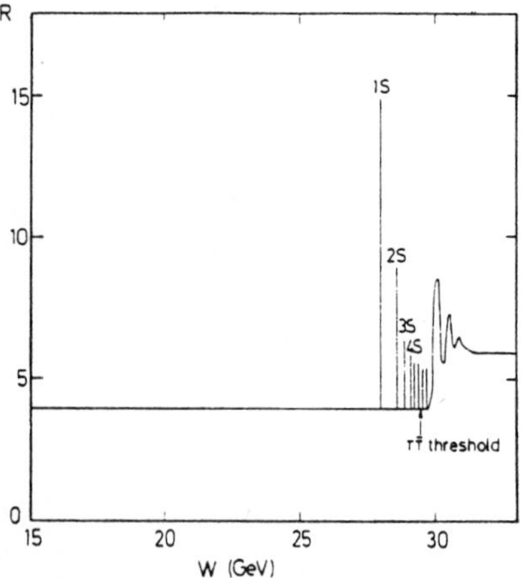


Fig.32. Theorist's view of the 30 GeV region before PETRA came into operation

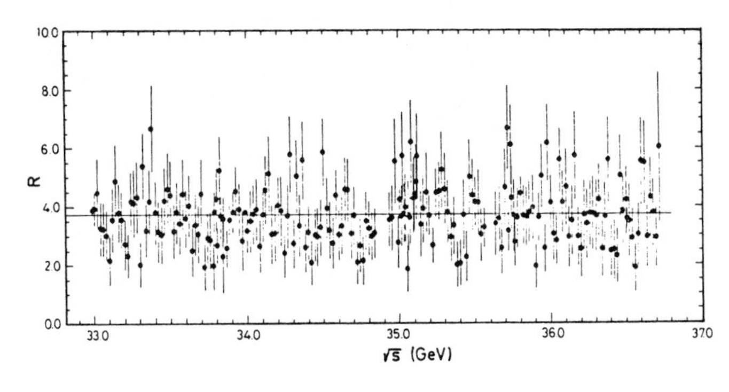


Fig.33. Normalized hadronic cross section R above 33 GeV cms energy (MARK-J)

- Decay of heavy quarkonia (fig.34a)

  As positronium can decay into three photons, quarkonium can decay by emitting three gluons[44]. One-gluon decay is inhibited by colour, two-gluon decay by C parity (like the photon, the gluon has C = -1). The three gluons would fragment to form jets of hadrons.

- Gluon bremsstrahlung (fig.34b)

  Analogous to the well-known bremsstrahlung of electrons in QED gluon-bremsstrahlung from quarks is predicted in QCD[45]. The process is proportional to $\alpha_s$ and will in first order result in a three-jet structure of hadronic final states in $e^+e^-$ annihilation.

## 1.4.2 Decay of Heavy Quarkonia

Bound vector states of heavy quarks (quarkonia) can decay through the one-photon channel. The corresponding term in $e^+e^-$ annihilation, a quark loop correction to the one-photon channel, is often called vacuum polarisation. Its contribution to the hadronic decay channel is given by

$$\Gamma_{q\bar{q}} = \Gamma_{ee} \, \Sigma \, Q_q^2 = \Gamma_{ee} \, R$$

a) Decay of Heavy Quarkonia

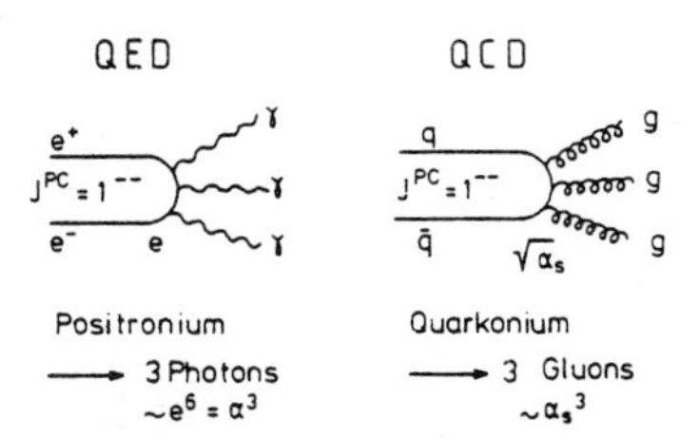


b) Gluon Bremsstrahlung

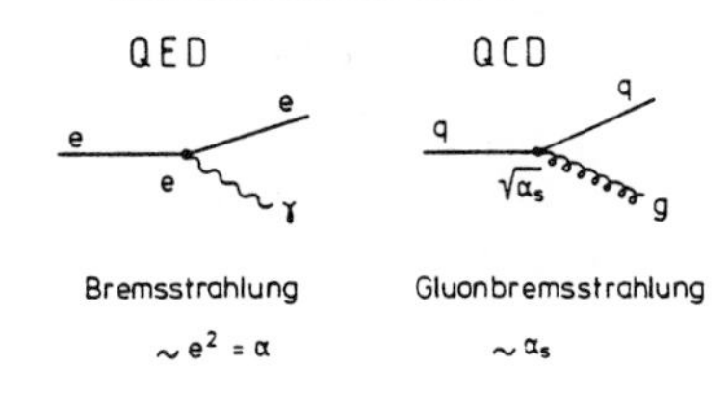


Fig.34a,b. Possible sources of gluon jets. a) Decay of heavy quarkonia. b) Gluon bremsstrahlung

where

$$\Gamma_{ee} = 16\,\pi \, \frac{\alpha^2\, Q_q^2}{M^2}\, |\psi(0)|^2$$

is known from QED.

Quantum chromo dynamics is the only theory that offers a prediction for the direct decay of quarkonia into hadrons. In this approach, the decay width is given by the coupling of the quark pair to more than 2 gluons. The lowest-order contribution (3 gluons) is then (e.g. ref.46)

$$\Gamma_{ggg} = \frac{160(\pi^2-9)\alpha_s^3}{81 M^2}\, |\psi(0)|^2$$

where $\alpha_s$ is the strong coupling constant.

The total width (for given $\Gamma_{ee}$)

$$\Gamma = \Gamma_{q\bar{q}} + \Gamma_{ggg} = R\,\Gamma_{ee}\,\left(1 + \frac{\Gamma_{ggg}}{\Gamma_{q\bar{q}}}\right)$$

is independent of the wave function of the quarks, since $|\psi(0)|$ drops out in

$$\frac{\Gamma_{ggg}}{\Gamma_{q\bar{q}}} = \frac{10(\pi^2-9)\alpha_s^3}{81\,\pi\,\alpha^2\,Q_q^2\,R}\ .$$

If we apply this formula to the $c\bar{c}$ ground state $J/\psi$ with $\Gamma_h = 57$ keV, $\Gamma_{ee} = 4.8$ keV, $R = 2.3$ and $Q_c = 2/3$ we get

$$\alpha_s = 0.19\ (J/\psi).$$

If we scale this value from $J/\psi$ to $\Upsilon$ taking the leading-order formula

$$\alpha_s(E) = \frac{12\,\pi}{(33-2N_f)\,\ln E^2/\Lambda^2}\ ;\quad N_f = \text{number of flavours}$$

we get

$$\alpha_s \simeq 0.17\ (\Upsilon).$$

With $R = 4$, $Q_b = -1/3$ and $\Gamma_{ee} = 1.2$ keV this yields

$$\Gamma_{tot} = \Gamma + 3\,\Gamma_{ee} \simeq 40 \text{ keV}$$

in fair agreement with the measured value of $36^{+8}_{-6}$ keV[33].

Since these numbers look like good agreement with theory, a word of caution is in order: Other determinations of $\alpha_s$ from $J/\psi$ radiative decays and charmonium models yield values of $\alpha_s \simeq 0.3./.0.4$ for J/ψ energies[46]. There is no generally accepted explanation for this discrepancy; higher-order radiative corrections and difficulties in defining $\alpha_s$ in the potential picture may cause the trouble.

Though, whatever the absolute value of $\alpha_s$ is, the J/ψ and Υ resonance parameters given above fit into a consistent picture with the right (small)energy dependence predicted by QCD.

### 1.4.3 Υ Decay Topology

In first order of QCD a $q\bar{q}$ bound state couples to three gluons. Once the energy of the $q\bar{q}$ state is high enough a fragmentation of these three gluons into jets will become the preferred decay mode[44].

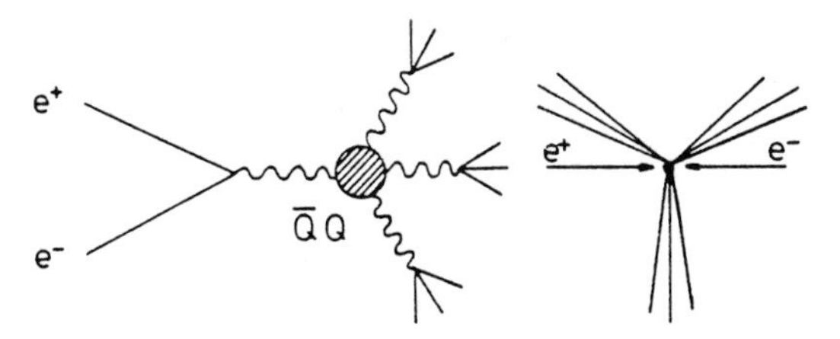


The observation of a three jet structure in the Υ decay would therefore be a decisive test on the existence of gluons and the validity of QCD[48,49]. This conjecture leads to the following predictions:

(1) Topological quantities like sphericity and thrust change drastically as one passes through the resonance.

(2) A three-jet structure would of course lead to a planar configuration of the events.

(3) Eventually three separated jets may be visible.

Although a possible observation of (1) and (2) may be indicative only (3) could be really decisive. Unfortunately it turns out that an asymmetric partition of energy among the three jets is preferred[48] which leads to a nearly back-to-back structure of the events instead of a symmetric three-star structure (fig.35). In addition at the present stage of theory and analysis any interpretation of the data suffers from the following problems:

- The fragmentation of gluons is not known. Phenomenological models have to be used when comparing the data.
- At the Υ resonance the energy of the proposed gluon jets is still very low (about 3 GeV/jet).
- Resonance events can only be separated statistically from the continuum.

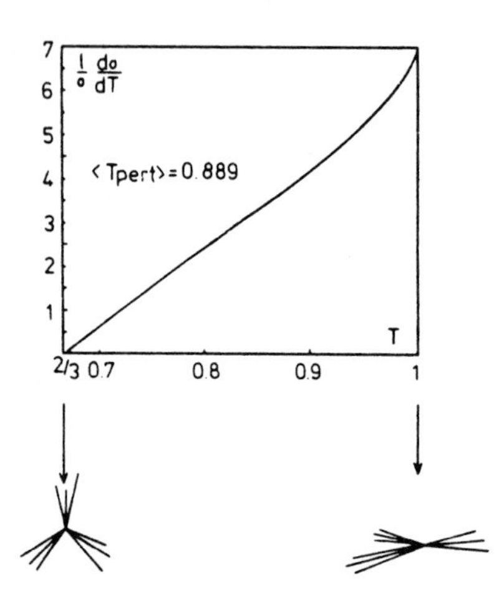


Fig.35. Relative cross section for symmetric and asymmetric three-gluon events

*Measruements*

Data are available from many groups at DORIS[28,50,51,52] and CESR[33]. I will restrict myself to a short discussion of the PLUTO results. The analysis proceeds in three steps.

- Isolate the direct decay mode.
- Define models.
- Compare the data with these models.

Data were compared with three models:

(1) a phase-space model,
(2) the two-jet Feynman-Field model, and
(3) a three-gluon jet model.

In this latter model the gluons fragment like quarks. The two ingredients of the model are:

- A three-gluon matrix element[48,53] for the production of three massless gluons through an intermediate virtual photon.
- A fragmentation of the gluons with limited $p_T$. The mean $p_T$ is adjusted to fit the two-jet data at 9.4 GeV below the resonance (at comparable jet energies). The charged multiplicity and neutral energy is adjusted to the $\Upsilon$ data.

Fig.36 shows the mean observed thrust in the $\Upsilon$ region compared to lower-energy data. $\Upsilon$ data are given before and after subtraction of the continuum value ($\Upsilon$ direct). We observe a drastic change of topology as one passes through the resonance (see also insert in fig.36). The predictions for the two-jet model and phase space are indicated in the figure. The three-gluon model[54] (not shown in the figure) coincides exactly with the $\Upsilon$ direct data point.

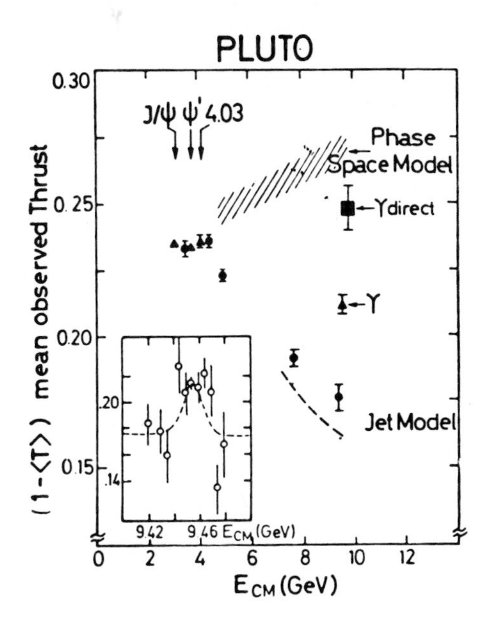


Fig.36. Observed mean thrust for charged particles including the Ypsilon. Data are compared to different model predictions: phase space and to jets. The three-gluon prediction coincides with the $\Upsilon$ direct data point

The thrust distribution on and off resonance is shown in fig.37. The $\Upsilon$ data are well reproduced by the three-gluon model whereas the phase-space models fail to describe the data correctly. From this and similar analyses we can certainly exclude the two-jet decay mode as a major contribution to the $\Upsilon$ decay. Data are in excellent agreement with the three-gluon model. The phase-space description is strongly disfavoured.

*Three Jets?*

In the proposed three-jet structure of $\Upsilon$ decays the mean jet energy would be ~ 3 GeV. We know from the corresponding two-jet data at 6 GeV that a jet struc-

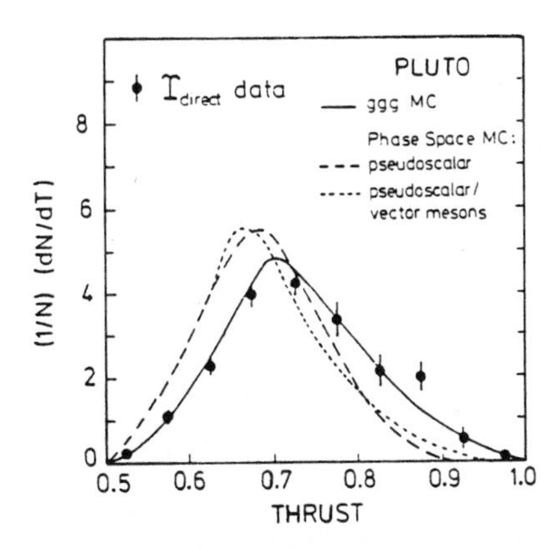


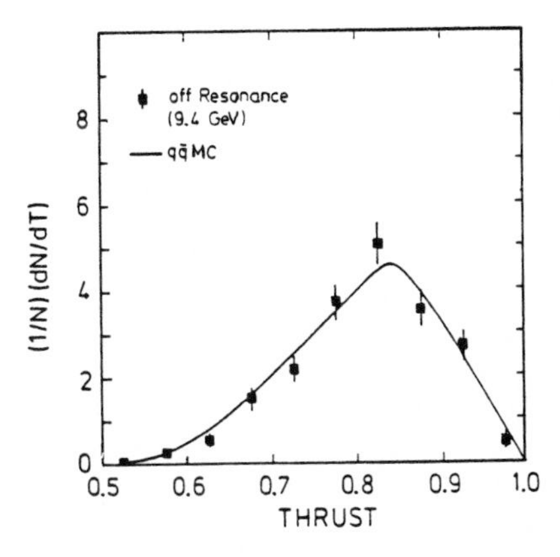


Fig.37. Trust distribution for $\Upsilon$ resonance decays and hadronic events near the resonance at 9.4 GeV. Data are compared with three gluon, two quark and various phase-space models

ture is very hard to defect. The identification of a three-jet structure is even more difficult since the average relative jet angles are smaller than in the two-jet case. Without going through the details of the many attempts to reveal this structure let me just summarize: Although data are well consistent with QCD no positive evidence has been found that 3 jets exist in $\Upsilon$ decays.

*Angular Distribution*

The alignment of the fastest jet with respect to the beam axis strongly depends on the gluon spin. If we accept the 3-gluon hypothesis for $\Upsilon$ decays, we can therefore perform a sensitive test on the spin of the gluons.

Fig.38 shows the angular distribution of the sphericity axis in direct $\Upsilon$ decays as measured by the LENA group[28)]. The data are compared with the predictions for spin 1 (full curve)[48)] and spin 0 (dashed curve) (ref.55).

The data are in good agreement with the vector gluon prediction. A fit to the data with $1 + \alpha \cos^2\Theta$ yields $\alpha = 0.7 \pm 0.3$ (LENA) and $\alpha = 0.83 \pm 0.23$ (PLUTO)[52,56)], which excludes scalar gluons.

*Conclusions*

The event shape encountered in Ypsilon decays strongly deviates from the two-jet structure found in the continuum. All topological quantities studied are in agreement with a three-gluon jet model. A simple phase-space model cannot explain the details of the data: QCD is the only model that offers a satisfactory explanation of all experimental aspects of Ypsilon decays. If we accept the three-gluon picture, we can exclude scalar gluons.

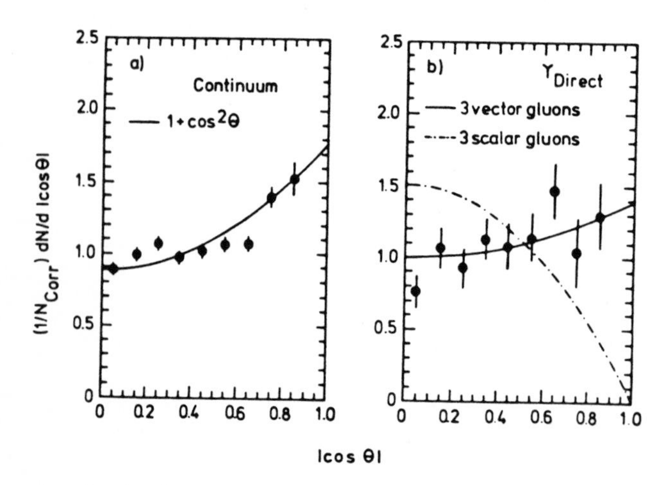


Fig.38a,b. Angular distribution of the thrust axis (LENA group). a) Near the resonance in the $q\bar{q}$ continuum. Data are compared with the expectation for spin 1/2 quarks. b) For resonance decays. Data are compared to the vector (-) and scalar gluon (-·-) prediction

### 1.4.4 QCD in High-Energy Jets

*Quark-Parton Model (QPM)*

The simple quark-parton model has led us throughout most of the preceeding discussions. In this picture electrons and positrons annihilate into a pair of quarks which fragment into jets. The basic assumptions of this model are:

- Quarks behave like leptons with fractional charges at the $\gamma q\bar{q}$ vertex, i.e. the process $e^+e^- \to q\bar{q}$ can be calculated from QED.
- The hadronic nature of quarks is introduced ad hoc: the quarks fragment with limited $p_T$ and thus produce jets.

The predictions of this model are:

$$R = 3 \sum_q Q_q^2 \qquad q = \text{quark flavours}$$

$$\langle p_T \rangle \simeq \text{const} \simeq 320 \text{ MeV}$$

$$\langle S \rangle \simeq \frac{\langle n(E) \rangle \langle p_T \rangle}{E^2} \sim \frac{\ln E}{E^2}$$

$$\langle 1-T \rangle \simeq \frac{\langle n(E) \rangle \langle p_T \rangle}{E} \sim \frac{\ln E}{E}$$

where we assume a logarithmic increase of the mean multiplicity n(E) with energy.

*First-Order Perturbative QCD*

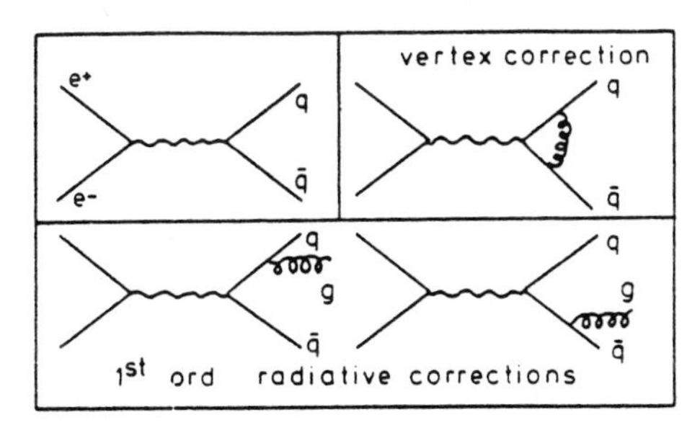


QCD tells us that quark-pair production is only the $0^{th}$ order in a perturbation expansion. In $1^{st}$ order radiative gluon corrections and a vertex correction term have to be added[45]. Like in QED the infrared divergences in the $q\bar{q}g$ terms cancel against those arising from the interference between the vertex correction and the Born term.

The process of gluon emission is governed by the strong coupling constant $\alpha_s$. Once $\alpha_s$ is known, QCD predicts quantitatively where the quark-parton model is a valid approximation and how large the radiative terms are. In this order of QCD the problem of fragmentation is of course not solved. Again both the quark and the gluon fragmentation have to be introduced heuristically. Thus the QPM assumptions change in the following way:

- Including first-order gluonic corrections the annihilation process $e^+e^- \to$ hadrons can be calculated in QCD with only one free parameter, $\alpha_s$.
- Quarks fragment like in the QPM. The fragmentation of gluons is probably softer due to the three-gluon vertex which exists in a non-abelian gauge theory like QCD[57].

*Characteristics of Gluon Emission*

The QCD cross section for gluon emission in the process

$$e^+e^- \to q\bar{q}g$$

in first-order $\alpha_s$ and integrated over all orientations of the $q\bar{q}g$ plane with respect to the beams reads[45]:

$$\frac{1}{\sigma_{q\bar{q}}}\frac{d\sigma(q\bar{q}g)}{dx_q\, dx_{\bar{q}}} = \frac{2}{3}\frac{\alpha_s}{\pi}\frac{x_q^2 + x_{\bar{q}}^2}{(1-x_q)(1-x_{\bar{q}})} \qquad \text{(a)}$$

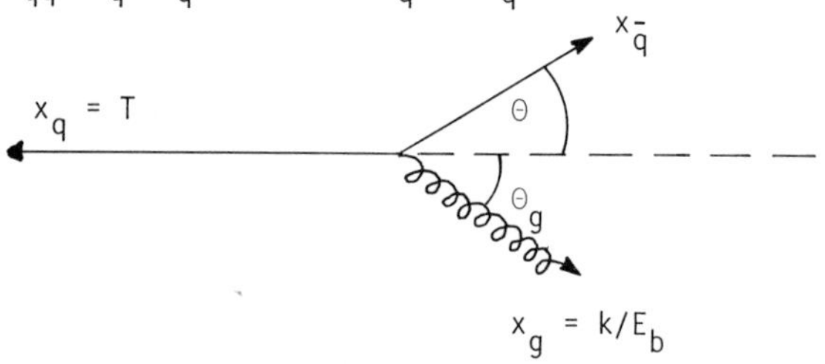

$x_{\bar{q}}$

$x_q = T$

$\Theta$

$\Theta_g$

$x_g = k/E_b$

where $x_i = E_i/E_b$ are the scaled energies of the quarks and the gluon (Feynman x; $E_b$ = beam energy).

For simplicity we assume that the quark q carries the largest momentum. We can easily see that without fragmentation the thrust of a $q\bar{q}g$ event is $T = x_q$. Since $x_q + x_{\bar{q}} + x_g = 2$ we can rewrite (a):

$$\frac{1}{\sigma_{q\bar{q}}}\frac{d\sigma(q\bar{q}g)}{dT\, dx_{\bar{q}}} = \frac{2}{3}\frac{\alpha_s}{\pi}\frac{x_q^2 + x_{\bar{q}}^2}{(1-T)(1-x_{\bar{q}})}$$

$$= \frac{2}{3}\frac{\alpha_s}{\pi}\frac{x_q^2 + x_{\bar{q}}^2}{(1-T)(T-1+x_g)} \qquad \text{(b)}$$

The expression has a singularity for $T \to 1$ when the two quarks are lined up (colinear singularity). In the small angle limit $T \to 1$ and for small gluon momenta ($x_g \ll 1$) (b) reads:

$$\frac{1}{\sigma_{q\bar{q}}} \frac{d\sigma(qqg)}{dT\, dx_{\bar{q}}} = \frac{2}{3}\frac{\alpha_s}{\pi} \frac{T^2 + (2-T-x_g)^2}{(1-T)(T-1+x_g)}$$

$$= \frac{4}{3}\frac{\alpha_s}{\pi}\frac{1}{1-T}\frac{E_b}{k} \qquad \text{(c)}$$

Formula (c) demonstrates the 1/k behaviour of the gluon brems-spectrum. A soft gluon singularity occurs for $k \to 0$.

*Transverse Momentum*

If we integrate over T we get

$$\frac{1}{\sigma_{q\bar{q}}}\frac{d\sigma}{dx_T} = \frac{1}{\sigma_{q\bar{q}}} \int dT \frac{d\sigma(q\bar{q}g)}{dx_T\, dT} \qquad \text{(d)}$$

where we have introduced

$$x_T = x_{\bar{q}} \sin\Theta \, .$$

(d) diverges for $x_T \to 0$. If we take moments, however,

$$\langle x_T^n \rangle = \frac{1}{\sigma_{q\bar{q}}} \int dx_T \int dT\, x_T^n \frac{d\sigma(q\bar{q}g)}{dT\, dx_T}$$

we get finite solutions of the type

$$\langle x_T^n \rangle = \alpha_s(E) \cdot f(x_q, x_{\bar{q}}) \, .$$

$f(x_q, x_{\bar{q}})$ does not depend on the beam energy. Therefore we get

$$\langle x_T^n \rangle \sim \alpha_s(E) \sim 1/\ln (E/\Lambda)^2 ,$$

i.e. $\langle x_T^n \rangle$ varies only slowly with energy. Consequently the moments of the transverse momenta increase with powers of the energy

$$\langle p_T^n \rangle \sim E^n/\ln (E/\Lambda)^2 \, .$$

In particular we get a linear or quadratic increase of the first two moments

$$\langle p_T \rangle \sim E/\ln (E/\Lambda)^2$$

$$\langle p_T^2 \rangle \sim E^2/\ln (E/\Lambda)^2 \, .$$

This is a dramatic deviation from the QPM prediction[58,59,60,61].

*Fixed Angle Jets*

At high energies, the gluon emission with increasing $p_T$ will eventually win over the fragmentation. The broadening of the 'jet' consisting of a quark and a gluon in $1^{st}$-order QCD is then given by an opening angle

$$\langle\Theta\rangle \simeq \frac{\langle p_T\rangle}{E_b} \sim \alpha_s(E) \sim 1/\ln(E/\Lambda)^2.$$

Thus the width of the jet will only depend slightly on energy, a phenomenon being denoted as: 'fixed angle jets' [62].

*Simple Tests of QCD*

From the above considerations a number of conceptually simple tests can be derived (table 4):

| Quantity | QPM ($q\bar{q}$) | + $1^{st}$-order QCD ($q\bar{q}g$) |
|---|---|---|
| R | $3\,\Sigma\,Q_q^2$ | $3\,\Sigma\,Q_q^2\,\frac{\alpha_s}{\pi}$ |
| $\langle p_T\rangle$ | const. | $\sim \alpha_s\,E \sim E/\ln\,(E/\Lambda)^2$ |
| $\langle p_T^n\rangle$ | const. | $\sim \alpha_s\,E^n \sim E^n/\ln(E/\Lambda)^2$ |
| $\langle S\rangle$ | $\sim \ln\,E/E^2$ | $\sim \alpha_s \quad \sim 1/\ln\,(E/\Lambda)^2$ |
| $\langle 1-T\rangle$ | $\sim \ln\,E/E$ | $\sim \alpha_s \quad \sim 1/\ln\,(E/\Lambda)^2$ |

Table 4: Additional contributions expected in $1^{st}$-order QCD compared to the simple QPM predictions

## 1.4.5 Experimental Tests of QCD at High-Energies

### *a) Total Hadronic Cross Section*

The most direct and transparent method to test QCD and to determine the strong coupling constant $\alpha_s$ would be a precision measurement of R. The theoretical value is known up to second order in QCD [63]

$$R = 3\cdot\sum_q e_q^2\,\left(1 + \frac{\alpha_s}{\pi} + C_2\,\left(\frac{\alpha_s}{\pi}\right)^2 + \ldots\right)$$

where $C_2$ depends on the renormalisation scheme chosen. $C_2$ has been calculated in different schemes and turns out to be of O(1). (The precise numbers are known but not essential for this discussion.)

Unfortunately the experimental determination of R implies an absolute measurement of the hadronic cross section. The systematic error on this number is still about 10%, i.e. of the same order as the increase in R expected from QCD. The mean value of R in the energy range around 30 GeV, R = 3.97 $\pm$ 0.06 (stat.)$\pm$ 0.4 (syst.)[64] is in good agreement with QCD for $\alpha_s \simeq$ 0.2 to 0.3. A decisive test, however, would require an accuracy of at least 5%.

*b) Rising Transverse Momentum*

The mean transverse and parallel momentum with respect to the thrust axis as a function of energy is shown in figs. 22, 39. The TASSO group[65] has determined the thrust axis from charged particles only, whereas the PLUTO data[66] are evaluated with respect to the common thrust axis of charged and neutral momentum. In both figures, the mean transverse momentum $\langle p_T \rangle$ and the mean parallel momentum $\langle p_L \rangle$ are calculated from charged tracks only. Apart from the gross features of the

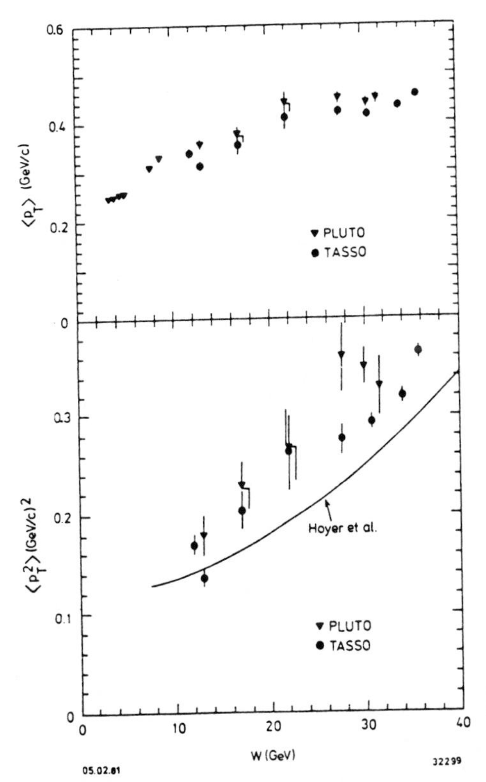


Fig.39. Energy dependence of $\langle p_\perp \rangle$ and $\langle p_\perp^2 \rangle$. The full curve is a QCD prediction made prior to the data (ref.58)

data - bounded $<p_T>$ and rising $<p_L>$ - both figures indicate a slight increase of $<p_T>$ with energy. This increase becomes dramatic if we plot the second moment $<p_T^2>$, which gives more weight to high momentum particles.

Only a small part of this increase can be accounted for by detector effects. This can be checked by comparing the data with the Feynman-Field model[67] for uds quarks, which has no inherent energy dependence of $<p_T^2>$. The TASSO data (fig.39) are corrected for this effect. The uds expectation would then be a constant $<p_T^2>$, in apparent contradiction to the data.

An increase in $p_T$ is expected from new heavy quark production. In fact, including detector effects and c and b quarks, the energy dependence is quite strong, accounting for roughly half of the increase in the PLUTO data (dashed line in fig.22). On the other hand including QCD effects the data are well reproduced in fig.39. The full line in fig.39 shows a QCD prediction which was made prior to the data and does not include heavy quarks. The trend of the data is well reproduced.

In summary, $<p_T^2>$ increases roughly by a factor of 2 between 10 and 30 GeV. This effect cannot fully be explained by detector effects and new quark flavours. Fig.40 shows a comparison of the $p_T^2$ distributions at low and high PETRA energies. To account for the data the parameter $\sigma_q$ in

$$d\sigma/dp_T^2 \sim e^{-p_T^2/2\sigma_q^2}$$

which describes the transverse spread of the momentum distribution in the Feynman-Field model has to be increased from $\sigma_q$ = 0.3 GeV at 13 and 17 GeV to $\sigma_q$ = 0.45 GeV in the high-energy range.

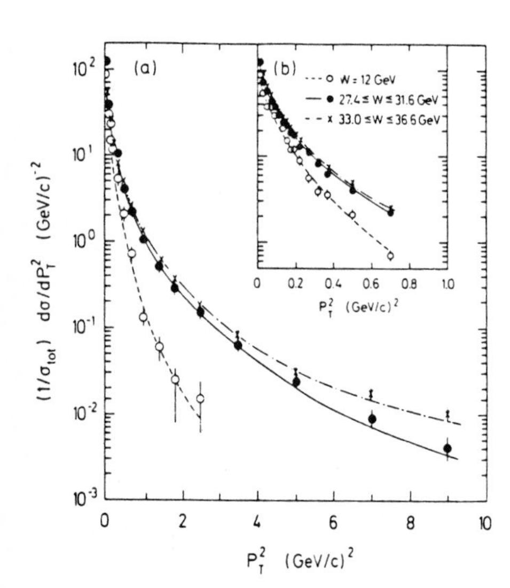


Fig.40. Distribution of $p_\perp^2$ of charged particles at different energies

### *c) Planar Event Structure*

Gluon bremsstrahlung would induce a planar event structure (compare section IV.3). Several methods have been used to investigate event shapes: the generalized sphericity[24] and the Q-parameters[68,69], the triplicity[70], the energy flow in terms of minor, major and oblateness[71]. I will only describe one of them, which has been applied by many groups:

*Generalized Spericity*

Let us look at the expression

$$T_{\alpha\beta} = \sum_i (\delta_{\alpha\beta}(p^i)^2 - p^i_\alpha\, p^i_\beta) \quad \alpha,\beta = 1,2,3$$

defined in analogy to the inertia tensor. If we diagonalize this expression we obtain the (normalized) eigenvalues $\lambda_k$ which correspond to the three main axes of the event in momentum space (fig.41). If we order these eigenvalues such that

$$\lambda_1 \geq \lambda_2 \geq \lambda_3$$

then $\lambda_3$ closely resembles our well-known sphericity definition

$$\lambda_3 = \frac{\sum_i (p^i_1)^2 + (p^i_2)^2}{\sum (p^i)^2} = \frac{\sum (p^i_T)^2}{\sum (p^i)^2} \; ; \; S = 3\lambda_3/(\lambda_1 + \lambda_2 + \lambda_3)$$

The physical meaning of $\lambda_3$ is again best understood from an analogy with the inertia tensor. $\lambda_3$ points in the direction of the smallest inertia moment in momentum space. To measure the flatness of events we have to study the other two eigenvalues, in particular $\lambda_1$, which points in the direction of the smallest extent of the event in momentum space.

It is convenient to define the following quantities[52,68] (fig.42):

$$Q_k = 1 - \frac{2\lambda_k}{\lambda_1 + \lambda_2 + \lambda_3} = \frac{\sum_i (p^i_k)^2}{\sum_i (p^i)^2}$$

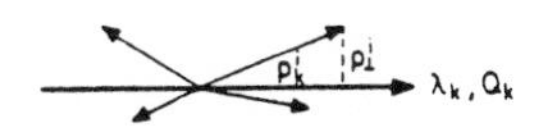


$\lambda_k \sim \sum_i (p^i_\perp)^2$

$Q_k \sim \sum_i (p^i_k)^2$

Fig.41. The spericity tensor in analogy to the inertia tensor. Definition of $\lambda_k, Q_k$

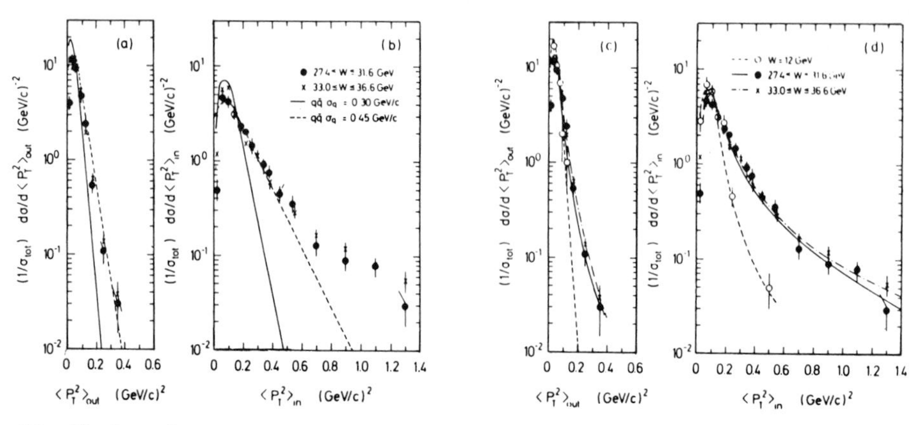

Fig.42. Mean transverse momentum spread in and normal to the event plane

$<p_{in}>$ *and* $<p_{out}>$

We recall the eigenvalues of the sphericity tensor $\lambda_k$ with $\lambda_1 \leq \lambda_2 \leq \lambda_3$ and introduce the corresponding eigenvectors $\vec{n}_1$, $\vec{n}_2$ and $\vec{n}_3$.

In a disklike structure, the normal to the event plane is given by $\vec{n}_1$. The vector $\vec{n}_2$ lies in the disk and is normal to the sphericity vector $\vec{n}_3$. The following quantities are defined[65]:

$$<p^2_{out}> = <\vec{p}\,\vec{n}_1)^2>$$

measures the mean momentum component pointing out of the plane, whereas

$$<p^2_{in}> = <\vec{p}\,\vec{n}_2)^2>$$

gives the mean momentum component in the plane perpendicular to the main event axis $\vec{n}_3$.

Fig.42 shows as an example the experimental distributions obtained by the TASSO collaboration[65]. We notice that, due to the method, the $<p^2_{out}>$ distributions are narrower than the ones for $<p^2_{in}>$. The width of the distributions increases with energy. Whereas $<p^2_{out}>$ and $<p^2_{in}>$ at low energies are well described by the conventional $q\bar{q}$ model (udscb), this model fails to describe the high-energy data. Increasing $\sigma_q$ to 450 MeV as suggested by the $p_T$ distribution (fig.40) gives a reasonable account of the data up to $<p^2_T> \simeq 0.8$ GeV (fig.42 a,b, dashed line). It does not reproduce the high momentum tail, however: this tail can be well explained by QCD (fig.42 c and d).

*Aplanarity*

The normalized eigenvalues

$$Q_k = \frac{\Sigma (p_k^i)^2}{\Sigma (p^i)^2}$$

defined above which measure the total momentum component along the axis $\vec{n}_k$ can be used to define a new variable

$$A = 3/2\ Q_1 = 3/2\ \langle p_{out}^2 \rangle / \langle p^2 \rangle,$$

the aplanarity[69]. In terms of $Q_k$ the sphericity can be written as

$$S = 3/2\ (Q_1+Q_2) = 3/2\ \langle p_T^2 \rangle / \langle p^2 \rangle.$$

Since $Q_1 + Q_2 + Q_3 = 1$, events can be plotted in a Dalitz-like triangle. Fig.43 shows the result for the low- and high-energy data separately[65]. At both energies, most events fall into the jet corner at low sphericity. In the high energy bin however, events congregate along the triangle edge of low aplanarity.

Fig.44 shows a similar analysis of the CELLO group[72]. If one eliminates the two-jet data with $S < 0.25$ the remaining aplanarity distribution is shown in fig.45. The two energy bins at 22 and 34 GeV result from high statistic runs in 1981. At both energies the simple $q\bar{q}$ model cannot explain the data (dashed curve). The discrepancy is larger in the high-energy bin. Including QCD effects (full curve) the measured distributions can be well described.

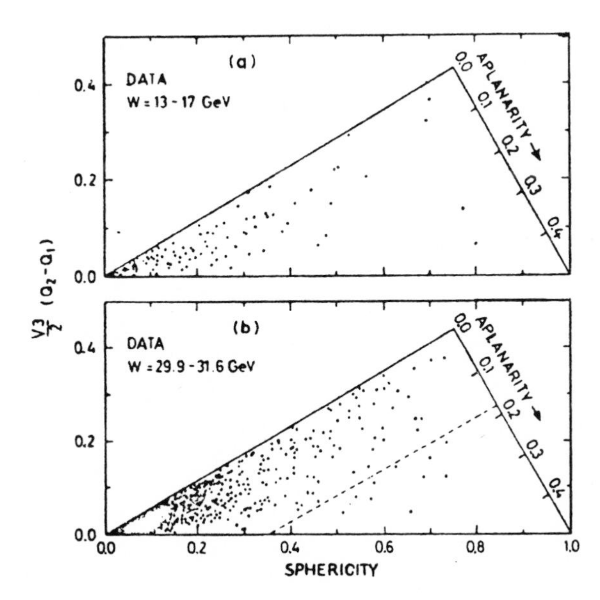


Fig.43a,b. Distribution of events in a scatter plot of aplanarity and sphericity. Data are shown for
a) low energies (13-17 GeV)
b) high energies (29.9-31.6 GeV)

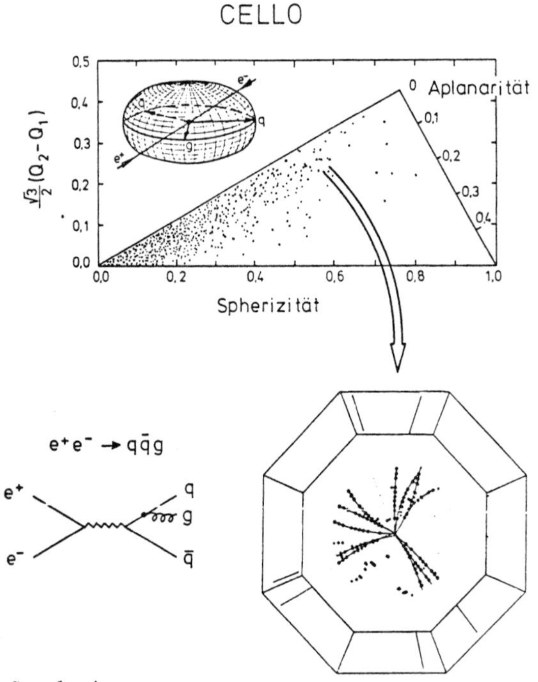


Fig.44. Generalized sphericity analysis of the CELLO data at about 34 GeV. The figure illustrates the selection of a hard gluon bremsstrahlung event (3-jet event)

*Conclusion*

In conclusion, all data indicate an increasing number of events with planar structure which cannot be accounted for by a simple rise in $\langle p_T^2 \rangle$.

## *d) Three-Jet Events*

We have seen that the number of planar events is well compatible with the expectation of QCD. If $q\bar{q}g$ is the source of these events, there should be some hard gluon bremsstrahlung events with a visible three-jet structure. (At the $\Upsilon$ there was no chance to observe 3 jets, due to the low energy. At high PETRA energies about 10 GeV per jet should be sufficient to see this structure.) Fig.44 shows an example of such an event in the CELLO data. Fig.46 shows a corresponding example of 3-jet events from JADE.

The differential cross section for gluon bremsstrahlung is given by[73,74]

$$\frac{1}{\sigma_0}\frac{d\sigma}{dx_1 dx_2} = \frac{2\alpha_s}{3\pi}\frac{x_1^2 + x_2^2}{(1-x_1)(1-x_2)} \qquad (1)$$

q̄, $x_2$, q, $x_1$, g, $x_3$

$$x_1 = E_1/E_{beam}$$

$$\Sigma\, x_i = 2 \, .$$

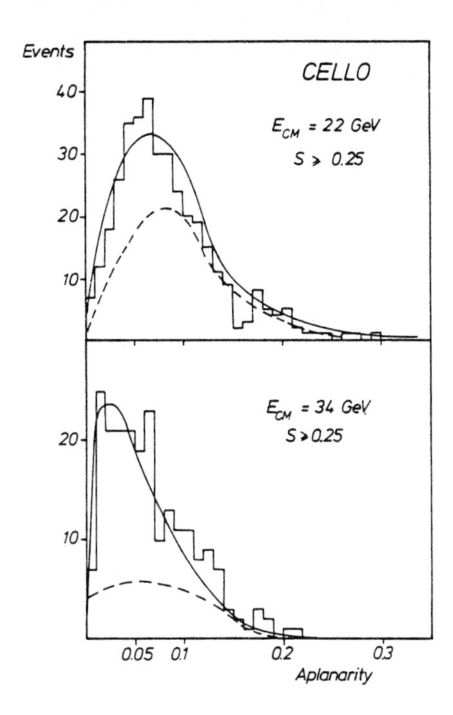


Fig.45. Aplanarity distribution at two different energies compared to $q\bar{q}$ (--) and $q\bar{q}$ + $q\bar{q}g$ (—) model predictions. $\alpha_s$ = 0.20 at 22 GeV and $\alpha_s$ = 0.16 at 34 GeV

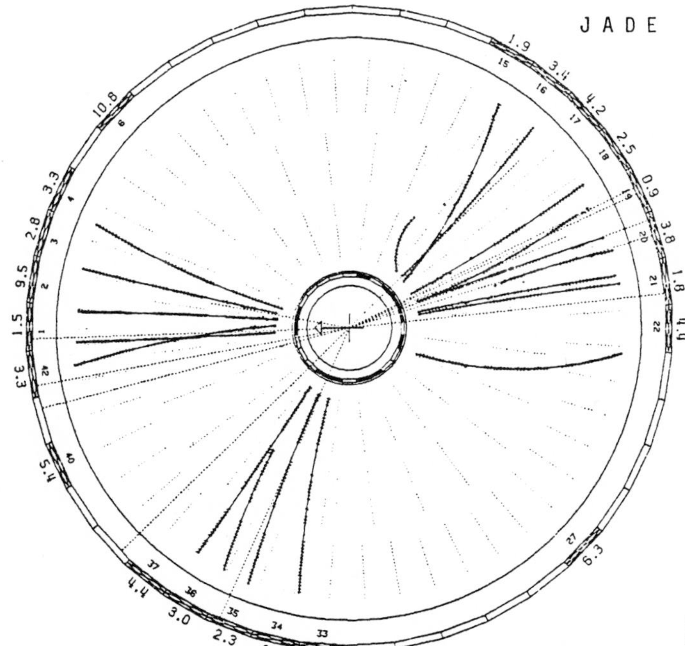


Fig.46. 3-jet event seen in the JADE detector

*e) Detailed Comparison with QCD*

This formula holds for the vector gluons(spin 1 gluons) of QCD. To test the sensitivity of this distribution one may compare the data with a 'scalar gluon'(spin 0 gluon) model which yields[73]

$$\frac{1}{\sigma_0}\frac{d\sigma}{dx_1\,dx_2} = \frac{\alpha_s}{3\pi}\frac{x_3^2}{(1-x_1)(1-x_2)} \cdot$$

The available statistics are not yet sufficient to check this double differential cross section in detail.

However, the differential cross section with respect to the fastest parton $\frac{1}{\sigma_0}\frac{d\sigma}{dx_1}$ has been tested by several groups[72,75,76,77].

The experimental procedure implies several steps:

- select an appropriate sample of three-jet events
- determine the energy fractions in the three jets and choose the most energetic one
- correct for detector efficiency and $q\bar{q}$ 'background' in the three-jet sample using MC methods.

As an example, the CELLO data at 34 GeV will be discussed[72]. All events are submitted to a cluster algorithm[78], in which adjacent particles (neutral and charged) are lumped together to form clusters of particles. These clusters are called jets if their energy exceeds 2 GeV. In events with 3 jets the most energetic one is selected.

Fig.47 shows the differential cross section with respect to the energy fraction of this most energetic jet corrected for efficiency and 2-jet background. The data are compared to vector gluon and scalar gluon predictions. Excellent agreement is obtained with QCD (spin 1) whereas spin 0 can be safely ruled out.

*f) Determination of $\alpha_s$*

In the preceeding section, $\alpha_s$ is determined by the absolute normalization of the distribution. Generally speaking, $\alpha_s$ can be determined by the number of 3-jet over 2-jet events. Different methods have been used by the PETRA groups; the most important ones are Q-plot analysis, cluster analysis, energy flow and energy-energy correlations. Table 5 summarizes the results of the 5 PETRA and the MARK II groups.

The mean value at 30 GeV is (all errors added quadratically)

$\alpha_s \simeq 0.17 \pm 0.01 \pm 0.02$ ($1^{st}$-order QCD)

$\Lambda \simeq 0.2$ GeV ($1^{st}$-order QCD)

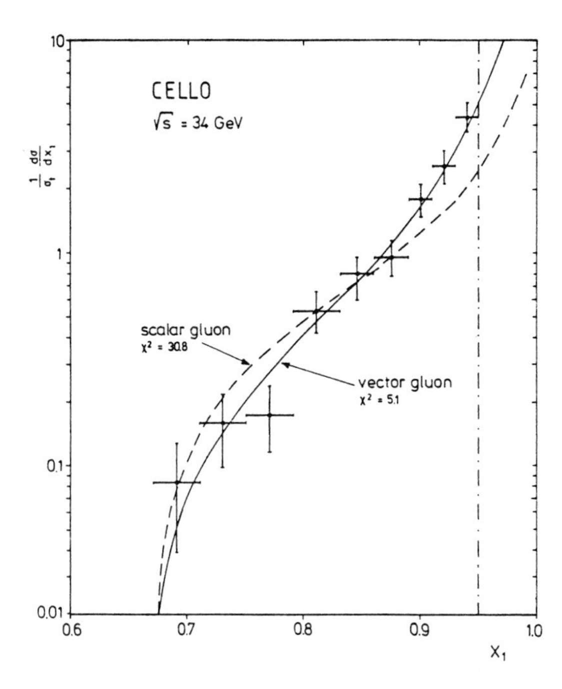


Fig.47. Differential cross section of three-jet events with respect to the energy fraction $x_1$ carried by the most energetic parton. Data are compared with the QCD prediction of vector gluons (—) and a scalar gluon model (---)

where the second error estimates the systematic uncertainties, e.g. in the details of the fragmentation process and the $p_\perp$ distribution in particular.

An additional uncertainty comes into the game if higher-order corrections are applied. Different groups numerically disagree in the amount of these corrections by ~ 30%[84].

| Collaboration | Ref. | $\alpha_s$ ± (stat.) ± (syst.) | Comment |
|---|---|---|---|
| CELLO | 72 | 0.16 ± 0.02 ± 0.03 | 34 GeV |
| | | 0.19 ± 0.04 ± 0.04 | 22 GeV |
| JADE | 79 | 0.18 ± 0.03 ± 0.03 | |
| MARK-J | 83,77 | 0.17 ± 0.02 | |
| PLUTO | 75,80, 77 | 0.16 ± 0.02 | |
| TASSO | 81 | 0.17 ± 0.02 ± 0.03 | |
| MARK II | 82 | 0.18 ± 0.015 ± 0.03 | from energy correlation asymmetry |

Table 5: Experimental values of $\alpha_s$ from gluon bremsstrahlung around 30 GeV cms energy

## 1.5 Summary

### 1.5.1 Quarks

(i) The naive quark-parton model (QPM) is strongly supported by

- deep inelastic scattering
- $e^+e^-$ annihilation into hadrons

Hadronic reactions like high $p_\perp$ phenomena and Drell-Yan processes (which were not mentioned in detail) can be fairly described.

(ii) We learn from deep inelastic scattering that the nucleon is composed of spin 1/2 pointlike particles:

- three-valence quarks
- $q\bar{q}$ pairs of sea quarks
- and gluons which take 55% of the total momentum.

The momentum distribution of valence and sea quarks is well known experimentally.

(iii) From $e^+e^-$ annihilation into hadrons we know that at least 5 flavours of point-like coloured spin 1/2 quarks exist (see table 2). The sixth quark t, which would establish a nice quark-lepton symmetry, has not been found so far. This poses a limit of > 18 GeV on the top quark mass.

### 1.5.2 Gluons

(i) Deep inelastic scattering
Scaling violations are consistent with QCD. The suggested value for the QCD scale parameter is

$\Lambda \simeq 100 - 300$ MeV.

The determination of $\Lambda$ is difficult since important higher-order twist contributions may be present in the data.

(ii) $e^+e^-$ annihilation
$\Upsilon$ decay data are in agreement with QCD. In first order we get

$\alpha_s \simeq 0.18$ at $\simeq 10$ GeV

$\Lambda \simeq 0.1$ GeV .

There is however no proof for planar events. Higher-order corrections to $\alpha_s$ may be important.

Event shapes in high-energy $e^+e^-$ annihilations show clear deviations from a naive QPM two-jet structure.

All features of the data are in quantitative agreement with QCD predictions yielding

$\alpha_s \simeq 0.17$ at $\simeq$ 30 GeV

$\Lambda \simeq 0.2$ GeV

Systematic uncertainties are 20 - 30% both from experiment (fragmentation) and theory (higher-order corrections).

(iii) So far, QCD has been overwhelmingly successful. The next crucial test will be the existence of the three-gluon vertex, e.g. through evidence for glueballs.

*Acknowledgements*

I am indebted to my colleagues in Karlsruhe who helped me to prepare this talk. In particular I am grateful to D.C.Fries, R.Maschuw and H.Müller for critically reading and G.Grundel for carefully typing the manuscript. I want to thank the organizers, in particular D.C.Fries and B.Zeitnitz, for their kind invitation. I very much appreciated the lively atmosphere of this school.

References

1) M. Gell-Mann Phys.Lett. 8 (1964) 214
G. Zweig, CERN report 8182/Th.401 (1964)

2) J.J. Aubert et al.,Phys.Lett. 33 (1974) 1404
J.-E. Augustin et al.,Phys.Rev.Lett. 33 (1974) 1406

3) S.W.Herb et al., Phys.Rev.Lett. 39 (1977) 252
W.R. Innes et al., Phys.Rev.Lett. 39 (1977) 1240

4) H. Fritzsch, M. Gell-Mann and H. Leutwyler, Phys.Lett. 47B (1973) 365
D.J. Gross and F. Wilczek, Phys.Rev.Lett. 30 (1973) 1343
H. Politzer, Phys.Rev.Lett. 30 (1973) 1346

5) D.P. Barber et al., Phys.Rev.Lett. 43 (1979) 830
TASSO-Collaboration, R. Brandelik et al.Phys.Lett. 86B (1979) 243
PLUTO-Collaboration, Ch. Berger et al.Phys.Lett. 86B (1979) 418

6) Review article: e.g.
K. Heinloth, DESY-report 72/5 (1972)

7) J.D. Bjorken, Phys.Rev. 179 (1969) 1547

8) E.D. Bloom et al., Phys.Rev.Lett. 23 (1969) 930
M. Breidenbach et al., Phys.Rev.Lett. 23 (1969) 935

9) Excellent reading: F.E. Close, Introduction to Quarks and Partons, Academic Press, London, 1979

10) J.G.H. de Groot et al., Phys.Lett 82B (1979) 292,
Phys.Lett. 82B (1979) 456
Z.Physik C1 (1979) 143
J.G.H. de Groot (CDHS Coll.) Int.Conf. on High-Energy Physics, Madison (1980)

11) M.D. Mestayer, Thesis, SLAC Report 214 (1978)

12) G. Callan and D.J. Gross, Phys.Rev.Lett. 22 (1969) 156

13) J. Steinberger, Int.School of Subnuclear Physics, Erice, Italy (1980)
CERN-EP/80-222 (1980)

14) J.J. Aubert et al., XX Int.Conf. on High-Energy Physics, Madison, Wisconsin (1980)

15) Recent review article: P.Söding and G. Wolf, DESY-Report 81/013 (1981)

16) J.D. Fox et al., Phys.Rev.Lett. 33 (1974) 1504
E.M. Riordan et al., Phys.Lett. 52B (1974) 249
W.B. Atwood et al., Phys.Lett 64B (1976) 479
H.L.Anderson et al., Phys.Rev.Lett. 38 (1977) 1450

17) J.B. Kogut, L. Susskind, Phys.Rev. D9 (1974) 697, 3391

18) D.J. Gross, F. Wilczek, Phys.Rev.Lett. 30 (1973) 1343, Phys.Rev. D8 (1973) 3633, D9 (1974) 980

19) Recent review articles: G. Flügge, KfK-Bericht 2995 (1980)
G. Wolf, DESY-report 80/13 (1980); B. Wiik, DESY-report 80/129 (1980)
P. Duinker, DESY-report 81/012 (1981); H. Meyer, University of Wuppertal preprint WUB 81/16

20) CELLO-Collaboration, H.-J. Behrend et al.Phys.Scripta 23 (1981) 610

21) G. Hanson et al., Phys.Rev.Lett. 35 (1975) 1609
G. Hanson, XIII Rencontre de Moriond, Les Arcs (1978) and SLAC-PUB-2118 (1978)

22) PLUTO Coll., Ch. Berger et al., Phys.Lett. 78B (1978) 176

23) J.D. Bjorken and S.J. Brodsky, Phys.Rev. D1 (1970) 1416

24) E. Fahri, Phys.Rev.Lett. 39 (1977) 1587. See also S. Brandt,
Ch. Peyrou, R. Sosnowski and A. Wroblewski, Phys.Lett. 12 (1964) 57
S. Brandt and H. Dahmen, Z.Physik C1 (1979) 61

25) PLUTO Coll., Ch. Berger et al., Phys.Lett. 81B (1978) 410, 86B (1979) 413, and 86B (1979) 418

26) TASSO Coll., R. Brandelik et al., Phys.Lett. 83B (1979) 261, 86B (1979) 243, and 89B (1980) 418

27) D. Barber et al., Phys.Rev.Lett. 42 (1979) 1110, 1113 and
D. Barber et al., MIT-LNS Report 107 (August 1979)

28) LENA Coll., Niczyporuk et al., DESY-report 81/008 (1981)
LENA Coll., Niczyporuk et al., Phys.Rev.Lett. 46 (1981) 92

29) S. Okubo, Phys.Lett. 5 (1963) 163
V. Zweig, CERN report 8419/Th 412 (1964)
J. Iizuka, Proc.Theor.Phys.Suppl. 37-38, p.21 (1966)

30) R. Partridge et al., Phys.Rev.Lett. 44 (1980) 712

31) R. Partridge et al., Phys.Rev.Lett. 45 (1980) 1150

32) T.M. Himel et al., Phys.Rev.Lett.45 (1980)1146

33) D. Schamberger, 1981 Int.Symp. on Lepton and Photon Interactions at High Energies, Bonn 1981

34) G. Goldhaber et al., Phys.Rev.Lett. 37 (1976) 255
I. Peruzzi et al., Phys.Rev.Lett. 37 (1976) 569

35) DASP Coll., R. Brandelik et al., Phys.Lett. 70B (1977) 132 and 80B (1979) 412

36) W. Lockmann et al., Phys.Lett. 85B (1979) 443
D. Drijard et al., Phys.Lett. 85B (1979) 452
K. Giboni et al., Phys.Lett. 85B (1979) 437
C. Baltay et al., Phys.Rev.Lett. 42 (1979) 1721
G.S. Abrams et al., Phys.Rev.Lett. 44 (1980) 10

37) C. Angelini et al., Phys.Lett 80B (1979) 428, 84B (1979) 150
J. Sandweiss et al., Phys.Rev.Lett.44 (1980) 1104
L. Voyvodic, 1979 Int.Symp. on Leptons and Photons at High Energies,FNAL, 1979

38) PLUTO Coll., Ch. Berger et al., Phys.Lett. 76B (1978) 243
C.W. Darden et al., Phys.Lett. 76B (1978) 246

39) D. Andrews et al., Phys.Rev.Lett. 45 (1980) 219
G. Finocchiaro et al., Phys.Rev.Lett. 45 (1980) 223

40) A. Silverman, 1981 Int.Symp. on Lepton and Photon Interactions at High Energies, Bonn, 1981

41) P.W. Higgs, Phys.Rev.Lett. 12 (1964) 132
Further reading e.g.:
Ch. Quigg, NATO Advanced Study Institute, St.Croix, 1980

42) JADE Coll., W. Bartel et al., Phys.Lett. 100B (1981) 364
CELLO Coll., H.-J. Behrend et al., Phys.Lett. (1981)
MARK-J Coll., D.P.Barber et al., MIT-LNS Report 115 (1981)

43) G. Flügge, Z. Physik C1 (1979) 121 and Proc.of the Int. Conf. on HEP, Geneva 1979
J. Kirkby, 1979 Int. Symp. on Lepton and Photon Interactions at High Energies, FNAL, 1979

44) T. Appelquist, H.D. Politzer, Phys.Rev. D12 (1975) 1404, Phys.Rev.Lett. 34 (1975) 43

45) A.M. Polyakov, Int.Symp. on Lepton and Photon Interactions at High Energies, Stanford, 1975
J. Ellis, M.K. Gaillard and G. Ross, Nucl. Phys. B111 (1976) 253
T.A. de Grand, Y.J. Ng., S.H.H. Tye, Phys.Rev. D16 (1977) 3251

46) M. Krammer,H. Krasemann, Int. Universitätswochen, Schladming, 1979

47) R. Barbieri et al., Nucl. Phys. B154 (1979) 535

48) K. Koller, H. Krasemann, T.F. Walsh, Z. Physik C1 (1979) 71
K. Koller, T.F. Walsh, Phys.Lett. 72B (1977) 227, 73B (1978) 504,and Nucl. Phys. B140 (1978) 449

49) T.A. de Grand et al., Phys.Rev. D16 (1977) 3251
S. Brodsky et al., Phys.Lett. 73B (1978) 203
H. Fritzsch, K.H. Streng, Phys.Lett. 74B (1978) 90

50) F.H. Heimlich et al., Phys.Lett. 86B (1979) 399

51) C.W. Darden et al., Int.Conf. on High-Energy Physics, Tokyo, 1978

52) PLUTO Coll., Ch. Berger et al., Phys.Lett. 82B (1979) 449
PLUTO Coll., Ch. Berger et al., Z. für Physik C8 (1981) 101

53) A. Ore, J.L. Powell, Phys.Rev. 75 (1949) 1696

54) Compare also: K. Hagiwara, Nucl.Phys. B137 (1978) 164

55) K. Koller, H. Krasemann, Phys.Lett 88B (1979) 119

56) H. Meyer, 1979 Int.Symp. on Lepton and Photon Interactions at High Energies, FNAL, 1979

57) I.I.Y. Bigi, R.F. Walsh, Phys.Lett. 82B (1979) 267

58) C.L. Basham et al., Phys.Rev. D17 (1978) 2298
F. Steiner, DESY 78/59 (1978)

59) G. Kramer, G. Schierholz, Phys.Lett. 82B (1979) 108

60) P. Hoyer, P. Osland, H.G. Sander, T.F. Walsh, P.M. Zerwas, DESY 79/21, to be published

61) A. de Rujula, J. Ellis, E.G. Floratos and M.K. Gaillard, Nucl.Phys. B138 (1978) 387

62) G. Sterman and S. Weinberg, Phys.Rev.Lett. 39 (1977) 1436

63) M. Dine and J. Sapirstein, Phys.Rev.Lett. 43 (1979) 668
K.G. Chetryrkin et al., Phys.Lett. 85B (1979) 277
W. Celmaster and R.J. Gonsalves, Phys.Rev.Lett. 44 (1979) 560

64) D. Cords, XX Int.Conf. on High-Energy Physics, Madison 1980

65) TASSO Coll., R. Brandelik et al., Phys.Lett. 86B (1979) 243
TASSO Coll., R. Brandelik et al., Phys.Lett. (1981)

66) PLUTO Coll., Ch. Berger et al., Phys.Lett. 86B (1979) 418

67) R.D. Field and R.P. Feynman, Phys.Rev. D15 (1977) 2590, Nucl.Phys. B136 (1978) 1

68) G. Alexander, XIX Int. Conf. on High-Energy Physics, Tokyo, 1979

69) S.L. Wu, G. Zobernig, Z.Physik C2 (1979) 107

70) S. Brandt and H. Dahmen, Z.Physik C1 (1979) 61

71) D.P. Barber et al., Phys.Rev.Lett. 43 (1979) 830

72) CELLO Coll., H.-J. Behrend et al., DESY report 81/080 (1981)

73) J.Ellis et al., Nucl.Phys. B111 (1976) 253

74) T.A. de Grand et al., Phys.Rev. D16 (1977) 3251

75) PLUTO Coll., Ch. Berger et al., Phys.Lett. 97B (1980) 459

76) TASSO Coll., R. Brandelik et al., Phys.Lett. 97B (1980) 437

77) W. Braunschweig, Int.Symp. on Lepton and Photon Interactions at High Energies, Bonn, 1981

78) H.J. Daum et al., Z.Physik C8 (1981) 167
J.Dorfan, Z.Physik C7 (1981) 349
K. Lanius et al., Z.Physik C8 (1981) 251

79) JADE Coll., W. Bartel et al., Phys.Lett. 91B (1980) 142

80) PLUTO Coll., Ch. Berger et al., Phys.Lett. 99B (1980) 292

81) TASSO Coll., R. Brandelik et al., Phys.Lett 94B (1980) 437

82) R.Hollebeek, Int.Symp. on Lepton and Photon Interactions at High Energies, Bonn, 1981

83) MARK J Coll., D.P. Barber et al., Phys.Rev.Lett. 43 (1979) 830
H. Newman, XX Int.Conf. on High-Energy Physics, Wisconsin, 1980

84) A. Ali et al., Phys.Lett. 93B (1980) 155, Nucl.Phys.B168 (1980) 409
R.K. Ellis et al., Phys.Rev.Lett. 45 (1980) 1226 and Nucl.Phys. B178 (1981) 421
Fabricius et al., Phys.Lett. 94B (1981) 431
Vermaseren et al., CERN rep. Th 3002

# 2. The Pauli Principle and QCD for Quarks and Nucleons in Hardons and Nuclei

F. E. Close

With 3 Figures

## 2.1 Introduction

Quarks are constituents of nucleons, which in turn are the building blocks of nuclei. Now that high-energy physicists believe that they have a (correct?) theory of quark forces, at least within nucleons, the obvious question is, can we exploit this theory to build insight into the nature of nuclear forces?

Other speakers here will address themselves more directly to this problem. I shall detail our knowledge of how the quarks build up baryons[1] and see what hints or problems emerge for the extension to nuclear physics.

The crucial property possessed by quarks, but not by nucleons, is that they carry "colour". Any pair of quarks in a nucleon is always antisymmetric under interchange of their colour, whereas nucleons are always symmetric in colour (trivially so because they have no net colour!). Thus the Pauli principle places constraints on quark clusters that are in a sense the precise opposite of those on nucleon clusters. The structure of nucleons and nuclei are intimately affected by this profound subtlety, as we shall see.

I shall begin by telling you about colour and how it plays a role in building up hadrons from quarks. The similarity between colour and electrical charge is manifested by the similarity between the respective quantum field theories of quantum chromodynamics[2] (QCD) and quantum electrodynamics (QED). In turn, nuclear forces may be to QCD what molecular forces are to QED.

This will lead us to compare and contrast quark clusters and nucleon clusters (i.e. hadrons and nuclei). The role of colour and the Pauli principle is dramatically seen by comparing the magnetic moments of clusters of three nucleons ($H^3$ and $He^3$) with those of three quarks (N,P). This is very intimately connected with symmetries: magnetic moments probe not just the internal charge but the way that these charges are correlated with spin - this is where symmetry, notably the Pauli principle, plays its role.

This will bring us to study the magnetic moments of the octet of baryons. There are some 20% deviations between theory and experiment[3]. Is this a problem?

In nuclear physics there are also such discrepancies - should the particle physicists expect tc do any better? In nuclei the induced exchange of pions is regarded as responsible (a); the induced gluon exchange (b) may play an analogous role for baryons (the gluon has no flavour, but it does have spin, and this is probed in the magnetic moment). Whether or not

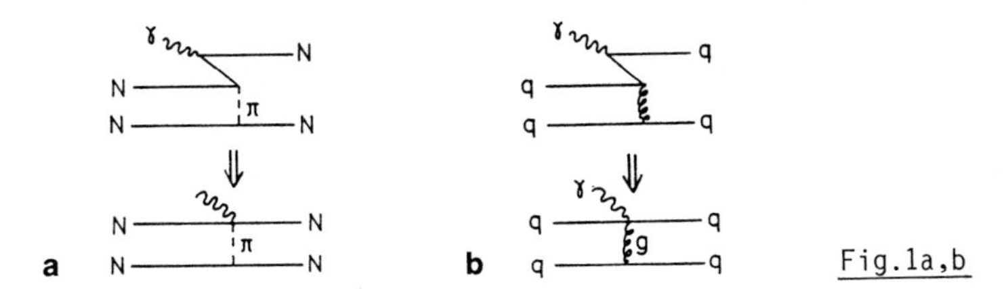


Fig.1a,b

this is important is moot, but the deviations from the naive picture do exhibit some systematic pattern which I will spell out. The problem (not solved) is what is the cause of this pattern.

Next we will study the empirical clues that hadron spectroscopy gives on the nature of the forces between quarks at long range. For heavy quarks a potential picture is probably reasonable. This will be discussed by Professors Heller(4) and Ono(5). Can one continue to apply potential models to light quarks? A priori this seems a hopeless task, yet many have done so and a significant measure of success has resulted over several years. If potential models supplemented by QCD give a good phenomenology of hadron spectroscopy, then we may hope to extend them to the realm of nuclear forces. This is the reason why I shall discuss potential models for light quarks here. QCD generates a single gluon exchange between quarks which perturbs the Hamiltonian and has consequences for hadron masses. For S-wave quark clusters this is very successful, but for P-waves there are some unsolved problems. Isgur and Karl(6) have shown us how an excellent fit to P-wave baryons can be obtained within QCD if one throws away some of the effects that QCD predicts should be there(7). I shall give a possible rationale for this and discuss some of the problems - theoretical and empirical - that must be solved before QCD can be reasonably expected to give a handle on nuclear forces.

## 2.2 Onium Atoms and Onium Hadrons

In QED there is one type of charge, described by positive or negative numbers. Opposites attract and orbit about one another forming neutral atoms. In positronium or hydrogen the excitation energies are O (eV).

Mesons are formed from quark and antiquark ("quarkonium"). Their radii are of order $10^6$ smaller than atoms; hence electromagnetic effects will be correspondingly

$10^6$ times larger. This brings us to O (MeV) - the size of the $\pi^o\pi^+$ mass difference which is indeed due to the differing electromagnetic interactions between their quarks (the quarks in $\pi^+$ and $\pi^o$ have different charges, hence lead to a net different strength of electromagnetic force within).

The excitation energies of quarkonia levels are of order hundreds of MeV. Hyperfine splittings and spin-orbit splittings are of order tens to hundreds of MeV. This shows that the forces acting between quarks in these situations are order 10-100 times stronger than electromagnetic, which agrees with the more quantitative remarks about QCD in other talks here.

Thus far we might conclude that interquark forces are merely "superstrong QED". However things are more complicated as evidenced by the fact that baryons exist. First of all this shows that two quarks can be attracted by a third - "two quarks can act like an antiquark". Second, the $\Omega^-$ is built from three strange quarks in the ground state, whereas the Pauli exclusion principle allows at most two identical spin-$\frac{1}{2}$ quarks to occupy that configuration.

*Colour*

If quarks possess a property called colour, any quark being able to carry any one of three colours (say red, yellow, blue), then the $\Omega^-$ (and any baryon) can be built from distinguishable quarks:

$$\Omega^-\left(S_R^\uparrow S_Y^\uparrow S_B^\uparrow\right) .$$

If quarks carry colour but leptons do not, then it is natural to speculate that colour may be the property that is the source of the strong interquark forces - absent for leptons.

Electric charges obey the rule "like repel, unlike attract" and cluster to net uncharged systems. Colours obey a similar rule: "Like colours repel, unlike (can) attract". If the 3 colours form the basis of an SU(3) group, then they cluster to form "white" systems - viz. the singlets of SU(3). Given a random soup of coloured quarks, the attractions gather them into white clusters, at which point the colour forces are saturated. The residual forces among these clusters are the nuclear forces whose origin will be mentioned later.

If quark (Q) and antiquark ($\bar{Q}$) are the $\underline{3}$ and $\underline{\bar{3}}$ of colour SU(3), then combining up to three together gives SU(3) multiplets of dimensions as follows[1]

$$QQ = \underset{\sim}{3} \times \underset{\sim}{3} = \underset{\sim}{6} + \underset{\sim}{\bar{3}}$$

$$Q\bar{Q} = \underset{\sim}{3} \times \underset{\sim}{\bar{3}} = \underset{\sim}{8} + \underset{\sim}{1}$$

(note the $\underset{\sim}{\bar{3}}$ in the QQ system - this is the explanation of the qualitative remark earlier that "two quarks act like an antiquark" - the "acting like" is in colour-space). The $Q\bar{Q}$ contains a singlet - the physical mesons.

$$QQ\bar{Q} \;=\; \underset{\sim}{15} + \underset{\sim}{6} + \underset{\sim}{3} + \underset{\sim}{3}$$

$$QQQ \;=\; \underset{\sim}{10} + \underset{\sim}{8} + \underset{\sim}{8} + \underset{\sim}{1} \quad .$$

Note the singlet in QQQ - the physical baryons.

For clusters of three or less only $Q\bar{Q}$ and QQQ contain colour singlets and, moreover, these are the only states realised physically. Thus are we led to hypothesise that only colour singlets can exist free in the laboratory; in particular the quarks will not exist as free particles.

Beyond threefold clusters we have $Q^{3N}$ (nuclei) $(Q\bar{Q})^N$ - multimesons and $Q^{3N}(Q\bar{Q})^M$ - baryon-meson systems. A major question is whether nontrivial realisations of these systems occur - $Q^6$ for example can occur in colour singlet either as $(Q^3)_{\underset{\sim}{1}}\,(Q^3)_{\underset{\sim}{1}}$ or $(Q^3)_{\underset{\sim}{8}}\,(Q^3)_{\underset{\sim}{8}}$. Do such "hidden colour" nuclei exist?

So far we have supposed only that there is a non-abelian SU(3) group of transformations based on the three colours. We can go further and build a <u>theory</u> of forces by postulating local invariance under phase transformations involving coloured fields. This is the SU(3) analogue of the way that QED emerges as a "gauge theory" of U(1) charge. The resulting "non-abelian gauge theory" of SU(3) colour is named quantum chromodynamics, QCD, the similarity in name bearing witness to the similarity in mathematical structure.

Local U(1) gauge invariance for electrical charge necessarily generates the existence of a massless photon - the carrier of the electromagnetic force. Local gauge invariance for SU(3) colour similarly generates eight (= $3^2$-1) massless gluons - the carriers of the (strong) colour force.

The gluons belong to an octet of SU(3) colour and thus themselves carry colour. They will be attracted to one another by the same colour forces that act on coloured quarks. If the law is that colours cluster to singlets, then the following states involving gluons should exist:

$$(\underset{\sim}{8} \times \underset{\sim}{8})_{\underset{\sim}{1}} \equiv (G^2)_{\underset{\sim}{1}}, \qquad (\underset{\sim}{8} \times \underset{\sim}{8} \times \underset{\sim}{8})_{\underset{\sim}{1}} \equiv (G^3)_{\underset{\sim}{1}} \qquad \text{"gluonium" or "glueballs"}$$

$$[\underset{\sim}{8} \times (\underset{\sim}{3} \times \underset{\sim}{\bar{3}})_{\underset{\sim}{8}}]_{\underset{\sim}{1}} \equiv (GQ\bar{Q})_{\underset{\sim}{1}}, \qquad [\underset{\sim}{8} \times (\underset{\sim}{3} \times \underset{\sim}{3} \times \underset{\sim}{3})_{\underset{\sim}{8}}]_{\underset{\sim}{1}} \equiv (GQ^3)_{\underset{\sim}{1}} \qquad \text{"hermaphrodites".}$$

There is much interest among particle physicists as to the properties and possible ways of detecting such states. Gluonia could play a role in nuclear forces if they are exchanged between nucleons.

## 2.3 QED, QCD; Molecular and Nuclear Forces

Parallels between atomic and nuclear systems vis à vis QED and QCD are exhibited in the summary table, which is self-explanatory.

| | | QED electric charge | QCD colour |
|---|---|---|---|
| Feel the force | Have manifest charge | $e^-$ $Z^+$, $Na^+$ $Cl^-$ | quarks, GLUONS |
| | contain it hidden | atoms, molecules | nucleons, nuclei |
| Don't feel force | totally Neutral | $\nu^0$, PHOTON | $\binom{\nu^0}{e^-}$ leptons |

Note that the force carriers, gluons and photon, do not occupy corresponding positions: the gluons carry colour, the photon does not carry charge. This is the reason why the electromagnetic and colour forces have different spatial behaviours.

Now let's look inside atoms and hadrons.

Inside atoms there is a Coulomb potential, $\alpha/r$, and relativistic corrections due to one (transverse) photon exchange which generates the famous Fermi-Breit Hamiltonian. This latter splits the energy levels of states with parallel or antiparallel spins: e.g. the $^3S_1$ and $^1S_0$ levels in hydrogen - the $^3S_1$ is pushed up in energy while the $^1S_0$ is pulled down.

Inside hadrons the single gluon exchange generates an analogous $\alpha_s/r$ and Fermi-Breit interaction. This will shift the $^3S_1$ $q\bar{q}$ energy up relative to $^1S_0$ - as empirically seen - vector mesons are more massive than their pseudoscalar counterparts.

In QED the $\alpha/r$ potential is the whole story, and so the constituents can be liberated if energy is supplied. In QCD the coloured gluons self-interact en route, and a rising (confining) potential $r^N$ obtains. (This is often taken to be linear or harmonic oscillator for convenience. There is no reason to believe that it is a simple power - indeed $r^2 + r^6 + \ldots + r^\infty$ could be the story - it seems that there are problems at long range if a simple power is extrapolated - see Heller's lectures.)

The forces between "neutral" clusters are the next topic for comparison. These may be classified in three types: constituent exchange (covalent), imbalanced charge distribution leading to residual "van der Waals" forces, and ionic forces. These are summarised in the diagram, which is again self-explanatory.

| | Covalent | Van der Waals | Ions |
|---|---|---|---|
| Atoms<br>Molecules | electron exchange | 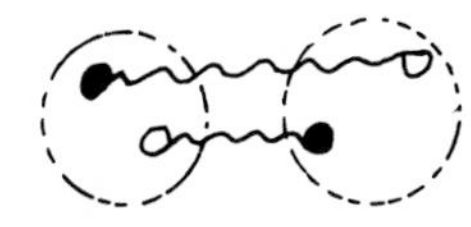 |  |
| Hadrons<br>Nuclei | quark exchange<br>at low energies (macroscopic" > 1 fm.)<br>colour-QCD hidden |  | Not > 1 fm.<br>Maybe in dense quark matter<br>(Baym Lectures) |

The colour SU(3) symmetry is the only exact internal symmetry that occurs at quark level. It was once thought that flavour was a broken symmetry. (Do not confuse SU(3) colour with the old, misidentified, SU(3) of flavour. We now know that there are at least five flavours u,d,s,c,b - there may be even more. There are only three colours - that is why three quarks cluster to form baryons). We now summarise some features of quark masses and how they are manifested in hadron spectroscopy.

## 2.4 "Constituent" Masses for Quarks

A quark and antiquark rotate around one another and build up a series of energy levels as in atomic physics: $^1S_0$ $^3S_1$; $^1P_1$, $^3P_0$ $^3P_1$ $^3P_2$ etc. Each configuration corresponds to what we call "mesons". At the $^1S_0$ level we find the lightest meson - the pion. Being light, it plays an important role in nuclear physics, but it is not fundamental: heavier mesons can be exchanged. Ultimately it is the q and $\bar{q}$ that are being exchanged by the nucleons - the nuclear forces are analogous to covalent forces.

There are three pions $\pi^-\pi^o\pi^+$, and this is because there is more than one flavour of quark. Similarly there are $\rho^-\rho^o\rho^+$ in the vector mesons. From up(u) and down(d) quarks and their corresponding antiquarks $(\bar{u},\bar{d})$ we can build these three charged states and a fourth (observed) neutral partner from the combinations $\bar{u}d$, $\frac{u\bar{u} \pm d\bar{d}}{\sqrt{2}}$, $\bar{d}u$. The quarks all have spin $\frac{1}{2}$ and they have fractional charges:

$$e_u = \frac{2}{3} e_{proton}$$

$$e_d = -\frac{1}{3} e_{proton}$$

(see Flügge's lectures for evidence in deep-inelastic scattering experiments).

The proton and neutron are then the ground states of three-quark systems:

P(uud) ; N(ddu) .

As the nucleons have masses of order 1 GeV, so do the up and down quarks act "as if" they had masses of order 1/3 GeV. (This is often called their "constituent" mass; the relation with their "real" mass, probably of order only a few MeV, comes about via confinement: the nucleons consists of quarks plus field energy, and it is this combination that carries the 1/3 GeV. Astonishingly, the nucleon's static properties appear to be quite successfully described by pretending that the nucleon is composed of three independent quarks each of 1/3 GeV.)

For example if quarks have canonical g factors then

$$\mu_q = \frac{e_q}{m_q} \frac{\hbar}{2c}$$

and the proton g = 2.79 is fitted if

$$m_u = m_p/2.79 = 336 \text{ MeV}$$

(the down quark being order 5 MeV heavier to fit the heavier neutron mass).

In addition to the above hadrons there are also strange particles whose existence requires there to be a strange quark. Strange mesons and baryons tend to be of order 100-200 MeV heavier than their non-strange counterparts, which suggests that the strange quark is correspondingly that much heavier than the non-strange.

This also is consistent with the magnetic moment of the strange $\Lambda$ baryon. From this we can infer the magnetic moment of the strange quark. It turns out that[8]

$$\mu_s = -0.614$$

(where $\mu_u \equiv 2.79$).

Now the charges are $\frac{e_s}{e_u} = \frac{-1/3}{+2/3}$

and so

$$\mu_s = -\frac{1}{2}\mu_u \frac{m_u}{m_s}$$

with the result that $m_s \simeq 510$ MeV. This is very reasonable.

And so it goes. Charmed hadrons suggest that charmed quarks weigh of order 1500 MeV. Their spectroscopy is discussed by Ono here and so I shall say no more about them. We shall see later that these masses also give an excellent description of the hyperfine splitting phenomenology in QCD.

I will not list all of the hadron states - they can be obtained from the data tables[9]. However, there is a newly discovered meson $J^{PC} = 3^{--}$ mass 1870 MeV which completes the $3^{--}$ nonet containing

| | |
|---|---|
| $g(u\bar{u} - d\bar{d})$ | 1700 MeV |
| $\omega_3(u\bar{u} + d\bar{d})$ | 1700 MeV |
| $K^*(u\bar{s})$ | 1780 MeV |
| $\phi_3(s\bar{s})$ | 1870 MeV |

(more discussion can be found in my Lisbon Conference talk[10]). This gives us the beginning of a "strangeonium" spectroscopy to be compared with the extant charmonium and bottonium. The $\phi(1020)$, $f'(1514)$, $\phi_3(1870)$ are non-controversial $1^{--}$, $2^{++}$, $3^{--}$ states. It is very probable that E(1420) is $1^{++}$ dominantly $s\bar{s}$. There are claims from DCI that $\phi'(1670)$ may be the radial excitation of $\phi(1020)$. If that is so, then the strangeonium, charmonium and bottonium spectra appear to be very similar. The $\vec{S}\cdot\vec{S}$ and $\vec{L}\cdot\vec{S}$ splittings within a level are flavour (mass) dependent, but the separation between 1S, 1P, 2S, etc. are, to first approximation, the same.

This "mass independence" yields important restrictions on the form of the (long-range) potential. What is the relation between the power dependence of $v(r) \sim r^N$ and the mass dependence of energy levels? Specifically, under what circumstances does mass independence obtain?

A simple potential is mass independent if

$$v(r) \sim \log r \quad .$$

A sum of power is approximately mass independent

$$v(r) \sim Ar^{N_1} + Br^{N_2} \quad .$$

If $N_1 < 0$, $N_2 > 0$. In particular if $N_1 = -1$ (one-gluon exchange) then, for suitable constant B

$$v(r) \sim \frac{\alpha_s}{r} + Br \ (\text{or } r^2)$$

there will be a region of r where this potential is approximately logarithmic. This question is studied in detail by Ono[5], to whom I refer for further discussion. I shall merely give a short explanation of where this result comes from and then continue with the main theme, namely symmetries and correlations in baryons.

## 2.5 Mass Dependence of Potential Energies

At short range the single gluon-exchange generates a $\alpha/r$ Coulomb-like potential. At long range, multiple exchanges of non-abelian gluons generate a rising potential $r^N$. This could be scalar or vector (or otherwise) - for the present I will only be concerned with the magnitude of N - whether the potential is linear, harmonic oscillator or otherwise.

To determine N we can scale the Schrödinger equation - what happens if m in

$$[\nabla^2 + 2mAr^N]\ \psi(r) \quad = \quad 2mE\ \psi(r) \tag{5.1}$$

is replaced by another M (where A is some dimensionless quantity)? The energy and distance scale has been altered. Thus $r \rightarrow \lambda r$ and we have new eigenvalues $\varepsilon$ given by the solution of

$$[\nabla^2 + 2M\ \lambda^{N+2}\ r^N A]\ \psi(\lambda r) \quad = \quad 2M\,\varepsilon\,\lambda^2\ \psi(\lambda r) \quad . \tag{5.2}$$

The immediate way to find these $\varepsilon$ is to choose $\lambda(N)$ such that the left-hand side is preserved; thus choose

$$\lambda \sim \left(\frac{m}{M}\right)^{1/N+2} . \tag{5.3}$$

Then the right hand rescales as

$$M\,\varepsilon\,\lambda^2 \sim mE$$

or

$$\frac{\varepsilon}{E} \sim \left(\frac{m}{M}\right)^{N/N+2} \quad . \tag{5.4}$$

Thus energies scale as $(m)^{-N/N+2}$ when masses are changed. Hence

$$\left.\begin{array}{lll} r^2 & \leftrightarrow & m^{-1/2} \\ r & \leftrightarrow & m^{-1/3} \end{array}\right\} \text{ fall with increasing m} \tag{5.5}$$

$$r^{-1} \quad \leftrightarrow \quad m \qquad \text{rises with increasing m} \quad .$$

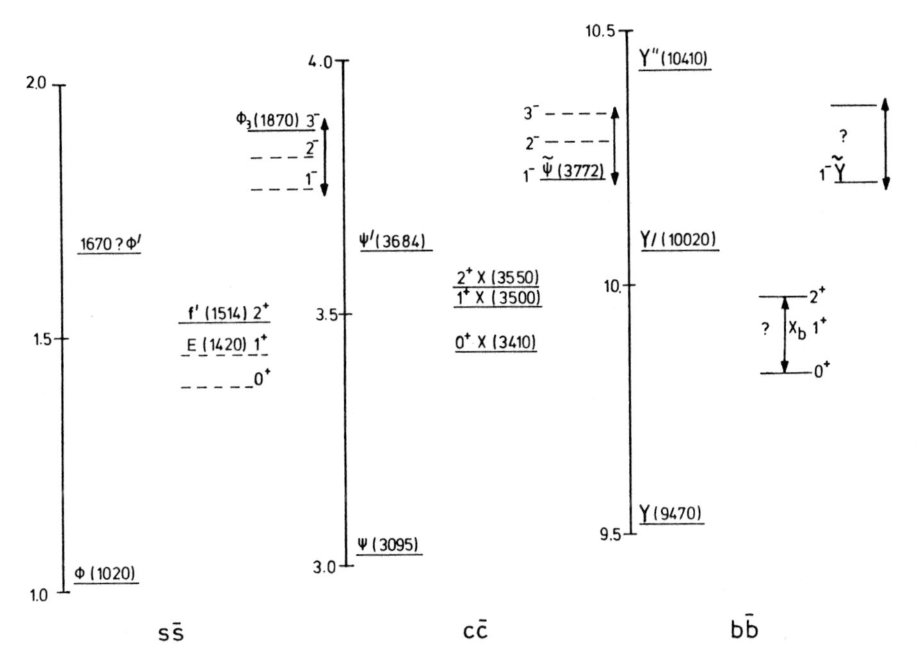


Strangeonium, Charmonium and Bottonium Spectra Compared

Fig.2

The (empirical) m independence obtains for a log r potential or by combining $r^{-1}$ (rise) with $r^{N>1}$ (fall). We have reason to believe that the $r^{-1}$ arises from one-gluon exchange and so the m independence comes because the quarks probe the distance where 1/r is turning into r or $r^2$. For very massive quarks, at shorter distances, an increasingly Coulombic potential should be felt.

So the moral is that it is not inconsistent to suppose that at large r a harmonic oscillator potential applies. It may be linear, but $r^2$ has Gaussian wavefunctions and is easy to analyse analytically, so a lot of papers assume $r^2$. Thus whenever a particular choice is required, I shall choose S.H.O. and, hopefully, not be far wrong.

## 2.6 Symmetries and Correlations in Baryons

To have three quarks in colour singlet:

$$1 \equiv \frac{1}{\sqrt{6}}\Big[(RB - BR)Y + (YR - RY)B + (BY - YB)R\Big] \qquad (6.1)$$

any pair is in the $\bar{\underset{\sim}{3}}$ and is antisymmetric. Note that $\underset{\sim}{3} \times \underset{\sim}{3} = \underset{\sim}{6} + \bar{\underset{\sim}{3}}$. These are, explicitly

| $\bar{\underset{\sim}{3}}_{anti}$ | $\underset{\sim}{6}_{sym}$ |
|---|---|
| RB-BR | RB+BR |
| RY-YR | RY+YR |
| BY-YB | BY+YB |
| | RR |
| | BB |
| | YY |

(6.2)

Note well: *Any Pair is Colour Antisymmetric*

The Pauli principle requires total antisymmetry and therefore any pair must be: -

*Symmetric in all else*

("else" means "apart from colour").

This is an important difference from nuclear clusters where the nucleons have no colour (hence are trivially symmetric in colour!). Hence for nucleons Pauli says

*Nucleons are Antisymmetric in Pairs* (6.3)

and for quarks

*Quarks are Symmetric in Pairs* (6.4)

If we forget about colour (colour has taken care of the antisymmetry and won't affect us again), then

(i) Two quarks can couple their spins as follows

$$\left\{\begin{array}{ll} S = 1: & \text{symmetric} \\ S = 0: & \text{antisymmetric} \end{array}\right\} \qquad (6.5)$$

(ii) Two u,d, quarks similarly form isospin states

$$\left\{ \begin{array}{ll} I = 1 : & \text{symmetric} \\ I = 0 : & \text{antisymmetric} \end{array} \right\} \quad (6.6)$$

(iii) In the ground state L = 0 for all quarks; hence the orbital state is trivially symmetric. Thus for pairs in L = 0 we have

$$\left\{ \begin{array}{ll} S = 1 \text{ and } I = 1 & \text{correlate} \\ S = 0 \text{ and } I = 0 & \text{correlate} \end{array} \right\} \quad . \quad (6.7)$$

Thus the $\Sigma^0$ and $\Lambda^0$ which are distinguished by their ud being I = 1 or 0 respectively also have the ud pair in spin = 1 or 0 respectively: -

$$\left\{ \begin{array}{lcl} \Sigma^0 (ud)_{I=1} s & \leftrightarrow & (ud)_{S=1} S \\ \Lambda^0 (ud)_{I=0} s & \leftrightarrow & (ud)_{S=0} S \end{array} \right\} \quad . \quad (6.8)$$

Thus the spin of the $\Lambda^0$ is carried entirely by the strange quark.

These correlations between flavour and spin cause interesting flavour-dependent effects to arise as a product of spin-dependent forces (e.g. the $\vec{S}\cdot\vec{S}$ force from the magnetic contact interaction between quarks). Thus can the u and d flavours have different momentum distributions in the proton as revealed in deep-inelastic scattering (Flügge's lectures).

This is also the source of the $\Sigma$-$\Lambda$ mass difference. The $\vec{S}\cdot\vec{S}$ interaction acts between all possible pairs; thus

$$\Sigma^0 \left[ (ud)_1 s \right] : \qquad \langle \vec{S}\cdot\vec{S} \rangle_1 + \langle \vec{S}\cdot\vec{S} \rangle_{s,1} \quad (6.9)$$

$$\Lambda^0 \left[ (ud)_0 s \right] : \qquad \langle \vec{S}\cdot\vec{S} \rangle_0 \quad (6.10)$$

(note $\langle \vec{S}\cdot\vec{S} \rangle$ between a spinless diquark and anything vanishes; hence the absence of $\langle S\cdot S \rangle_{s,o}$).

Now

$$\langle \vec{S}\cdot\vec{S} \rangle_0 = -3 \langle \vec{S}\cdot\vec{S} \rangle_1 \quad , \quad (6.11)$$

(see p. 91 of ref. 1). Further, if $m_s = m_{u,d}$ the $\Sigma$ and $\Lambda$ become mass degenerate and so in this limit

$$\langle \vec{S}\cdot\vec{S} \rangle_{s,1} = -4 \langle \vec{S}\cdot\vec{S} \rangle_1 \quad . \quad (6.12)$$

For unequal masses of u and s the magnetic interaction scales as the inverse mass. Hence finally

$$\Sigma^0 \sim \langle\vec{\xi}\cdot\vec{\xi}\rangle_1 \left\{1-4\frac{m_u}{m_s}\right\} \tag{6.13}$$

$$\Lambda^0 \sim \langle\vec{\xi}\cdot\vec{\xi}\rangle_0 \left\{-3 \right\} \quad . \tag{6.14}$$

Then with $m_s > m_u$ we find $m_\Sigma > m_\Lambda$ as observed. Increasing $m_s/m_u$ enhances the effect (e.g. for the charmed analogues $\Sigma_c[(ud)c]$ and $\Lambda_c[(ud)c]$ the splitting will be larger - again observed).

This is very nice, but when a quark is excited to P wave we find that there are $\Sigma^*$ and $\Lambda^*$ with total quark spin 3/2 (J = L + S = 5/2 in particular) and here

$$m_{\Sigma^*} < m_{\Lambda^*} \quad .$$

Why is this reversed relative to the ground state? Let's look again at the symmetry.

Recall that the total effect is that a pair must be symmetrical (after colour). If s = 3/2 then any pair has S = 1 i.e. symmetrical. Notice what this does.

| $\Sigma^*$ | $\Lambda^*$ |
|---|---|
| $(ud)_{sym}$ | $(ud)_{anti}$ |
| $S_{ud}$ sym | $S_{ud}$ sym |
| $\therefore L_{ud}$ sym | $\therefore L_{ud}$ anti |
| $\therefore L_{ud}^{\Sigma^*} = 0$ | $\therefore L_{ud}^{\Lambda^*} = 1$ |
| $L_s = 1$ | $L_s = 0$ |

Thus in $\Sigma^*$ it is the (heavy) s quark that is excited, whereas for $\Lambda^*$ it is the (lighter) u,d. Recall our scaling of the Schrödinger equation in section 5 - energy excitations fall with increasing mass. Thus the ud is excited more than s, hence $\Lambda^*$ heavier than $\Sigma^*$ ! (This was first noted in the S.H.O. model in ref. 6.)

I want to defer discussion of the P-wave baryons until later, as there are some interesting questions there. Let me complete the study of symmetries in S-wave baryons by looking at magnetic moments. The charge of a baryon is just the sum of its constituent quark charges. The magnetic moment is an intimate probe of the correlations between the charges and spins of the constituents. Being wise, today we can say that the neutron magnetic moment was the first clue that the nucleons are not elementary particles. Conversely the fact that quarks appear to have $g \simeq 2$ suggests that they <u>are</u> elementary (or that new dynamics is at work if composite).

## 2.7 Magnetic Moments: Spin Flavour Correlation

A very beautiful demonstration of symmetry at work is the magnetic moment of two similar sets of systems of three, viz.

$$\left\{ \begin{matrix} N & ; & P \\ ddu & ; & uud \end{matrix} \right\} \qquad \mu_P/\mu_N = -3/2$$

and the nuclei

$$\left\{ \begin{matrix} H^3 & ; & He^3 \\ NNP & ; & PPN \end{matrix} \right\} \qquad \mu_{He}/\mu_H = -2/3 \; .$$

The Pauli principle for nucleons requires $He^4$ to have <u>no</u> magnetic moment:

$$\mu[He^4; P^{\uparrow}P^{\downarrow}N^{\uparrow}N^{\downarrow}] = 0 \; .$$

Then

$$He^3 \equiv He^4 - N$$

$$H^3 \equiv He^4 - P$$

and so

$$\frac{\mu_{He^3}}{\mu_{H^3}} = \frac{\mu_N}{\mu_P} \; .$$

To get at this result in a way that will bring best comparison with the nucleon-three quark example, let's study the $He^3$ directly.

$He^3$ = ppn: pp are flavour symmetric, hence spin antisymmetric i.e. S = 0.

Thus

$$[He^3]^{\uparrow} \equiv (pp)_0 \, n^{\uparrow} \tag{7.1}$$

and so the pp do not contribute to its magnetic moment. The magnetic moment (up to mass scale factors) is

$$\mu_{He^3} = 0 + \mu_N \tag{7.2}$$

Similarly,

$$\mu_{H^3} = 0 + \mu_p \quad . \tag{7.3}$$

Now let's study the nucleons in analogous manner.

The proton contains uu flavour symmetric and colour antisymmetric; thus the spin of the "like" pair is symmetric (S = 1) in contrast to the nuclear example where this pair had S = 0. Thus coupling spin 1 and spin $\frac{1}{2}$ together, the Clebsches yield

$$p^{\uparrow} = \frac{1}{\sqrt{3}} (uu)_0 d^{\uparrow} + \sqrt{\frac{2}{3}} (uu)_1 d^{\downarrow} \tag{7.4}$$

(contrast eq. 7.1), and (up to mass factors)

$$\mu_p = \frac{1}{3} (0+d) + \frac{2}{3} (2u-d) \quad . \tag{7.5}$$

Suppose that $\mu_{u,d} \propto e_{u,d}$ then

$$\mu_u = -2 \mu_d \tag{7.6}$$

so

$$\frac{\mu_p}{\mu_N} = \frac{4u-d}{4d-u} = -\frac{3}{2} \tag{7.7}$$

(the neutron follows from proton by replacing $u \leftrightarrow d$).

I cannot overstress the crucial, hidden role that colour played here in getting the flavour-spin correlation right.

We can extend this discussion to the full baryon octet. Six of these states contain two identical quark flavours (which by symmetry necessitates that this pair have total spin S = 1):

$$\left.\begin{array}{l} [\,P\ (uu)_1 d\,] \\ [\,N\ (dd)_1 u\,] \\ [\Sigma^+\ (uu)_1 s\,] \\ [\Sigma^-\ (dd)_1 s\,] \\ [\Xi^0\ (ss)_1 u\,] \\ [\Xi^-\ (ss)_1 d\,] \end{array}\right\} \tag{7.8}$$

The remaining pair are $\Sigma^0$ and $\Lambda^0$, both uds. In the former, the (ud) have I = 1 and hence S = 1. For the $\Lambda^0$, on the other hand, the (ud) have I = 0 and hence S = 0.

Thus the $\Lambda^0$ $[(ud)_0 s]$ is analogous to the $He^3$ nuclear example. The magnetic moment is carried entirely by the third quark, namely s. The data yield[8]

$$\mu_\Lambda \rightarrow \mu_s \simeq \frac{3}{5} \mu_d \quad . \tag{7.9}$$

The strange and down quarks have the same charge (- 1/3) and so the datum fits with

$$m_d = \frac{3}{5} m_s \tag{7.10}$$

as already noted.

If we approximate $m_u = \frac{1}{3} m_p$ (thus the proton would have $g = 3$) then we can do a quick computation of baryon magnetic moments where the individual contributions to the g factors are

$$\left.\begin{aligned} u &= 2 \\ d &= -1 \quad (\text{ratio of } e_d/e_u) \\ s &= -3/5 \quad (\text{ratio of } m_d/m_s) \end{aligned}\right\} \quad . \tag{7.11}$$

From the general spin structure of eq. (7.4)

$$B^\uparrow = \frac{1}{\sqrt{3}} (q_1 q_2)_{\Rightarrow} q_3^\uparrow + \sqrt{\frac{2}{3}} (q_1 q_2)_{\Uparrow} q_3^\downarrow \tag{7.4 bis}$$

we have

$$\mu = \frac{2}{3} (q_1 + q_2) - \frac{1}{3} q_3 \tag{7.12}$$

into which the (7.11) are to be substituted as required. The resulting pattern is as follows

| | Prediction | Data[9,11] | |
|---|---|---|---|
| $P[(uu)_1 d]$ | $\frac{1}{3}(4u-d) = 3$ | 2.79 | (7.13) |
| $N[(dd)_1 u]$ | $\frac{1}{3}(4d-u) = -2$ | -1.9 | (7.14) |
| $\Sigma^+[(uu)_1 s]$ | $\frac{1}{3}(4u-s) = 2.8$ | $2.33 \pm 0.13$ | (7.15) |
| $\Sigma^-[(dd)_1 s]$ | $\frac{1}{3}(4d-s) = -1.1$ | $-1.41 \pm 0.25$ | (7.16) |
| $\Xi^0[(ss)_1 u]$ | $\frac{1}{3}(4s-u) = -1.5$ | $-1.25 \pm 0.02$ | (7.17) |
| $\Xi^-[(ss)_1 d]$ | $\frac{1}{3}(4s-d) = -0.5$ | $-0.75 \pm 0.06$ | (7.18) |

The trend is exceptionally well described. There are undeniably 20% effects not fully accounted for.

Encouraged by this success, we might look further at this problem since, after all, there are exchange effects in nuclei that cause 20% deviations from the naive additive approach analogous to that which we have used for quarks.

We can form contributions of u and d quarks from (7.13) and (7.14) for nucleon, (7.15, 16) for $\Sigma$, (7.17, 18) for $\Xi$, and the data yield

$$(u\text{-}d)_N = 2.9$$

$$(u\text{-}d)_\Sigma = 1.7 \pm 0.15$$

$$(u\text{-}d)_\Xi = 0.9 \pm 0.12 .$$

As we go to systems with more strange quarks, the u,d quarks act as if their effective mass increases (by a factor of three??). There is a systematic trend but extremely dramatic.

Now let's study the strange quark. We can do this by supposing that the environmental dependence for u and d flavours is the same. Then

$$(s)_\Xi \equiv \frac{1}{4}(\Xi^0 + 2\Xi^-) = -0.69 \pm 0.03$$

$$(s)_\Sigma \quad (\Sigma^+ + 2\Sigma^-) = -0.5 \pm 0.5$$

$$(s)_\Lambda = -0.6$$

So the strange quark gives its "canonical" contribution to baryons containing either one or two strange quarks.

## 2.8 Fermi-Breit Interaction

The vector current of a spin-$\frac{1}{2}$ particle with anomalous magnetic moment $\kappa$ and mass m interacts with a vector field $A_\mu$

$$H_I = e\,\bar{\psi}(p_2)\,[\gamma_\mu + \frac{\kappa}{2m}\sigma_{\mu\nu}q^\nu]\,\psi(p_1)\,A^\mu \tag{8.1}$$

where $q \equiv p_2 - p_1$, (and we assume it pointlike i.e. $F_{1,2}(q^2) = 1$).

The magnetic fields (B) and electric field (E) are derived from $A_\mu$, and for large m the interaction becomes

$$H_I = -\mu_1 \sigma_1 \cdot B_2 - (2\mu_1 - \frac{e_1}{2m_1})\, \sigma_1 \cdot E_2 \times p_1/2m_1 \tag{8.2}$$

If $\underline{B}$, $\underline{E}$ are generated by a pointlike particle at $r_2$ then

$$B_2 = \mu_2\, \sigma_2\, \delta(r_1 - r_2) - \frac{\mu_2}{r^3}\left[\sigma_2 - 3\hat{r}\, \sigma_2 \cdot \hat{r}\right] - \frac{1}{r^3}\left[\frac{e_2}{m_2}\, r \times p_2\right] \tag{8.3}$$

$$E_2 = e_2 \hat{r}/r^2 , \tag{8.4}$$

and substituting into (8.2) generates the Fermi-Breit interaction upon summing over particles 1 and 2 [12]

$$\begin{aligned} & e_1 e_2/r - \frac{8\pi}{3}\mu_1\mu_2\, \underline{\sigma}_1 \cdot \underline{\sigma}_2\ \delta(\underline{r}_1 - \underline{r}_2) \\ & + \frac{\mu_1\mu_2}{r^3}\left[\underline{\sigma}_1 \cdot \underline{\sigma}_2 - 3\underline{\sigma}_1 \cdot \hat{r}\ \ \underline{\sigma}_2 \cdot \hat{r}\right] \\ & + \frac{1}{r^3}\left[\mu_1 \frac{e_2}{m_2}\, \underline{\sigma}_1 \cdot \underline{r} \times \underline{p}_2 + (1 \leftrightarrow 2)\right] \\ & + \left[-(\mu_1 - \frac{e_1}{4m_1})\, \underline{\sigma}_1 \cdot \frac{e_2 \underline{r}}{r^3} \times \frac{\underline{p}_1}{m_1} + (1 \leftrightarrow 2)\right] \end{aligned} \tag{8.5}$$

In QCD there is an analogous interaction that arises from one-gluon exchange in contrast to the one-photon exchange in QED. The charges $e_{1,2}$ are replaced by a 3 x 3 matrix $\lambda$ that takes account of the colour carried by the gluon.

The magnetic contact interaction $\sim \sigma_1 \cdot \sigma_2/m_1 m_2$ is well manifested in the data. For S-wave hadrons $H_I$ reduces to the following mass operator[13]

$$M = \sum_i m_i + c^2 \sum_{ij} \frac{\sigma_i \cdot \sigma_j}{m_i m_j}$$

where $m_i$ are the phenomenological masses. If we choose

$$m = m_{u,d} = 350 \text{ MeV} ; \quad \frac{m_s}{m_d} = \frac{3}{2} ; \quad \frac{4c^2}{m^2} = 50 \text{ MeV}$$

then the S-wave baryons and relative splittings of vector and pseudoscalar mesons are excellently reproduced:

| S Baryons | N | Δ | Σ | Λ | Σ* | Ξ | Ξ* | Ω |
|---|---|---|---|---|---|---|---|---|
| Predicted | 930 | 1230 | 1178 | 1110 | 1377 | 1329 | 1529 | 1675 |
| Actual | 938 | 1232 | 1189 | 1115 | 1385 | 1320 | 1530 | 1672 |

| S Mesons | ρ-π | K*-K | D*-D | F*-F | B*-B |
|---|---|---|---|---|---|
| Predicted | 1 | 0.67 | 0.25 | 0.17 | 0.18 |
| Actual | 1 | 0.65 | 0.24 | 0.18 | ? |

With the discovery of charm it gives us four measures of $m_c/m_u$, namely

| | $\Lambda_c$-N | $\Sigma_c-\Lambda_c$ | $\frac{3D^*+D}{3\rho+\pi}$ | $\frac{D^*-D}{\rho-\pi}$ |
|---|---|---|---|---|
| $\frac{m_c}{m_u}$ | 4.7 | 5.6 | 5.0 | 4.2 |

This is not perfect, but the overall conclusion is clear: the naive quark model plus QCD contact interaction gives a good description of S-wave hadron masses.

## 2.9 Spin-Orbit Forces and P-Wave Baryons

If we look at the constituent mass dependence of the various pieces in $H_{FB}$ we can see how the atomic and quark systems differ.

Hydrogen has a light electron mass m and massive nucleus M.

The contact interaction is killed by the large M. However, it is the only contributor to the S-wave system and manifests itself due to the lack of competition. Its small size, due to $M^{-1}$, is the source of its naming "hyperfine" structure. For P,D ... waves the terms proportional to $g^2/mM$ and $2g/mM$ are hyperfine and dominated by the $(2g-1)/m^2$ - so-called fine structure, [$\mu \equiv ge/2m$ in eq. 8.5].

The resulting $\underset{\sim}{L}\cdot\underset{\sim}{S}$ splittings are O(M/m) bigger than the $\underset{\sim}{S}\cdot\underset{\sim}{S}$.

In quark systems the masses of the various constituents are comparable to one another. The S wave $\underset{\sim}{S}\cdot\underset{\sim}{S}$ and P wave $\underset{\sim}{L}\cdot\underset{\sim}{S}$ should therefore be comparable. Experimentally the situation is even more extreme - the $\underset{\sim}{L}\cdot\underset{\sim}{S}$ appear to be much smaller than the $\underset{\sim}{S}\cdot\underset{\sim}{S}$ effects! For example, the states with different J but same L=1 and spin for the P-wave baryons are almost degenerate - contrast the 300 MeV $\underset{\sim}{S}\cdot\underset{\sim}{S}$ splitting between the S-wave $\Delta$ and nucleon or the $\underset{\sim}{S}\cdot\underset{\sim}{S}$ effects in the P-wave baryons which again are of order hundreds of MeV.

So there is a further important difference between $H_{FB}$ in QED and QCD. In QCD the potential is not simply 1/r but contains a long-distance rising component. This presumably arises from multi-gluon exchange and could be effectively scalar, pseudoscalar, vector, axial or tensor in nature.

The $\underset{\sim}{S}\cdot\underset{\sim}{S}$ $\delta(r)$ contact interaction arises only for 1/r vector potential, in general $\delta(r) \to \nabla^2 V(r)$. The absence of long-range $\underset{\sim}{S}\cdot\underset{\sim}{S}$ effects in baryons and mesons shows that the long range potential is not (magnetic) vector. Electric-vector may be possible, I think. It has been supposed that the long-range potential has an important scalar component. The reason is that the Thomas precession in a scalar potential has opposite sign to the spin-orbit effects in a vector potential and so tends to cancel them. Thus can the small $\underset{\sim}{L}\cdot\underset{\sim}{S}$ be accommodated - the naive expectation is reduced by the competition from the long-range scalar potential.

To be more specific let's quote the spin-dependent terms for vector V(r) and scalar S(r) potentials, particles mass $m_{1,2}$. I have assumed that they have a g factor. Setting g = 1 yields the naive vector results; set g = 0 and the scalar potential form results[14].

Vector:

$$V + g^2\left\{ \frac{\underline{\sigma}_1\cdot\underline{\sigma}_2}{6m_1m_2}\nabla^2 V - \frac{1}{4m_1m_2}\left(V'' - \frac{V'}{r}\right)\left\{\frac{\underline{\sigma}_1\cdot\vec{r}\;\underline{\sigma}_2\cdot r}{r^2} - \frac{1}{3}\underline{\sigma}_1\cdot\underline{\sigma}_2\right\}\right\}$$

$$+ g\frac{V'}{m_1m_2r}\left[(\underline{r}\times\underline{p}_1)\cdot\underline{\sigma}_2 - (\underline{r}\times\underline{p}_2)\cdot\underline{\sigma}_1\right]$$

$$+ \frac{V'}{r}(2g-1)\left\{\frac{\underline{r}\times\underline{p}_1\cdot\underline{\sigma}_1}{m_1^2} - \frac{\underline{r}\times\underline{p}_2\cdot\underline{\sigma}_2}{m_2^2}\right\} \tag{9.1}$$

Scalar:

$$S + \frac{S'}{r}\left(\frac{\underline{r}\times\underline{p}_2\cdot\underline{\sigma}_2}{4m_2^2} - \frac{\underline{r}\times\underline{p}_1\cdot\underline{\sigma}_1}{4m_1^2}\right) \tag{9.2}$$

Notice that the scalar form corresponds to the g = 0 limit of a vector interaction. This is because the scalar does not transmit the "magnetic" field, only the "electric" - or Thomas precession.

For simplicity consider the case of equal masses. We can gather the spin-orbit and Thomas precession terms into the combination proportional to[7]

$$\underbrace{-(4g-1)\ [(\underline{\sigma}_1+\underline{\sigma}_2)\cdot\underline{r}\times(\underline{p}_1-\underline{p}_2)]}_{\text{"}L\cdot S^{+}\text{"}} + \underbrace{(\underline{\sigma}_1-\underline{\sigma}_2)\cdot\underline{r}\times(\underline{p}_1+\underline{p}_2)}_{\text{"}K\cdot S^{-}\text{"}} \tag{9.3}$$

Note how $g = 1$ (vector) and $g = 0$ (scalar) contributions will cancel against one another in the first term but add in the $\underset{\sim}{K}\cdot\underset{\sim}{S}$ contributions. Herein lies the phenomenological problem.

For mesons this causes no trouble. The Hamiltonian yields the mass of the system in its rest frame and for a meson at rest $\underset{\sim}{K} \equiv \underset{\sim}{P}_1 + \underset{\sim}{P}_2 = 0$. So only the $\underset{\sim}{L}\cdot\underset{\sim}{S}$ pieces contribute; the relative sign between vector and scalar enables these to cancel against each other, minimising the magnitude of the spin-orbit splittings.

For baryons, which are three-quark systems, the baryon rest frame is $\underset{\sim}{P}_1 + \underset{\sim}{P}_2 = -\underset{\sim}{P}_3 \neq 0$ in general, and so the terms $\underset{\sim}{K}.\underset{\sim}{S}.$ do not vanish. The cavalier cancellation between scalar and vector does not work - these $\underset{\sim}{K}.\underset{\sim}{S}.$ add together. In the literature these terms have often been "neglected" (which is a euphemism for "ignored").

The reason why people have ignored the $\underset{\sim}{K}.\underset{\sim}{S}.$ term is that there is a feeling that perhaps they shouldn't be there (augmented probably by their detrimental effect on phenomenology)[15,6].

They do not occur in the list of permissible forms made for two-body (nuclear physics) interactions by Eisenbud and Wigner[16], and indeed, these terms violate the symmetry conditions that two-body operators must satisfy: they are not invariant under Galilean transformations.

However, there is an extra $\underset{\sim}{S}\cdot\underset{\sim}{L}$ type of term arising from the Wigner rotation of spins in boosting from the baryon rest frame to the two-quark rest frame. This is discussed in detail in refs. 7,12,17, and the effect is that K is replaced by K

$$\underset{\sim}{\tilde{K}} = (\underset{\sim}{r}_1 - \underset{\sim}{r}_2) \times [\underset{\sim}{P}_1 + \underset{\sim}{P}_2 - \frac{m_1+m_2}{M_{tot}}\underset{\sim}{P}] \tag{9.4}$$

where $\underset{\sim}{P} = \underset{\sim}{P}_1 + \underset{\sim}{P}_2 + \underset{\sim}{P}_3$.

For a baryon this involves the coordinates of all three quarks, and so these terms are often, misleadingly, referred to as "three-body terms". They are <u>not</u>:

they are two-body interactions whose physical origin involves the boost from the two-body to overall rest frame. This latter "knows" about the third quark - hence the occurence of the third coordinate.

The $\underset{\sim}{L}\cdot\underset{\sim}{S}$ terms give contributions proportional to (g-c) and the $\underset{\sim}{K}\cdot\underset{\sim}{S}$ give contributions proportional to $(\frac{g+3c}{4})$ where g arises from the potential $V_g$ and c from the (confining) scalar potential. The expressions for g and c depend on both the form of the shell-model orbitals used and the nature of the confining potential used.

Isgur and Karl[6] used a harmonic confining potential and shell-model orbitals which are eigenfunctions of this potential. Their expressions for g and c are

$$g = \frac{4\alpha_s\omega^2}{3\sqrt{2\pi}m\omega} \quad ; \quad c = \frac{\omega^2}{2m} \tag{9.5}$$

where $\omega$ is the oscillator frequence (520 MeV), m = 420 MeV, and $\alpha_s$ = 0.97. These give

$$g = 300 \text{ MeV} , \quad C = 215 \text{ MeV} \tag{9.6}$$

which are individually large but cancel in $\underset{\sim}{L}\cdot\underset{\sim}{S}$. Thus

$$\begin{aligned} \Delta D_{33} - \Delta S_{31} &= 0\ (\underset{\sim}{L}\cdot\underset{\sim}{S}) + \frac{g+3c}{4}\ (\underset{\sim}{K}\cdot\underset{\sim}{S}) \\ \Lambda D_{03} - \Lambda S_{01} &= (g-c)\ (\underset{\sim}{L}\cdot\underset{\sim}{S}) - (\frac{g+3c}{4})\ (\underset{\sim}{K}\cdot\underset{\sim}{S}) , \end{aligned} \tag{9.7}$$

and the $\underset{\sim}{L}\cdot\underset{\sim}{S}$ alone yields zero and + 125 MeV for these.

Ignoring the $\underset{\sim}{K}\cdot\underset{\sim}{S}$ has given a reasonable result, but $\underset{\sim}{K}\cdot\underset{\sim}{S}$ has to be included and gives + 236 MeV for the $\Delta$ splitting (three times too big?) and - 108 MeV for the $\Lambda$ states - large and of the wrong sign.

At a theoretical level the $\underset{\sim}{K}\cdot\underset{\sim}{S}$ terms have caused confusion. Their presence is only understood completely in the weak binding limit. The relationship between these terms and Wigner rotation has been discussed in ref. 7, 12, where the delicacy in definition of centre of mass coordinate is spelled out.

A more transparent approach is that of Brodsky and Primack[17]. They show that the Foldy-Wouthuysen Hamiltonian for the interaction of a composite system of spin-$\frac{1}{2}$ constituents and external electromagnetic field is not the sum of the individual constituent Hamiltonians. The Wigner rotation associated with the overall motion of the centre of mass generates an additional "non-additive" contribution

$$\Delta H = \frac{\underset{\sim}{\sigma}_1}{4M} \frac{\underset{\sim}{\sigma}_1}{m_1} - \frac{\underset{\sim}{\sigma}_2}{m_2} \cdot \left[ e_2 \underset{\sim}{E}_1 \times \underset{\sim}{P}_1 - e_1 \underset{\sim}{E}_2 \times \underset{\sim}{P}_2 \right] . \tag{9.8}$$

If the "external" field felt by particle 2 comes from particle 1 and vice versa, then

$$e_2 \underset{\sim}{E}_1 = -e_1 \underset{\sim}{E}_2 \sim - \underset{\sim}{r}/r^3$$

since $|\underset{\sim}{E}_{1,2}| \propto e_{1,2}$, but in relatively opposing directions. Thus

$$\Delta H = - \frac{e_1 e_2}{4M} \left( \frac{\underset{\sim}{\sigma}_1}{m_1} - \frac{\underset{\sim}{\sigma}_2}{m_2} \right) \cdot \frac{\underset{\sim}{r}}{r_3} \times (\underset{\sim}{P}_1 + \underset{\sim}{P}_2) \tag{9.9}$$

and identically kills the $\underset{\sim}{K}\cdot\underset{\sim}{S}$ term for mesons. For a three-body system the analysis is more involved. The net effect is to replace $\underset{\sim}{K}$ by $\underset{\sim}{\tilde{K}}$ as at eq. (9.4).

## 2.10 What Kills the $\underset{\sim}{K}\cdot\underset{\sim}{S}$ Contributions?

The Hamiltonian that we have been discussing involves only two-body interactions. This is too simple in general, e.g. at the next order in gluon-exchange there is the set of diagrams

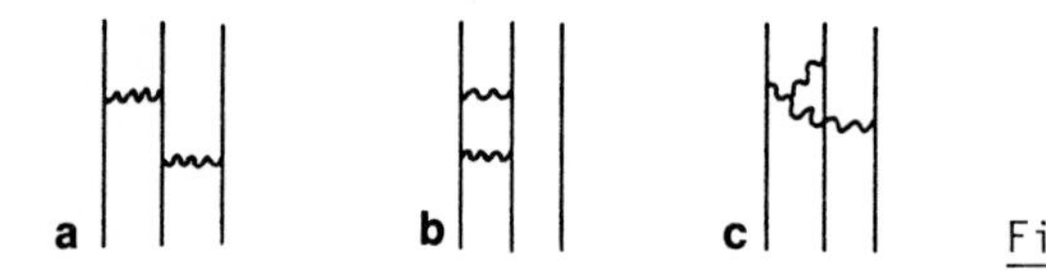


Fig.3a-c

Figure 3a and b are indeed iterations of two-body terms, but c is manifestly a non-reducible three-body contribution. Order by order in perturbation theory there are more topologies that are two-body reducible than the contrary. Thus it seems prima facie that the dynamical three-body contributions do not conspire to kill the unwanted $\underset{\sim}{K}\cdot\underset{\sim}{S}$ terms (however, for heavy quarks in a bag we have learned from Heller(4) that three-body forces are dominant. It is possible therefore that there is some residuum for light quarks - this needs more study, but in its absence we cannot yet claim to understand baryon spectroscopy in QCD).

There is another way of killing the $\underset{\sim}{K}\cdot\underset{\sim}{S}$ contributions within the two-body framework. Suppose that the short-range force is dominantly magnetic and the long-range confinement is electric vector plus a scalar.

The short-range ($H_{FB}$) behaviour is proportional to g factors as follows

i) $\alpha_s g^2 \; \underset{\sim}{S}\cdot\underset{\sim}{S}$

ii) $\alpha_s g \; \underset{\sim}{L}\cdot\underset{\sim}{S}$

iii) $\alpha_s 1 \; \underset{\sim}{K}\cdot\underset{\sim}{S}$

Thus a smaller value of $\alpha_s$ is needed to fit the hyperfine splitting because $g^2$ helps enhance it. The $\underset{\sim}{L}\cdot\underset{\sim}{S}$ will be relatively smaller (g versus $g^2$) and the $\underset{\sim}{K}\cdot\underset{\sim}{S}$ terms smaller still. These are all in line with what we want but are not enough - the long-range scalar potential still generates unacceptably large effects.

If the long-range potential had no relativistic spin dependence, then we would be home and dry. The electric spin-dependent contribution with g = 1 for a vector potential is equal and opposite to a scalar potentials Thomas precession (g = 0). For this particular combination of long-range confining forces the spin-orbit (Thomas) effects cancel.

But this is pure phenomenology. There is no reason why this particular conspiracy should occur.

## 2.11 Conclusions

We have seen how symmetries - notably the Pauli principle - shape the hadrons and their properties. The presence of colour causes quark pairs to be overall symmetric in other quantum numbers - in contrast to overall antisymmetry for nucleons. This has important consequences for baryon magnetic moments in contrast to those of tritium and helium-three.

I have illustrated some of the effects that symmetry places on spectroscopy and the impact that QCD may have there. While the contact interaction between pairs of quarks colour magnetic moments does appear to be manifested empirically, the spin-orbit and Thomas precession effects are still not totally understood, at least for baryons. This is not to belittle the considerable progress, inspired by QCD, that has taken place in hadron spectroscopy. But it is important to realise that the baryons are still not totally understood. To the extent that we believe QCD to be the theory of quark forces, then nuclear forces should in principle be derivable from QCD. Until we understand fully the way that baryons are built via QCD, then, derivation of nuclear forces seems some way off.

Ultimately we may have to face a fact of life. Physics tends to deal with approximations. QCD perturbation theory is applicable at very short distances and well describes quark forces there. Conventional nuclear physics models apply well at distances greater than 1 fm. Whether these two realms of approximation/models can be fused is a question for the future, but an exciting one to contemplate.

## References


1. F.E. Close, Introduction to Quarks and Partons, Academic Press (1979)
2. S.J. Brodsky, these proceedings
3. This is discussed by N. Isgur, Proc. YVI Rencontre de Moriond, CNRS, France (1981)
4. L. Heller, these proceedings
5. S. Ono, these proceedings
6. N. Isgur and G. Karl, Phys.Lett. 72B (1977) 109
   N. Isgur and G. Karl, Phys.Rev. D18 (1978) 4187
7. F.E. Close and R.H. Dalitz, in Low and Intermediate K-N physics, (ed. E. Ferrari and G. Violini), pub. D. Reidel (1981)
8. L. Schachinger et al., Phys.Rev.Lett. 41 (1978) 1348
9. Particle Data Group, Rev.Mod.Phys. 52, April 1980
10. F.E. Close in Proc. of European Physical Society Meeting, Lisbon (1981) (ed. J. Dias de Deus) and Rutherford Lab. report RL-81-066
11. O.E. Overseth, Proc. "Baryon 1980", Univ. of Toronto, p. 289 (ed. N. Isgur)
    R. Handler, Proc. XX Internat. Conf. on HEP, Madison (1980), p. 539 (ed. L. Durand and L.G. Pondrom)
12. F.E. Close and H. Osborn, Phys.Rev. D2 (1970) 2127
13. A. de Rujula, H. Georgi and S.L. Glashow, Phys.Rev. D12 (1975) 147
14. D. Gromes, Nucl.Phys. B130 (1971) 18
15. L.J. Reinders, J. Phys. G4 (1978) 1241 but see also
    D. Gromes and I. Stamatescu, Nucl.Phys. B112 (1976) 213
16. L. Eisenbud and E.P. Wigner, Proc. Nat. Acad. Sci. 27 (1941) 281
17. S.J. Brodsky and J.R. Primack, Ann Phys. 52 (1969) 315

# 3. Quantum Chromodynamics at Nuclear Dimensions*

S.J. Brodsky

With 26 Figures

## 3.1 Introduction

An historic and central goal of physics has been the determination of the fundamental theory of the nuclear force. Incredibly, it appears that the search may well be over: Quantum Chromodynamics [1] (QCD), the $SU(3)_{color}$ non-Abelian gauge theory of quarks and gluons, appears to be the theory of the strong and nuclear interactions in the same sense that quantum electrodynamics accounts for electrodynamic interactions. QCD solves many crucial problems: the meson and baryon spectra, quark statistics, the structure of the weak and electromagnetic currents of hadrons, the scale invariance of interactions at short distance, and, most likely, color (i.e., quark and gluon) confinement at large distances. Many different and diverse tests [2] have confirmed the basic features of QCD although the fact that the tests of quark and gluon interactions must be done within the confines of hadrons, as well as various technical difficulties, have prevented truly quantitative confirmation of the theory. The structure of the theory satisfies all prerequisites of elegance and beauty.

Despite the evidence that QCD — or something close to it — gives a correct description of the structure of hadrons and their interactions, it seems paradoxical that the theory has thus far had very little impact in nuclear physics. One reason for this is that the application of QCD to distances larger than 1 fm involves coherent, non-perturbative dynamics which is beyond present calculational techniques. For example, in QCD the nuclear force can evidently be ascribed to quark interchange and gluon exchange processes. These, however, are as complicated to analyze from a fundamental point of view as is the analogous covalent bond in molecular physics. Since a detailed description of quark-quark interactions and the structure of hadronic wavefunctions is not yet well understood in QCD, it is evident that a quantitative first-principle description of the nuclear force will require a great deal of theoretical effort.

Another reason for the limited impact of QCD in nuclear physics has been the conventional assumption that nuclear interactions can for the most part be analyzed in terms of an effective meson-nucleon field theory or potential model in isolation from the details of short-distance quark and gluon structure of hadrons. However, in these lectures, I will argue that this view is untenable: in fact, there is no "correspondence principle" which yields traditional nuclear physics as a rigorous large-distance or non-relativistic limit of QCD dynamics. On the other hand, the distinctions between standard nuclear physics dynamics and QCD at nuclear dimensions are extremely interesting and illuminating for both particle and nuclear physics. For example:

(1) Meson and nucleon degrees of freedom are insufficient to describe nuclei in QCD: mixed color configurations appear as Fock components of ground state nuclei and as excited multiquark nuclear states. In fact, the hidden color wavefunction components contribute to basic properties of nuclei including magnetic and quadrupole moments, charge distributions, etc.

---

*Work supported by the Department of Energy under contract number DE-AC03-76SF00515.

(2) The usual impulse approximation formula for elastic form factors of nuclei,

$$F_A(Q^2) \cong F_N(Q^2)\, F_A^{Body}(Q^2),$$

which is conventionally used to separate nucleon size effects from nuclear dynamics, is incorrect in QCD because of off-shell and recoil effects. An alternative, QCD-based formula is discussed in Section VIII. We also shall show (see Sec. IV) that even so-called static properties such as the nuclear magnetic moment which are derived in the limit $Q^2 \to 0$ receive non-trivial recoil contributions.

(3) Since quarks are the ultimate carriers of the electromagnetic current in QCD, the identification of specific nucleon-anti-nucleon pair production terms in the analysis of the electromagnetic structure of nuclei cannot be justified.

(4) Conventional effective meson-nucleon field theories with nucleons coupled to isovector $\rho$ mesons violate unitarity in tree-graph (Born) approximation. Since such theories are not renormalizable, they have no predictive content in higher orders. A renormalizable theory requires tri-linear and quartic vector meson couplings and a spontaneous symmetry-breaking mechanism to provide meson masses.

The real conflict between quark and nuclear physics is at a very basic level: because of Lorentz invariance a conserved charge must be carried by a local (point-like) current; there is no consistent relativistic theory where fundamental constituent nucleon fields have an extended charge structure.

The plan of these lectures is as follows. In Section II we review the basic structure and features of QCD. Light-cone perturbation theory is then introduced in Section III. This method can be regarded as an elegant relativistic generalization of ordinary Schroedinger many-body theory, and it has many applications to nuclear physics problems. Sections III through VII are intended as a general introduction to QCD analysis and phenomenology with special emphasis on exclusive and inclusive large-momentum-transfer reactions, and the structure of hadronic wavefunctions.

The most dramatic and definitive area of application of QCD to nuclear physics is the short-distance structure of the nuclear force and large-momentum-transfer nuclear reactions. We will discuss these applications in detail in Section VIII. The importance of these predictions is not only the asymptotic large momentum behavior, but also the analytic constraints placed on nuclear amplitudes. For example, we give predictions for the power-law form of effective meson-nucleon couplings as dictated by the underlying renormalizable gauge theory. In Section IX we conclude with a list of experiments which could illuminate QCD dynamics within nuclei. The eventual goal is the complete synthesis of nuclear, hadronic and quark/gluon dynamics. Indeed, if QCD is correct, it must account for all the features and interactions of nuclei as well as mesons and baryons.

## 3.2 Basic Features of QCD

In quantum chromodynamics the fundamental degrees of freedom of hadrons and their interactions are the quanta of quark and gluon fields which obey an exact internal SU(3) (color) symmetry. The spin-1/2 quarks are in the fundamental (triplet) representation of $SU(3)_C$, the spin-1 gluons are in the adjoint (octet) representation, and hadrons are identified with singlet states; e.g., mesons $|M\rangle \sim \sum_{i=1}^{3} |q_i\bar{q}_i\rangle$ and baryons $|B\rangle \sim \sum \varepsilon_{ijk}|q_iq_jq_k\rangle$. In addition, gluonium (color-singlet bound states of 2 and 3 gluons) should exist. As we discuss in Section VI, new types of "hidden color" nuclear states are also predicted in QCD. The different types of quarks, u,d,s,c,b,...are distinguishable by their flavor label and mass.

It is well known that the general structure of QCD meshes remarkably with the facts of the hadronic world, especially quark-based spectroscopy (including the charm and beauty quark systems); current algebra; the dimensional counting parton-model structure of large-momentum-transfer reactions (up to computable logarithmic corrections to scale invariance). Experiments at large momentum transfer, both exclusive and inclusive, are consistent with the QCD postulate that the electromagnetic and weak currents of hadrons are carried by point-like spin-1/2 quarks which interact via a Dirac coupling to spin-1 gluons. The most important phenomenological evidence for QCD comes from inelastic lepton scattering, $e^+e^-$ annihilation processes, and those high-momentum-transfer exclusive and inclusive reactions where the structure of perturbative quark and gluon subprocesses can be studied in relative isolation from the bound-state dynamics of the hadrons. From the theoretical standpoint, the elegant structure of QCD makes it appear almost compelling as a fundamental theory of hadronic and nuclear phenomena, even though many crucial questions concerning quark and gluon confinement, and the effects of non-perturbative phenomena remain unanswered.[3]

A critical feature of QCD is asymptotic freedom, [4] i.e., the logarithmic decrease of the effective quark and gluon coupling constant $\alpha_s(Q^2)$ with momentum transfer which implies that the strong interactions become weak, and even calculable, in perturbative theory at short distance. The fact that the annihilation ratio

$$R_{e^+e^-}(s) = \frac{\sigma(e^+e^- \to \text{hadrons})}{\sigma(e^+e^- \to \mu^+\mu^-)} \qquad (1.1)$$

is empirically [5] close to the zeroth-order QCD prediction, $R^o = 3\sum_q e_q^2$ for energies above the heavy-quark thresholds, is a crucial check of asymptotic freedom and the color, charge, and spin assignments of the quark quanta in QCD. Critical features of QCD are also confirmed by the observed logarithmic breaking of scale invariance in deep-inelastic lepton-scattering [2] and the measurements of two-jet and three-jet structure of $e^+e^-$ annihilation final states. [5] The recent observations of jet structure [6] in two-photon reactions (consistent with $\gamma\gamma \to q\bar{q}$ subprocesses), and measurements [7] of the photon structure function also provide fundamental checks of predictions which are essentially unique to QCD. However, despite these successes, there is no direct experimental evidence for (near) scale-invariant quark-quark, quark-gluon, or gluon-gluon scattering amplitudes as predicted by QCD; the cross section for large-transverse-momentum hadron production in hadron-hadron collisions appears to reflect much more complicated dynamical mechanisms. On the other hand, as we discuss in Section IV, the fact that the proton form factor $G_M(Q^2)$ scales as $(Q^2)^{-2}$ reflects the fact that the minimum Fock state in the nucleon contains 3 quarks, and that the internal quark-quark interactions which control the nucleon wavefunction at short distances are consistent with scale invariance. [8,9] Thus far, experiments are not sufficiently sensitive to distinguish a logarithmically decreasing $\alpha_s(Q^2)$ from a constant; i.e., fixed point behavior. The sensitivity of the nucleon form factors to the form of $\alpha_s(Q^2)$ is discussed in Section VI.

Although there have been remarkable technical achievements in perturbative QCD calculations in the past few years, [1,2,10] there has also been the realization that precise and detailed comparisons with experiment require consideration of effects and phenomena not readily computable with present methods. There are, in fact, only a very few large-momentum-transfer processes which can be studied rigorously to all orders in perturbation theory such as $R_{e^+e^-}(s)$, [1] the meson form factors $F_M(Q^2)$ [11] (and $F_{\gamma\to M}(Q^2)$), the two-photon processes [12] $\gamma\gamma \to M\bar{M}$ at large momentum transfer, the photon structure function, [13] and the $Q^2$ evolution of the hadron structure functions. Although, in principle, these processes can be calculated to arbitrary orders in perturbation theory, in practice, there are serious complications involving the dependence of predictions made to finite order on the choice of renormalization scheme and the scale parameterization chosen for the argument of $\alpha_s$. [2,13] We shall discuss a new method [14] for avoiding these ambiguities in Section II. Aside from this, there is always the question of the radius of convergence of the perturbation expansion. Even for processes which can be calculated to arbitrary orders in $\alpha_s$, there are (presently) uncalculable power-law suppressed (higher twist) contributions [15] which must be included in detailed fits to experiment, especially at the edge of phase space. [16]

In the case of jet production, QCD-based predictions based on the elementary features of $e^+e^- \to q\bar{q}$ and $q\bar{q}g$, $\gamma\gamma \to q\bar{q}$, etc. must also take into account higher twist contributions, model-dependent non-perturbative effects intrinsic to hadron formation and decay, [5] and possibly dynamical effects due to quark confinement. [3] In the case of some exclusive processes such as the baryon form factor there are non-leading QCD contributions which are asymptotically suppressed by Sudakhov form factors. [9,10] The precise evaluation requires an all orders resumption of perturbation theory. QCD predictions for elastic hadron-hadron scattering are complicated by the presence of Landshoff [17] pinch singularity contributions which are only partially suppressed by Sudakhov form factors. [10] Despite these complications, we can still derive general properties for exclusive reactions such as hadron-helicity conservation [18] and the leading power-law behavior. [19]

An even more interesting (and perplexing) situation occurs for all inclusive high-momentum-transfer inclusive reactions involving hadronic initial states such as Drell-Yan massive lepton pair production, direct photon production, and large $p_T$ hadron production. As shown in Ref. 20, initial-state interactions violate the usual QCD factorization theorem order by order in perturbation theory and affect the normalization and transverse momentum dependence of the inclusive cross sections. In addition, final-state interactions also affect the associated multiplicity and transverse momentum dependence of the outgoing jets in deep-inelastic lepton scattering reactions. A detailed report on these effects is given in Ref. 20.

Perhaps the most serious complication to QCD phenomenology is the presence of higher twist subprocesses, since power-law suppressed contributions can often mimic (and thus confuse the identification) of the logarithmic modifications predicted for the leading twist contributions. [16] Examples of this for deep-inelastic structure functions and fragmentation distributions are discussed in [21] and [22] and Section V. In the case of three-jet production in $e^+e^-$ annihilation, higher twist terms give contributions [23] $dN/dk_\perp^2 \sim (k_\perp^2)^{-2}$ for the hadron transverse momentum distribution in quark and gluon jets. These hard components can complicate the separation of the $e^+e^- \to q\bar{q}g$ and $e^+e^- \to q\bar{q}$ subprocesses. In the case of hadron production at large transverse momentum, "direct-coupled" higher twist subprocesses such as $gq \to \pi q$ actually dominate [24] the leading twist $qq \to qq \to q\pi q$ subprocesses at large $x_T = 2p_T/\sqrt{s}$. Evidence for direct-coupled $\pi q \to \gamma^* q$ subprocesses in $\pi p \to \mu^+\mu^- x$ reactions is discussed in Section V and Ref. 22.

Present QCD phenomenology is also incomplete in the sense that although much attention is paid to the $Q^2$ evolution of hadron structure functions there is no real understanding of the basic x-dependent form of the quark and gluon distribution in hadrons, or how to relate them to other hadronic phenomena. The relation of the $x \sim 1$ behavior of structure functions to the exclusive fixed $W^2$, high $Q^2$ domain is only roughly understood. [25] The $x \sim 0$ behavior of structure functions and the connection to the photoabsorption cross section at fixed $Q^2$, high $\nu$, and nuclear shadowing phenomena is also not well understood. [26]

The main purpose of these lectures is to begin to extend QCD phenomenology by taking into account the physics of hadronic wavefunctions. [27] Our eventual goal is to obtain a parameterization of the wavefunctions which will bridge the gap between the non-perturbative and perturbative aspects of QCD. The lack of knowledge of hadronic matrix elements is the main difficulty in computing and normalizing dynamical higher twist contributions for many processes.

In Section III we emphasize the utility of a Fock-state representation of the meson and baryon wavefunctions as a means not only to parameterize the effects of bound-state dynamics in QCD phenomena, but also to interrelate exclusive, inclusive, and higher twist processes. It is particularly convenient to choose a momentum space Fock-state basis [19,27]

$$\psi_n(x_i, k_{\perp i}; \lambda_i) \quad ; \quad \sum_{i=1}^{n} x_i = 1 \quad , \quad \sum_{i=1}^{n} k_{\perp i} = 0 \quad ,$$

defined at equal "time" $\tau = t + z$ on the light cone. Here $x_i = (k^o + k^3)_i/(p^o + p^3)$, $\vec{k}_{\perp i}$, and $\lambda_i$ specify the longitudinal and transverse momenta and spin projection $S_z$ of each (on-mass-shell) quark and gluon in the n-particle Fock state ($n \geqq 2$ for mesons and $n \geqq 3$ for baryons). We also choose the light-cone gauge $A^+ = A^o + A^3 = 0$ so that only physical polarizations of the gluons occur. The color-singlet wave-functions are regulated so that they are finite in both the infrared and ultra-violet regimes. [28]

There are a number of reasons why this representation of hadrons in terms of the quark and gluon degrees of freedom is useful:

(1) In light-cone perturbation theory, the perturbative vacuum is also an eigenstate of the total QCD Hamiltonian on the light cone; perturbative calculations are enormously simplified by the absence of vacuum to pair production amplitudes.

(2) All form factors, charge radii, magnetic moments, etc. have exact expressions in terms of the $\psi_n$.

(3) The structure functions $G_q(x,Q)$ and $G_g(x,Q)$ (and more general multiparticle distributions) which control large-momentum-transfer (leading and higher twist) inclusive reactions, and the distribution amplitudes $\phi(x,Q)$ which control large-momentum-transfer exclusive reactions (and directly coupled inclusive reactions) are each specific, basic measures of the $\psi_n$. Examples of these calculations are schematically illustrated in Figs. 1 through 3.

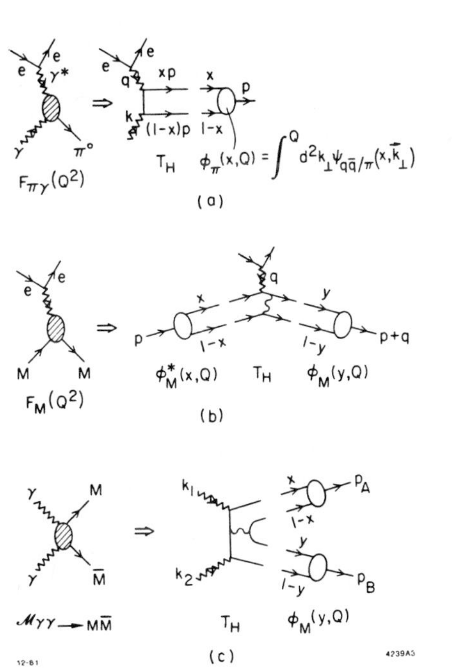


Fig. 1. Calculable large-momentum-transfer meson processes in QCD, and their connection to the meson Fock-state wavefunction $\psi_{q\bar{q}}$ and distributions amplitude $\phi(x,Q)$. Only a representative diagram for the hard scattering amplitude $T_H$ is shown. (a) The $\gamma \to \pi^0$ transition form factor (measurable in single tagged ee $\to$ ee $\pi^0$ experiments), (b) the meson form factor, (c) the $\gamma\gamma \to M\bar{M}$ scattering amplitude. Details are discussed in Sect. IV

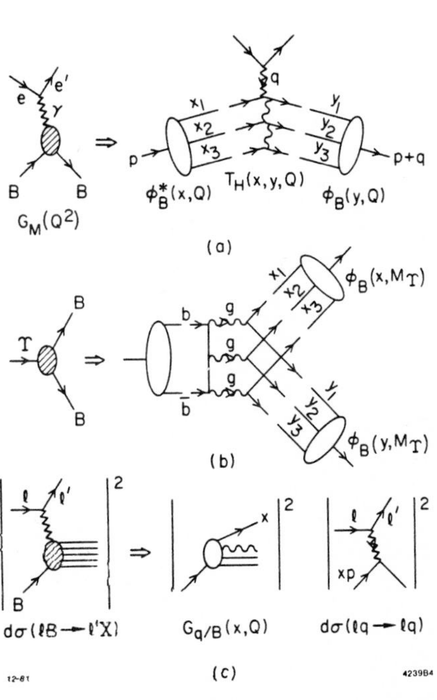


Fig. 2. Baryon processes at large momentum transfer in QCD and the connection to the baryon Fock-state wavefunction. (a) Baryon form factors, (b) heavy quarkonium decay $\Upsilon \to p\bar{p}$, (c) deep-inelastic lepton-baryon scattering. Only representative contributions are shown. The inclusive cross section and structure function $G_{q/B}(x,Q)$ is computed from the square of the baryon wavefunction summed over all contributing Fock states

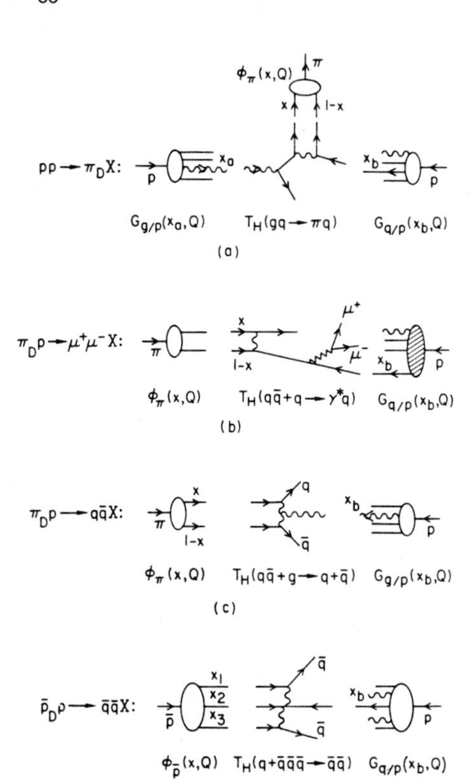


Fig. 3. Examples of QCD-computable higher twist "direct-coupled" subprocesses for inclusive reactions. The subscript D indicates that the hadronic wavefunction is involved directly in the high-momentum-transfer subprocesses. (a) Direct production of high $p_T$ mesons in hadron-hadron cross section. The predicted cross section is proportional to the meson form factor $F_M(p_T^2)$ times the leading twist cross section. (b) Higher twist contribution to meson-induced massive lepton pair production. The predicted cross section is equivalent to a contribution $F_L(x,Q^2) \sim C/Q^2$ to the longitudinal structure function of the meson. (c) Direct meson production of quark jets in meson-baryon collisions. All of the meson energy is used to produce jets at large transverse momentum. The cross section is proportional to $F_M(p_T^2)$ times the leading twist $qq \to qq$ cross section. (d) Direct production of anti-quark jets in $\bar{B}B$ collisions. The cross section is proportional to $G_M^B(p_T^2)$ times the leading twist $qq \to qq$ cross section. In each case the direct process dominates over the leading twist contribution in a large x kinematic region

(4) Other physical quantities such as decay amplitudes provide rigorous sum-rule or local constraints on the form of the valence components of meson and baryon wavefunctions. [2]

In the remainder of this section we will give a brief introduction to QCD and asymptotic freedom. We then discuss a new method to avoid scheme and scale ambiguities in perturbative QCD predictions. In Section III we give a detailed discussion of light-cone perturbation theory and the Fock-state expansion of hadronic wavefunctions. The QCD equation of motion is also discussed. In Section IV we discuss measures of the hadronic and nuclear wavefunctions (form factors, magnetic moments, etc.), and the QCD analysis of high-momentum-transfer exclusive processes. We also show how meson distribution amplitudes can be measured in $\gamma\gamma \to M\bar{M}$ reactions. The connection of the Fock-state basis to leading and higher twist contributions to deep-inelastic scattering is given in Section V. In Section VI we discuss how many different QCD processes are interrelated (as in Figs. 1 through 3) through the hadronic Fock states. We also discuss a novel type of QCD subprocess — direct-coupled hadron-induced reactions. [29] A new prediction for the proton form factor is also given. In Section VI we also introduce a simple phenomenology of hadron wavefunctions and discuss present constraints on the form and normalization of the valence meson and nucleon Fock states. An important conclusion is that the valence Fock state as defined at equal time or the light cone appears to have a significantly smaller radius than that of the physical hadron; [27] higher Fock states thus play an essential role in low-momentum-transfer phenomenology. Applications to quark jet diffraction excitation [30] and the hidden heavy-quark Fock-state structure of hadrons are also discussed. [31] The effects of initial- and final-state interactions on QCD inclusive reactions are discussed in Ref. 20.

### 3.2.1 The QCD Lagrangian

An essential feature of QCD is that $SU(3)_c$ is an exact local symmetry: rotations in color space can be made independently at any space-time point. The mathematical realization of this is the Yang-Mills non-Abelian gauge field theory. The QCD Lagrangian density is [1]

$$\mathscr{L}_{QCD} = \bar{\psi}(i\not{D} - m)\,\psi - \frac{1}{4}\,\mathrm{Tr}\,F^2_{\mu\nu} \tag{2.1}$$

$$iD^\mu = i\,\partial^\mu\, I + gA^\mu \tag{2.2}$$

$$F^{\mu\nu} = \partial^\mu A^\nu - \partial^\nu A^\mu + g[A^\mu, A^\nu]. \tag{2.3}$$

Here

$$\psi(x) = \begin{pmatrix} q_R(x) \\ q_Y(x) \\ q_B(x) \end{pmatrix}$$

is the color triplet of quark fields, and $A^\mu(x) = \sum_{a=1,8} \lambda_a A^\mu_a(x)$ is the color-octet gluon field summed over the $3 \times 3$ traceless matrices $\lambda_a$ satisfying $[\lambda_a, \lambda_b] = i\, f_{abc}\, \lambda_c$ and $\mathrm{Tr}[\lambda^a \lambda^b] = 2\delta^{ab}$. $\mathscr{L}_{QCD}$ is obviously a color singlet. Local gauge invariance and color symmetry follow from the invariance of $\mathscr{L}_{QCD}$ under the general gauge transformation

$$\psi(x) \to U(x)\;\psi(x) \tag{2.4}$$

$$A^\mu(x) \to U(x)\;A^\mu(x)\;U^{-1}(x) + \frac{i}{g}\,U(x)\left(\partial^\mu U^{-1}(x)\right) \tag{2.5}$$

where the unitary matrix $U(x) = \exp i \sum_a \lambda_a \theta_a(x)$ is an arbitrary function of space and time. Note that the field strength $F^{\mu\nu}(x) \to U(x)\; F^{\mu\nu}\; U^{-1}(x)$ is not invariant, since it is in the adjoint representation of $SU(3)_c$. The local gauge invariance of the Yang-Mills is an essential ingredient in proving the renormalizability and consistency of the theory. [1]

In general, a sum over quark flavors $i = u,d,s,c,b...$ is understood in $\mathscr{L}_{QCD}$. (In fact, the mass matrix $m_{ij}$ is not diagonal when the weak and electromagnetic interactions are taken into account. [32]) The fundamental origin of the quark flavors and their masses remains an outstanding problem in hadron physics.

In a sense QCD can be regarded as the non-Abelian generalization of QED:

$$\mathscr{L}_{QED} = \psi(x)(i\not{D} - m)\,\psi - \frac{1}{4}\,F^2_{\mu\nu} \tag{2.6}$$

where $iD^\mu = i\partial^\mu + eA^\mu$, $F^{\mu\nu} = \partial^\mu A^\nu - \partial^\nu A^\mu$. From the point of view of formal perturbation theory there are close similarities in the Feynman rules and treatment of ultraviolet renormalization and infrared divergences. The Feynman rules for QCD are given in Table 1. In the case of covariant gauges one must formally include "ghost" scalar particles in loops, or else unitarity of amplitudes involving the non-Abelian-couplings will be lost. In the case of axial gauges ($\eta^\mu A_\mu = 0$ where $\eta^\mu$ is a fixed 4-vector) there are no ghosts, but renormalization is somewhat more complicated. The color trace algebra for any Feynman diagram can be done almost automatically using the graphical rules given by CVITANOVIC. [33] The main algorithm is that as far as color is concerned, the gluon propagator ~ in SU(N) is equivalent to two quark lines $\rightleftarrows$ minus 1/N times the identity (to remove the U(N) singlet). The complete rules are given in Ref. 33.

Table 1. Feynman rules for quantum chromodynamics*

|  |  |
|---|---|
| Fermion Propagator | $\frac{i}{\not p - m + i\epsilon}\,\delta_{\alpha\beta}$ |
| Gluon Propagator | $-i\left[g_{\mu\nu} - (1-a)\frac{p_\mu p_\nu}{p^2+i\epsilon}\right]\frac{\delta_{ab}}{p^2+i\epsilon}$ |
| Ghost Propagator | $\frac{i\,\delta_{ab}}{p^2+i\epsilon}$ |
| Fermion Vertex | $ig\,\gamma_\mu\,\lambda^a_{\alpha\beta}$ |
| Triple Vertex | $g f_{abc}\left[g_{\mu\nu}(k-q)_\sigma + g_{\nu\sigma}(q-r)_\mu + g_{\sigma\mu}(r-k)_\nu\right]$ |
| Quartic Vertex | $-ig^2\left[f_{abe}f_{cde}(g_{\mu\sigma}g_{\nu\rho} - g_{\mu\rho}g_{\nu\sigma}) + f_{ace}f_{bde}(g_{\mu\nu}g_{\sigma\rho} - g_{\mu\rho}g_{\nu\sigma}) + f_{ade}f_{cbe}(g_{\mu\sigma}g_{\nu\rho} - g_{\mu\nu}g_{\sigma\rho})\right]$ |
| Ghost Vertex | $g f_{abc}\, r_\mu$ |

*From A. J. Buras, Ref. 1.

Although QCD and QED perturbation theory have many similarities, there are non-perturbative aspects of the non-Abelian theory which have no analog in electrodynamics, e.g., classical ("instanton") solutions to the pure gauge theory. These solutions can have profound consequences for the QCD vacuum state. [34] Furthermore, the absence of asymptotic color states implies that, at best, the perturbation rules are only valid in a far-off-shell short-distance regime.

Fortunately for many processes of experimental interest it is possible to prove factorization theories which separate the long-distance dynamics associated with the hadron wavefunction and color confinement from quark and gluon subprocesses which only involve short-distance propagation of color. [35] If this factorization can be proved to all orders in perturbation theory, it is reasonable to assume that the corresponding perturbative predictions are legitimate predictions of the complete theory. In the case of predictions dependent on hadronic fragmentation from quark or gluon jets one has to make an extra assumption that the essential effects of color confinement are restricted to large distances. [3]

### 3.2.2 QCD Perturbation Theory

As in QED, one can sum the effects of vacuum polarization into a "running" coupling constant ($\alpha_s = g^2/4\pi$)

$$\alpha_s(Q^2) = \frac{\alpha_s(Q_0^2)}{1 - \alpha_s(Q^2)\left[\pi(Q^2) - \pi(Q_0^2)\right]}$$

where $\pi(Q^2)$ can be computed (in some gauges) from the single-particle-irreducible contributions to the gluon propagator. Given the gluon propagator at any scale $Q_0^2$, one can use Eq. (2.7) to determine the effective interaction at the scale $Q^2$. To lowest order in perturbation theory the quark and gluon loop insertions give $\left[Q^2, Q_0^2 >> m_i^2,\ i = 1, 2 \ldots n_f\right]$

$$\pi(Q^2) - \pi(Q_0^2) = \frac{1}{4\pi} \log \frac{Q^2}{Q_0^2} \left[\frac{2}{3} n_f - 11\right] + \mathcal{O}(\alpha_s) \tag{2.8}$$

i.e., for $n_f < 33/2$, $\alpha_s(Q^2)$ decreases with $Q^2$, exactly opposite to QED. More generally, one can calculate the $Q^2$ dependence of $\alpha_s$ in higher orders

$$\frac{\partial}{\partial \log Q^2} \alpha_s(Q^2) \equiv \beta\left[\alpha_s(Q^2)\right] = \frac{-\beta_0}{4\pi} \alpha_s^2(Q^2) - \frac{\beta_1}{(4\pi)^2} \alpha_s^3(Q^2) + \ldots \tag{2.9}$$

where [1] $\beta_0 = 11 - 2/3\, n_f$, $\beta_1 = 102 - 38/3\, n_f$. The solution for $\alpha_s(Q^2)$ at large $Q^2$ to two-loop accuracy then has the form

$$\alpha_s(Q^2) = \frac{4\pi}{\beta_0 \log \frac{Q^2}{\Lambda^2} + \frac{\beta_1}{\beta_0} \log\log \frac{Q^2}{\Lambda^2}} \tag{2.10}$$

where $\Lambda$ is introduced as a constant of integration. The fact that $\alpha_s(Q^2)$ decreases at large momentum transfer [asymptotic freedom] is an extra-ordinary feature of QCD which in principle allows a systematic computation of short-distance processes. A graph of $\alpha_s(Q^2)$ showing the effect of the $\beta_1/\beta_0$ term is shown in Fig.4. It should be emphasized that perturbation theory does not determine the form of $\alpha_s$ at small $Q^2$ where its magnitude becomes large. As noted by PARISI and PETRONZIO, [36] consistent calculations of perturbative loops demand that $\alpha_s(Q^2)$ remains finite at all values of the loop integration. Thus far there is no direct experimental evidence that $\alpha_s(Q^2)$ decreases logarithmically.

If we choose $Q_0^2$ to be the ultimate ultraviolet cutoff scale of QCD, then $\alpha_s(Q_0^2) = \alpha_s^0$ is the "bare charge" of the theory. We can then identify $\alpha_s(Q^2)$ as the effective coupling constant which takes into account all vacuum polarization contributions of invariant mass $\mathcal{M}^2$: $Q^2 < \mathcal{M}^2 < Q_0^2$. Similarly, we can define the running quark mass $m(Q^2)$ which takes into account all self-energy insertions in the range $Q^2 < \mathcal{M}^2 < Q_0^2$.

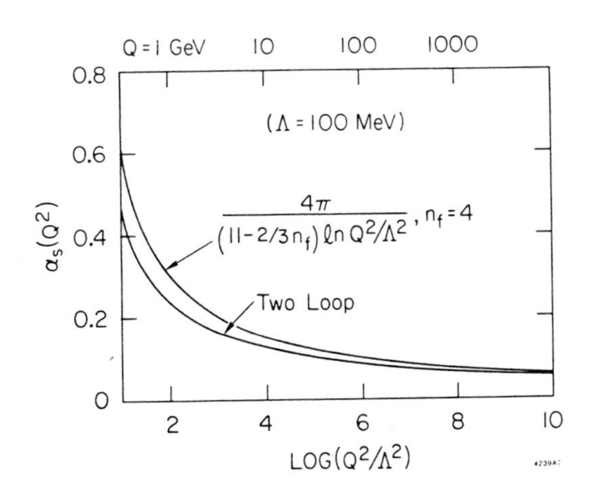


Fig. 4. The QCD coupling constant $\alpha_s(Q^2)$ for $n_f = 4$ to one- and two-loop accuracy. Empirical specifications of $\Lambda$ in a given scheme should use the two-loop formula [Eq. (2.10)]

Let us now define a cutoff Lagrangian $\mathscr{L}^{\kappa}_{QED}$ density for QCD by excluding all intermediate states with $\mathscr{M}^2 > \kappa^2$. The fact that the theory is renormalizable implies that

$$\mathscr{L}^{\kappa}_{QCD} = \bar{\psi}\left(i\not\partial + g(\kappa)\not A - m(\kappa)\right)\psi - \frac{1}{4}\,\mathrm{Tr}\,F^2 + \mathcal{O}\frac{1}{\kappa^2}\, m(\kappa)\;\bar{\psi}\,\sigma_{\mu\nu}F^{\mu\nu}\,\psi + \ldots \tag{2.11}$$

i.e., all effects of very high mass states $\mathscr{M}^2 > \kappa^2$ are completely contained in the effective coupling constant $g(\kappa)$, the quark running mass $m(\kappa)$, and "higher twist" power-law suppressed $1/\kappa^2$, $1/\kappa^4$, etc. terms. If $\kappa^2$ is taken at the ultimate cutoff scale $Q_0^2$, then $\mathscr{L}^{\kappa}_{QCD}$ is the bare Lagrangian. If $\kappa^2$ is chosen sufficiently large, then the higher twist terms are negligible in (2.11).

The classic perturbative calculation in QCD is that of the annihilation cross section $\sigma_{e^+e^-\to \text{hadrons}}$ which can be computed from the hadronic absorptive part of the forward $e^+e^- \to e^+e^-$ amplitude to order $\alpha^2$. Since there are no external color charges, there can be no gluon-mass infrared divergences or quark mass singularities. Thus the only relevant scale is $Q^2 = s = E^2_{c.m.}$, and we can compute perturbatively from $\mathscr{L}^{\kappa}_{QCD}$ with $\kappa^2 = Q^2$. The result to order $\alpha_s^2(Q^2)$ is

$$R_{e^+e^-}(Q^2) = 3\sum_q e_q^2\left[1 + \frac{\alpha_s^{\overline{MS}}(Q^2)}{\pi} + \frac{\alpha_s^2(Q^2)}{\pi^2}\,(B + A\,n_f) + \ldots\right] \tag{2.12}$$

where the $A\,n_f$ term arises from virtual quark loops. An essential and unique prediction of asymptotic freedom is that $\lim_{Q^2\to\infty} R(Q^2) = 3\sum e_q^2 = R^o$, the free quark prediction. The specific values of B and A in Eq. (2.12) depend on the method of implementing the ultraviolet cutoff. In the $\overline{MS}$ scheme (a particular dimensional regularization scheme) one finds [37] $B \cong 1.98$, $A \cong -0.115$. However, in analogy to QED, it is clear that the $A\,n_f$ term should be identified with the fermion loop vacuum polarization contribution to the running coupling constant in the $\alpha_s(\kappa)/\pi$ term; the particular numerical value for A is rather arbitrary, since we could have chosen any scale $\kappa^2 = f^2Q^2$ for the perturbation expansion. In QCD, $\alpha_s$ is essentially a function of $\beta_0 = 11 - 2/3\,n_f$. Thus we write $B + A\,n_f = -3/2\,\beta_0 A + C$, where $C = (33/2)A + B \cong 0.0825$ must be scheme independent (since to the order of interest the cutoff schemes can only differ by the definition of the scale constant $\Lambda^2$). We thus have the QCD prediction: [14]

$$R_{e^+e^-}(Q^2) = 3\sum_q e_q^2\left[1 + \frac{\alpha_s^{\overline{MS}}(f^2Q^2)}{\pi} + 0.0825\,\frac{\alpha_s^2}{\pi^2} + \ldots\right] \tag{2.13}$$

where $f = f_{\overline{MS}} = e^{3A} \cong 0.71$ in the $\overline{MS}$ scheme. Let us imagine that eventually measurements of $\sigma_{e^+e^-\to(Q^2)\,\text{hadrons}}$ will be sufficiently accurate that we can choose $R(Q^2)$ to define a "canonical" measurement of the QCD running coupling constant:

$$\alpha_s^R(Q^2) \equiv \pi\left[\frac{R(Q^2) - R^o}{R^o}\right]\left[1 - 0.0825\left(\frac{R - R^o}{R^o}\right)\right] = \alpha_s^{\overline{MS}}(f^2R^2)\quad . \tag{2.14}$$

Our goal is then to show that all observables in QCD which have a perturbative expansion in $\alpha_s$ can (in principle) be expressed in terms of $\alpha_s^R(Q^2)$ without any scheme or scale ambiguity. We will define the scale parameter $\Lambda = \Lambda^R$ using Eq. (2.10) for $\alpha_s^R$.

We thus propose the following prescription for making scheme- and scale-independent perturbative QCD predictions: [14] For any observable $\rho(Q^2)$ which has a perturbative expansion in $\alpha_s(Q^2)$ one can compute in a given renormalization scheme

$$\rho(Q^2) = \frac{\alpha_s(Q^2)}{\pi} + (A_\rho n_f + B_\rho)\frac{\alpha_s^2(Q^2)}{\pi^2} + \ldots \quad . \tag{2.15}$$

As in the case of $R(Q^2)$, we identify $(-3/2)\beta_0 A_\rho$ as the vacuum polarization correction to the running coupling constant in the $\alpha_s/\pi$ term. Thus

$$\rho(Q^2) = \frac{\alpha_s^R(\hat{Q}^2)}{\pi} + C_\rho\left(\frac{\alpha_s}{\pi}\right)^2 + \ldots \tag{2.16}$$

where

$$\hat{Q}_\rho^2 = e^{3A_\rho - 3A_R}\, Q^2 \tag{2.17}$$

and

$$C_\rho = \frac{33}{2} A_\rho + B_\rho \tag{2.18}$$

are scheme independent. The leading order prediction for $\rho(Q^2)$ can thus be written unambiguously in terms of $\alpha_s^R$. If $C_\rho\,\alpha_s/\pi$ is reasonably small, then we can expect that Eq. (2.16) gives a meaningful perturbative QCD prediction. An important task will be to carry out the above procedure to higher orders in $\alpha_s$.

As an example of the above method, let us consider the decay rate for pseudoscalar quarkonium states which is computed in terms of $Q\bar{Q} \to gg$ plus higher-order subprocesses. In the $\overline{MS}$ scheme [29] (C is a known color factor):

$$\frac{\Gamma(\eta_c \to \text{hadrons})}{\Gamma(\eta_c \to \gamma\gamma)} = C\left[\alpha_s^{\overline{MS}}\left(M_{\eta_c}^2\right)\right]^2 \left\{1 + \frac{\alpha_s^{\overline{MS}}}{\pi}\left(17.13 - \frac{8}{9} n_f\right) + \ldots\right\} \tag{2.19}$$

$$= C\left[\alpha_s^R\left(\left(.37M_{\eta_c}\right)^2\right)\right]^2 \left\{1 + 2.46\,\frac{\alpha_s^R}{\pi} + \ldots\right\}$$

i.e.: the effective scale in the vacuum polarization contributions is ~0.37 $M_{\eta_c}$ relative to the scale in $e^+e^- \to$ hadrons. If $\alpha_s \cong 0.2$, then the correction term in Eq. (2.19) gives only a 7% correction to the determination of $\alpha_s$. In the case of the hadronic decays of $J^{CP} = 1^{--}$ heavy quarkonium states, the correction to the $Q\bar{Q} \to 3g$ decay amplitude appears to be very large, so that the leading order expressions may not be meaningful. One finds [40]

$$\frac{\Gamma(\Upsilon \to \text{hadrons})}{(\Upsilon \to \mu^+\mu^-)} = \frac{10(\pi^2 - 9)}{81\,\pi\, e_b^2\,\alpha^2}\left[\alpha_s^R\left((.22M_\Upsilon)^2\right)\right]^3 \left\{1 - 13.94\,\frac{\alpha_s^R}{\pi} + \ldots\right\} . \tag{2.20}$$

For $\alpha_s \cong 0.2$, the correction term gives a correction of order 30% to the determination of $\alpha_s$. Note that even in QED, the radiative corrections to orthopositronium decay are very large:

$$\Gamma_{3\gamma} = \Gamma_{3\gamma}^{o}\left\{1 - 12.61\ (3)\ \frac{\alpha}{\pi} + \ldots\right\} , \tag{2.21}$$

so this appears to be an intrinsic problem to this type of decay process. Additionally, the QCD prediction for quarkonium decay is complicated by some uncertainties from relativistic and higher Fock-state components in the quarkonium wavefunction.

One of the most important predictions from QCD is the logarithmic variation of structure function moments, $M_n(Q^2) = \int_0^1 dx\, x^n F_3(x,Q)$. Using the above renormalization procedure we find [14]

$$\frac{d}{d \log Q^2} \log M_n(Q^2) = \frac{-\gamma_n}{8\pi} \alpha_s^R \left(f_n^2 Q^2\right) \left[1 - \frac{\alpha_s^R}{\pi} C_n + \ldots\right]$$

where the $\gamma_n$ are known anomalous dimensions (see Sect. IV). The coefficient $C_n$ varies from ~0.27 to 1.1 for non-singlets moments n = 2 to 10, thus giving reasonably small corrections to the lowest-order predictions. The monotonic decrease of $f_n$ with n reflects the fact that the momentum scale for gluon emission becomes increasingly restricted at large n ($\langle 1 - x\rangle \sim O(1/n)$) due to phase-space effects. [41] Further applications and discussions will be given in Ref. 14. We also note that in processes with several large-momentum-transfer scales, the effective argument for $\alpha_s^R$ in the leading order predictions can be very complicated. For example in the case of large $p_T$ jet production due to $qq \to qq$ scattering, the subprocess scattering amplitude involves $\alpha_s$ evaluated at the subprocess invariants $\hat{t}$ and $\hat{u}$, whereas the evolution of each hadronic structure function is sensitive to its respective x-dependent phase-space boundary as well as the quark momentum transfer.

### 3.3 Hadronic Wavefunctions in QCD [27]

Even though quark and gluon perturbative subprocesses are simple in QCD, the complete description of a physical hadronic process requires the consideration of many different coherent and incoherent amplitudes, as well as the effects of non-perturbative phenomena associated with the hadronic wavefunctions and color confinement. Despite this complexity, it is still possible to obtain predictions for many exclusive and inclusive reactions at large momentum transfer, provided we make the ansatz that the effect of non-perturbative dynamics is negligible in the short-distance and far-off-shell domain. (This assumption appears reasonable, since a linear confining potential V ~ r is neglibible compared to perturbative 1/r contributions.) For many large-momentum-transfer processes, such as deep-inelastic lepton-hadron scattering reactions and meson form factors, one can then rigorously isolate the long-distance confinement dynamics from the short-distance quark and gluon dynamics — at least to leading order in $1/Q^2$. [35] The essential QCD dynamics can thus be computed from (irreducible) quark and gluon subprocesses amplitudes as a perturbative expansion in an asymptotically small coupling constant $\alpha_s(Q^2)$.

An essential part of the QCD predictions are the hadronic wavefunctions which determine the probability amplitudes and distributions of the quark and gluons which enter the short-distance subprocesses. The hadronic wavefunctions provide the link between the long-distance non-perturbative and short-distance perturbative physics. Eventually, one can hope to compute the wavefunctions from the theory, e.g., from lattice or bag models, or directly from the QCD equations of motions, as we shall outline below. Knowledge of hadronic wavefunctions will also provide explicit connections between exclusive and inclusive processes, and will allow the normalization and specification of the power-law (higher twist) corrections to the leading impulse approximation results. As we shall discuss in Sect. VI, there are a number of novel QCD phenomena associated with hadronic wavefunctions, including the effects of intrinsic gluons, intrinsic heavy-quark Fock components, diffraction dissociation phenomena, and "direct" hadron processes where the valence Fock state of a hadron enters coherently into a short-distance quark-gluon subprocess.

The most convenient representation of a wavefunction in a relativistic field theory is to use a momentum space Fock-state basis defined at equal "time" $\tau = t + z$ on the light cone (see Fig.5a): [42]

$$\left\{ \psi_n(\vec{k}_{\perp i}, x_i ; \lambda_i) \right\} \tag{3.1}$$

Momentum conservation requires

$$\sum_{i=1}^{n} \vec{k}_{\perp i} = 0, \qquad \sum_{i=1}^{n} x_i = 1, \qquad 0 < x_i < 1 \quad . \tag{3.2}$$

The $\vec{k}_{\perp i}$ are the transverse momentum of the (on-mass-shell) constituents relative to the bound-state 3-momentum $\vec{P} = P^3$. The $x_i$ are the light-cone momentum fractions $\left(k^{\pm} = k^o \pm k^3,\ A \cdot B = 1/2(A^+B^- + A^-B^+) - \vec{A}_\perp \cdot \vec{B}_\perp\right)$

$$x_i = \frac{k_i^+}{p^+} = \frac{(k^o + k^3)_i}{p^o + p^3} \tag{3.3}$$

(In a frame where $P^3 \to \infty$, the $x_i$ are the longitudinal momentum fractions.) The mass shell condition is $k^2 = m^2$, or $k^- = (k_\perp^2 + m^2)/k^+$. As we shall see, the equal-$\tau$ formalism is equivalent to the usual Schroedinger equal-time theory in the non-relativistic limit.

A unique and remarkable advantage of quantizing a relativistic theory at equal $\tau$ is the fact that the perturbative vacuum state $|0\rangle$ is also an eigenstate of the full Hamiltonian. Matrix elements where particles are created out of the vacuum are excluded because of the fact that all particles must have $k_i^+ > 0$. Furthermore, the charge operator and the current $J^+ = J^o + J^3$ are diagonal in the Fock-state basis.

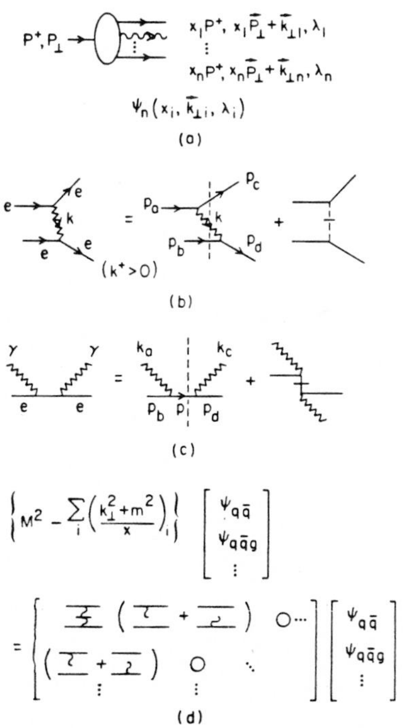

Fig. 5. (a) the n-particle Fock-state amplitude defined at equal $\tau$. The state is off the $p^-$ light-cone energy shell [see Eq. (3.12)]. (b,c) Examples of light-cone time-ordered perturbation theory calculations. The frame is chosen so that $k^+ > 0$. (d) QCD equation of motion for the meson wavefunction

It is particularly advantageous to choose the light-cone gauge $A^+ = A^o + A^3 = 0$, since unphysical degrees of freedom do not appear. A comparison between time-ordered and τ-ordered perturbation theory is given in Table II.

Thus at a given "time" we can define the (color-singlet) basis

$$|0\rangle \tag{3.4}$$

$$|q\bar{q}\rangle = a^{+}_{k^+,\vec{k}_\perp} b^{+}_{k^{+\prime},\vec{k}'_\perp} |0\rangle$$

$$\vdots$$

The pion state, for example, can be expanded as

$$|\pi\rangle = |q\bar{q}\rangle \Psi_{q\bar{q}} + |q\bar{q}g\rangle \Psi_{q\bar{q}g} + \ldots \tag{3.5}$$

where $\Psi_n = \langle n|\pi\rangle$ is the amplitude for finding the Fock state $|n\rangle$ in $|\pi\rangle$ at time τ. The full Fock-state wavefunction which describes the n-particle state of a hadron with 4-momentum $P^\mu = (P^+, P^-, \vec{P}_\perp)$ and constituents with momenta

$$k^\mu = \left(k^+, k^-, \vec{k}_\perp\right) = \left(xP^+, \frac{(x\vec{P}_\perp + \vec{k}_\perp)^2 + m^2}{x P^+}, x\vec{P}_\perp + \vec{k}_\perp\right) \tag{3.6}$$

Table 2. Time-ordered perturbation theory

| Equal t | Equal τ = t + z |
|---|---|
| $k^o = \sqrt{\vec{k}^2 + m^2}$ (particle mass shell) | $k^- = \frac{k_\perp^2 + m^2}{k^+}$ (particle mass shell) |
| $\sum \vec{k}$ conserved | $\sum \vec{k}_\perp, k^+$ conserved |
| $\mathcal{M}_{ab} = V_{ab} + \sum_c V_{ac} \frac{1}{\sum_a k^o - \sum_c k^o + i\epsilon} V_{cb}$ | $\mathcal{M}_{ab} = V_{ab} + \sum_c V_{ac} \frac{1}{\sum_a k^- - \sum_c k^- + i\epsilon} V_{cb}$ |
| n! time-ordered contributions | $k^+ > 0$ only |
| Fock states $\psi_n(\vec{k}_i)$ | Fock states $\psi_n(\vec{k}_{\perp i}, x_i)$ |
| $\sum_{i=1}^{n} \vec{k}_i = \vec{P} = 0$ | $x = \frac{k^+}{P^+}$, $\sum_{i=1}^{n} x_i = 1$, $\sum_{i=1}^{n} \vec{k}_{\perp i} = 0$ $(0 < x_i < 1)$ |
| $\mathcal{E} = P^o - \sum_{i=1}^{n} k_i^o = M - \sum_{i=1}^{n} \sqrt{k_i^2 + m_i^2}$ | $\mathcal{E} = P^+\left(P^- - \sum_{i=1}^{n} k_i^-\right) = M^2 - \sum_{i=1}^{n} \left(\frac{k_\perp^2 + m^2}{x}\right)_i$ |

and spin projection $\lambda_i$ is

$$\Psi_n = \psi_n\left(x_i, k_{\perp i}; \lambda_i\right) \prod_{\text{fermions}} \frac{u\left(x_i P^+, x_i \vec{P}_\perp + \vec{k}_{\perp i}\right)_{\lambda_i}}{\sqrt{x_i}} \cdot \prod_{\text{gluons}} \frac{\varepsilon\left(x_i P^+, x_i \vec{P}_\perp + \vec{k}_{\perp i}\right)_{\lambda_i}}{\sqrt{x_i}} . \tag{3.7}$$

Note that $\psi_n(x_i, \vec{k}_{\perp i}; \lambda_i)$ is independent of $P^+$, $\vec{P}_\perp$. The general normalization condition is

$$\sum_n \int \left[d^2 k_\perp\right] \int [dx] \left|\psi_n(x_i, \vec{k}_{\perp i}; \lambda_i)\right|^2 = 1 \tag{3.8}$$

where by momentum conservation

$$\left[d^2 k_\perp\right] = 16\,\pi^3 \delta^2 \left(\sum_{i=1}^n \vec{k}_{\perp i}\right) \prod_{i=1}^n \frac{d^2 k_{\perp i}}{16\,\pi^3} \tag{3.9}$$

and

$$[dx] = \delta\left(1 - \sum_{i=1}^n x_i\right) \prod_{i=1}^n dx_i . \tag{3.10}$$

In the non-relativistic limit the equal $\tau = t + z/c$ and equal time t theories coincide. For example, for the Fock-state wavefunction in the rest system we can identify

$$x = \frac{k^o + k^3}{M} \cong \frac{m}{M} + \frac{k^3}{M} , \tag{3.11}$$

and the off-shell light-cone energy is

$$\mathscr{E} = P^+ \left[P^- - \sum_{i=1}^n k^-\right] = M^2 - \sum_{i=1}^n \left(\frac{k_\perp^2 + m^2}{x}\right)_i$$
$$\cong 2M \left[\mathscr{E}_{NR} - \sum_{i=1}^n \left(\frac{k_\perp^2 + k_3^2}{2m}\right)_i\right] . \tag{3.12}$$

Thus, in the non-relativistic limit, the hydrogen atom wavefunction is

$$\psi_{1s} = \frac{C}{\left[k_\perp^2 + \left(m_e - xM\right)^2 + \alpha^2 m_e^2\right]^2} . \tag{3.13}$$

Light-cone perturbation theory rules can be derived by either evaluating standard equal-time time-ordered perturbation theory for an observer in a fast-moving Lorentz frame (the "infinite momentum" method), [43] or more directly, by quantizing at equal $\tau$. The LCPTh rules are: [19,44]

(1) For each Feynman diagram, assign particle 4-momentum $k^\mu$ such that $k^+, \vec{k}_\perp$ is conserved at each of the n vertices. (This is the analogue of 3-momentum conservation.) Since all particles are on the (positive energy) mass shell ($k^2 = m^2$) we have

$$k^- = \frac{k_\perp^2 + m^2}{k^+} > 0 . \tag{3.14}$$

(2) Construct all time orderings (up to n!) such that $k^+ > 0$ for all particles.

(3) For each intermediate state, assign a propagator

$$\frac{1}{\sum_{\text{initial}} k_i^- - \sum_{\text{intermediate}} k_i^- + i\varepsilon} \tag{3.15}$$

and a factor $1/k^+$ for each internal line. (This is the analogue of

$1/\left(\sum_{\text{initial}} E_i - \sum_{\text{intermediate}} E_i + i\varepsilon\right)$ and $1/(2E)$ in TOPTh.)

(4) For each loop, integrate

$$\int \frac{d^2 k_\perp}{2(2\pi)^3} \int_0^\infty dk^+$$

and sum over intermediate-state spins and polarization.

(5) The vertex factors depend on the theory. In the case of $g\phi^3$ interaction, assign a factor g at each vertex. In gauge theories the gluon-fermion vertices are

$$g\bar{u}\not{\varepsilon}u,\ -g\bar{v}\not{\varepsilon}v,\ g\bar{u}\not{\varepsilon}v,\ -g\bar{v}\not{\varepsilon}u \quad . \tag{3.17}$$

The trigluon and quartic-gluon vertices are given in Table 1.

(6) Finally, there are instantaneous gluon contributions in $A^+ = 0$ gauge:

$$\frac{\gamma^+ \ldots \gamma^+}{(k^+)^2} \tag{3.18}$$

(analogous to Coulomb interactions) and instantaneous fermion contributions $\gamma^+/2k^+$ (the remnant of backward-moving "Z-graph" fermion lines). For example, the electron-electron scattering diagrams of Fig.5b give

$$\mathcal{M}_{ee\to ee} = e^2 \frac{\bar{u}\gamma^\mu u\ \bar{u}\gamma^\nu u}{k^+ D} d_{\mu\nu} + e^2 \frac{u\gamma^+ u\ u\gamma^+ u}{(k^+)^2} \tag{3.19}$$

where the polarization sum is

$$d^{\mu\nu} = \sum_{\lambda=1,2} \varepsilon^\mu_\lambda \varepsilon^\nu_\lambda, \quad \varepsilon^+ = 0, \quad k\cdot\varepsilon = 0 \tag{3.20}$$

and the light-cone and energy denominator is

$$D = p_a^- - k^- - p_c^- + i\varepsilon \quad . \tag{3.21}$$

Similarly, the Compton scattering diagrams of Fig.5c give

$$\mathcal{M}_{\gamma e\to\gamma e} = e^2 \sum_{\lambda=1,2} \frac{\bar{u}\not{\varepsilon}_c u_\lambda\ \bar{u}_\lambda \not{\varepsilon}_a u}{p^+ D} + e^2 \frac{\bar{u}\not{\varepsilon}_c\ \gamma^+ \not{\varepsilon}_a u}{2p^+} \tag{3.22}$$

$$D = k_a^- + p_b^- - p^- + i\varepsilon \ .$$

(This is analogous to the decomposition of the Feynman propagator $(\not{p} - m + i\varepsilon)^{-1}$ into positive and negative frequency components.)

Calculations in light-cone perturbation theory are often surprisingly simple, since one can usually choose Lorentz frames for the external particles such that only a few time orderings need to be considered. All the variables have a direct physical interpretation. The formalism is also ideal for computing helicity amplitudes directly without trace projection techniques. A list of all the gluon-fermion vertices which are required as gauge theory calculations is given in Tables I and II of Ref. 19.

It is straightforward to implement ultraviolet renormalization in light-cone perturbation theory. We define truncated wavefunctions $\psi^\kappa$ and a truncated Hamiltonian $H^\kappa$ such that all intermediate states with $|\mathcal{E}| > \kappa^2$ are excluded. [45] Thus $\kappa^{-1}$ is analogous to the lattice spacing in lattice field theory. Since QCD is renormalizable, the effects of the neglected states are accounted for by the use of the running coupling constant $\alpha_s(\kappa^2)$ and running mass $m(\kappa^2)$, as long as $\kappa^2$ is sufficiently large compared to all physical mass thresholds. Completeness implies

$$\sum_{n,\lambda_i} \int \left[d^2k_\perp\right] \int [dx] \left|\psi_n^\kappa(x_i, k_{\perp i}; \lambda_i)\right|^2 = 1 - \mathcal{O}\left(\frac{m^2}{\kappa^2}\right) . \tag{3.23}$$

The equation of state for the meson or baryon wavefunction in QCD is a set of coupled multiparticle equations (see Fig.5d):

$$\left[M^2 - \sum_{i=1}^{n} \left(\frac{k_\perp^2 + m^2}{x}\right)_i\right] \psi_n^\kappa = \sum_{n'} V_{nn'}^\kappa \, \psi_{n'}^\kappa \tag{3.24}$$

where $M^2$ is the eigenvalue and $V_{nn'}$ is the set of diagonal (from instantaneous gluon and fermion exchange) and off-diagonal (from the 3-and 4-particle vertices) momentum-space matrix elements dictated by the QCD rules. Because of the $\kappa$ cutoff, the equations truncate at finite n,n'. In analogy to non-relativistic theory, one can imagine starting with a trial wavefunction for the lowest $|q\bar{q}\rangle$ or $|qqq\rangle$ valence state of a meson or baryon and iterating the equations of motion to determine the lowest eigenstate Fock-state wavefunctions and mass M. Invariance under changes in the cutoff scale provides an important check on the consistency of the results. Note that the general solution for the hadron wavefunction in QCD is expected to have Fock-state components with arbitrary numbers of gluons and quark-antiquark pairs.

The two-particle "valence" light-cone Fock-state wavefunction for mesons or positronium can also be related to the Bethe-Salpeter wavefunction evaluated at equal $\tau$:

$$\int \frac{dk^-}{2\pi} \psi_{BS}(k;p) = \frac{u(x_1,\vec{k}_\perp)}{\sqrt{x_1}} \frac{\bar{v}(x_2,-\vec{k}_\perp)}{\sqrt{x_2}} \psi(x_i,\vec{k}_\perp) \tag{3.25}$$

**+ negative energy components,**

where $\psi$ satisfies an exact bound-state equation [19]

$$\left[M^2 - \frac{k_\perp^2 + m_1^2}{x} - \frac{k_\perp^2 + m_2^2}{x_2}\right] \psi(x_i,\vec{k}_{\perp i}) = \int_0^1 dy \int \frac{d^2\ell_\perp}{16\,\pi^3} \tilde{K}\left(x_i,\vec{k}_{\perp i}; y_i,\vec{\ell}_{\perp i}; M^2\right) \psi(y_i,\vec{\ell}_{\perp i}) \quad . \tag{3.26}$$

The kernel $\tilde{K}$ is computed from the sum of all two-particle-irreducible contributions to the two-particle scattering amplitude. For example, the equation of motion for the $|e^+e^-\rangle$ Fock state of positronium reduces in the non-relativistic limit to $\left(k_\perp, \ell_\perp \sim \mathcal{O}(\alpha m),\ x = x_1 - x_2 \sim O(\alpha)\right)$ $M^2 = 4m^2 + 4m\varepsilon$

$$\left\{\varepsilon - \frac{k_\perp^2 + x^2 m^2}{m}\right\} \psi(x_i, k_\perp) \tag{3.27}$$

$$= (4x_1x_2) \int_{-1}^{1} dy\ m \int \frac{d^2\ell_\perp}{(2\pi)^3} \left[\frac{-e^2}{(\vec{k}_\perp - \vec{\ell}_\perp)^2 + (x-y)^2 m^2}\right] \psi(y_i, \ell_\perp) .$$

The non-relativistic solution is ($\beta = \alpha m/2$) [19]

$$\psi(x_i, k_\perp) = \sqrt{\frac{m\beta^3}{\pi}} \frac{64\pi\beta\, x_1 x_2}{\left[k_\perp^2 + (x_1 - x_2)^2 m^2 + \beta^2\right]^2} \begin{cases} \dfrac{u_\uparrow \bar{v}_\downarrow - u_\downarrow \bar{v}_\uparrow}{\sqrt{2\, x_1 x_2}} \\ \\ \dfrac{u_\uparrow \bar{v}_\uparrow}{\sqrt{x_1 x_2}} \end{cases} \tag{3.28}$$

for para and ortho states respectively.

More generally, we can make an (approximate) connection between the equal-time wavefunction of a composite system and the light-cone wavefunction by equating the off-shell propagator $\mathcal{E} = M^2 - \left(\sum_{i=1}^{n} k_i\right)^2$ in the two frames:

$$\mathcal{E} = \begin{cases} M^2 - \left(\sum\limits_{i=1}^{n} q^o_{(i)}\right)^2 , & \sum\limits_{i=1}^{n} q_i = 0 \text{ [C.M.]} \\ \\ M^2 - \sum\limits_{i=1}^{n} \left(\dfrac{k_\perp^2 + m^2}{x}\right)_i , & \sum \vec{k}_{\perp i} = 0, \quad \sum x_i = 1 \text{ [L.C.]} . \end{cases} \tag{3.29}$$

In addition we can identify

$$x_i = \frac{k_\perp^+}{p^+} \rightleftarrows \frac{\left(q^o + q^3\right)_i}{\sum\limits_{j=1}^{n} q^o_{(j)}} , \quad \vec{k}_{\perp i} \rightleftarrows \vec{q}_{\perp i} . \tag{3.30}$$

For a relativistic two-particle state with a wavefunction which is a function of the off-shell variable $\mathcal{E}$ only, then we can identify ($m_1 = m_2 = m$, $x = x_1 - x_2$) [27]

$$\psi_{L.C.}\left(\frac{k_\perp^2 + m^2}{1 - x^2} - m^2\right) \rightleftarrows \psi_{C.M.}(\vec{q}^{\,2}) . \tag{3.31}$$

In the non-relativistic limit, this corresponds to the identification $\vec{q}_\perp = \vec{k}_\perp$, $q_3^2 = x^2 m^2$.

## 3.4 Measures of Hadronic Wave Functions

### 3.4.1 Form Factors of Composite Systems

If we could solve the QCD equation of motion, Eq. (3.24), for the light-cone wave-functions $\psi_n$ of a hadron, then we could (in principle) calculate all of its electromagnetic properties. For example, to compute the elastic form factors $\langle p|J^\mu(0)|p+q\rangle$ of a hadron we choose the Lorentz frame [46]

$$p^\mu = (p^+,p^-,\vec{p}_\perp) = \left(P^+,\frac{M^2}{P^+},\vec{0}_\perp\right) \tag{4.1}$$

$$q^\mu = (q^+,q^-,\vec{q}_\perp) = \left(0,\frac{2p\cdot q}{P^+},\vec{q}_\perp\right)$$

where $p^2 = (p+q)^2 = M^2$ and $-q^2 = Q^2 = \vec{q}_\perp^{\,2}$. Then the only time ordering which contributes to the $\langle p|J^+|p+q\rangle$ matrix element is where the photon attaches directly to the $e_j\bar{u}_j\gamma^+u_j$ currents of the constituent quarks. The spin-averaged form factor is [46,19] (see Fig.6a)

$$F(Q^2) = \sum_n\sum_j e_j\int[dx]\left[d^2k_\perp\right]\sum_{\lambda_i}\psi_n^{*\kappa}(x_i,\vec{k}'_{\perp i};\lambda_i)\,\psi_n^\kappa(x_i,\vec{k}_{\perp i};\lambda_i) \tag{4.2}$$

where $\vec{k}_j^{\perp\prime} = \vec{k}_j^\perp + (1-x_j)\vec{q}^\perp$ for the struck quark and $\vec{k}_i^\perp - x_i\vec{q}^\perp$ $(i\neq j)$ for the spectator quarks. (The $-x_i\vec{q}^\perp$ terms occur because the arguments $\vec{k}^{\perp\prime}$ are calculated relative to the direction of the final-state hadron.) We choose $\kappa^2 \gg Q^2,M^2$. We note here the special advantage of light-cone perturbation theory: the current $J^+$ is diagonal in the Fock-state basis.

Because of Eq. (3.23) the form factor is normalized to 1 at zero momentum transfer. We can also compute the helicity flip form factors in the same manner. [19,47] For example, the anomalous moment $a = F_2(0)$ of any spin-1/2 system can be written [47]

$$\frac{a}{M} = -\sum_j e_j\int[dx]\left[d^2k_\perp\right]\psi_{p\uparrow}^{*\kappa}\sum_{i\neq j}x_j\left(\frac{\partial}{\partial k_i^1}+i\frac{\partial}{\partial k_i^2}\right)\psi_{p\downarrow}^\kappa \quad . \tag{4.3}$$

Explicit calculations of the electron anomalous moment in QED using this result are given in Ref. 47. We notice that in general all Fock states $\psi_n^\kappa$ contribute to the anomalous moment of a system, although states with $\kappa^2$ much larger than the mean off-shell energy $\langle\mathscr{E}\rangle$ are not expected to be important. The general result (4.3) also includes the effects of the Lorentz boost of the wavefunction from $p^\mu$ to $(p+q)^\mu$. In particular, the Wigner spin rotation contributes to $F_2(q^2)$ and the charge radius $F_1'(q^2)$ in the $q^2\to 0$ limit and can only be neglected in the limit of non-relativistic binding $\langle\mathscr{E}\rangle \ll M^2$. This effect gives non-trivial relativistic corrections [48] to nuclear magnetic moment calculations based on simple additivity $\vec{\mu} = \left\langle\sum_j\vec{\mu}_j\right\rangle$.

### 3.4.2 Form Factors of Mesons

Results such as Eqs. (4.2) and (4.3) are formally exact but useless unless we have complete knowledge of the hadronic or nuclear wave-function. However, by making use of the impulse approximation and the smallness of the QCD running coupling constant, we can calculate features of elastic and inelastic large-momentum-transfer processes [19] without explicit knowledge of the wavefunction. For example consider the $|q\bar{q}\rangle$ Fock state component contribution to the pion form factor. Choosing $\kappa^2 = Q^2$, we have

$$F_\pi(Q^2) = \int_0^1 dx\int^Q\frac{d^2k_\perp}{16\pi^3}\,\psi^{*Q}(x,\vec{k}_\perp)\,\psi^Q\!\left(x,\vec{k}_\perp+(1-x)\vec{q}_\perp\right) \tag{4.4}$$

$$+\ \text{higher Fock-state contributions} \quad .$$

The bound-state wavefunctions are peaked at low transverse momentum, i.e., small off-shell energy $\mathcal{E}$. Thus the leading contributions at large $Q^2$ come from the regimes (a) $\vec{k}_\perp^2 \ll \vec{q}_\perp^2$ and (b) $\left(\vec{k}_\perp + (1-x)\vec{q}_\perp\right)^2 \ll \vec{q}_\perp^2$. Thus

$$F_\pi^{(a)}(Q^2) \cong \int_0^1 dx\, \phi(x,Q)\, \psi^Q\left(x,(1-x)\vec{q}_\perp\right) \tag{4.5}$$

where [19]

$$\phi(x,Q) \equiv \int^Q \frac{d^2k}{16\pi^3}\, \psi^Q(x,k_\perp) \quad . \tag{4.6}$$

If we simply iterate the one-gluon exchange kernel $V_1$ in the equation of motion for $\psi$, then for $q_\perp^2 \gg \langle \ell_\perp^2 \rangle$

$$\begin{aligned}\psi^Q\left(x,(1-x)q_\perp\right) &\cong \int_0^1 dy \int^Q \frac{d^2\ell_\perp}{16\pi^3}\, \frac{V_1\left(x,(1-x)q_\perp;y,\ell_\perp\right)\psi^Q(y,\ell_\perp)}{-\,q_\perp^2(1-x)/x} \\ &\cong \int_0^1 dy\, \frac{V_1\left(x,(1-x)q_\perp;y,0_\perp\right)}{-\,q_\perp^2(1-x)/x}\, \phi(y,Q) \quad .\end{aligned} \tag{4.7}$$

Thus we can write the gluon exchange contribution to the form factor in the form [11,19] (see Fig.6b);

$$F_\pi(Q^2) = \int_0^1 dx\, dy\, \phi^*(y,Q)\, T_H(x,y;\,Q)\, \phi(y,Q) \tag{4.8}$$

where

$$T_H = \frac{16\pi\, C_F\, \alpha_s(Q^2)}{Q^2}\left[\frac{e_1}{(1-y)(1-x)} + \frac{e_2}{xy}\right] \tag{4.9}$$

is the "hard scattering amplitude" for scattering collinear constituents q and $\bar{q}$ from the initial to the final direction. The color factor is $C_F = (n_c^2-1)/2n_c =$

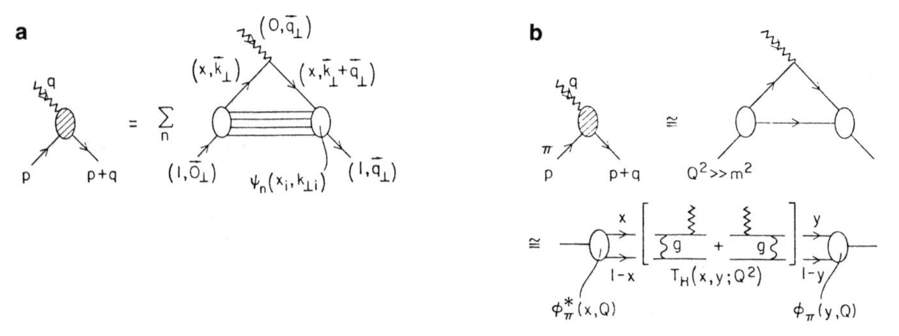


Fig.6. (a) Calculation of current matrix elements in light-cone perturbation theory. (b) Valence Fock-state contribution to the large-momentum-transfer meson form factor. $T_H$ is computed for zero-mass quarks q and $\bar{q}$ parallel to the pion momentum

4/3. The "distribution amplitude" $\phi(x,Q)$ is the amplitude for finding the $|q\bar{q}\rangle$ Fock state in the pion collinear up to the scale Q. (It is analogous to the wave-function at the origin in non-relativistic calculations.) The distribution amplitude enters universally in all large-momentum-transfer exclusive amplitudes and is a process-independent measure of the valence quark distribution in each hadron; its (logarithmic) dependence on $Q^2$ can be determined directly from the operator product expansion or the light cone or from an evolution equation, as we discuss below.

Thus the simplest estimate for the asymptotic behavior of the meson form factor is $F_\pi(Q^2) \sim \alpha_s(Q^2)/Q^2$. To see if this is correct we must examine the higher-order corrections: [19]

(1) Contributions from higher particle number Fock states $|q\bar{q}g\rangle$, $|q\bar{q}q\bar{q}\rangle$, etc. are power-law suppressed, since (in light-cone gauge) the numerator couplings cannot compensate the extra fall-off in $Q^2$ from the extra energy denominators.

(2) All infrared singularities and contributions from soft ($\ell_\perp \to 0$) gluons cancel in color-singlet matrix elements. [It is interesting to note that the quark (Sudakov) form factor falls faster at large $Q^2$ than $F_\pi(Q^2)$.]

(3) Vertex and vacuum polarization corrections to the $T_H$ are higher order in $\alpha_s(Q^2)$, since we choose $\kappa^2 = Q^2$. The effective argument of $\alpha_s$ in $T_H$ is $Q^2 = xyQ^2$ or $(1-x)(1-y)Q^2$ corresponding to the actual momentum transfer carried by the gluon.

(4) By definition, $\phi(x,\kappa^2)$ sums all (reducible) contributions from low-momentum-transfer gluon exchange in the $q\bar{q}$ wavefunction. Hard gluon contributions with $|\mathscr{E}| > \kappa^2$ and the irreducible (cross-graph, etc.) give contributions to $T_H$ which are higher-order $\alpha_s(Q^2)$. By analyzing the denominators in $T_H$ one can show that the natural $\mathscr{E}$ cutoff for $\phi(x,\kappa)$ which minimizes higher-order contributions is $\kappa^2 = Q_x^2 = Q^2 \min\left\{\frac{x}{1-x}, \frac{1-x}{x}\right\}$.

(5) Although $T_H$ is singular at $x \to 0,1$, the endpoint behavior of $\phi(x,Q^2) \sim x^\varepsilon$, $(1-x)^\varepsilon$ ($\varepsilon > 0$) is sufficient to render this region harmless.

### 3.4.3 The Meson Distribution Amplitude

The essential prediction of QCD for the pion form factor is the power-law behavior [8] $F_\pi \sim 1/Q^2$, with logarithmic corrections from the explicit powers of $\alpha_s(Q^2)$ in $T_H$ and the $Q^2$ dependence of the distribution amplitudes $\phi(x,Q^2)$.

The variation of $\phi$ with $Q^2$ comes from the upper limit of the $\vec{k}_\perp$ integration (since $\psi \sim 1/k_\perp^2$) and the renormalization scale dependence:

$$\psi^Q(x,\vec{k}_\perp) = \frac{Z_2(Q)}{Z_2(Q_0)}\,\psi^{Q_0}(x,\vec{k}_\perp) \tag{4.10}$$

due to the vertex and self-energy insertions. Thus

$$Q^2\frac{\partial}{\partial Q^2}\phi(x,Q) = \frac{Q^2}{16\pi^2}\psi^Q(x,\vec{q}_\perp) + \frac{d}{d\log Q^2}\log Z_2(Q^2)\,\phi(x,Q) \quad . \tag{4.11}$$

To order $\alpha_s(Q^2)$ we can compute $Q^2\psi$ from one-gluon exchange [as in Eq. (4.7)], and $d\log Z_2(Q^2)/d\log Q^2 = \alpha_s(Q^2)\gamma_F/4\pi$. Setting $\phi(x,Q) = x(1-x)$, $\tilde{\phi}(x,Q) = x_1x_2\tilde{\phi}$, we obtain an "evolutional equation" [19]

$$x_1x_2Q^2\frac{\partial}{\partial\log Q^2}\tilde{\phi}(x_i,Q) = \frac{\alpha_s(Q^2)}{4\pi}\int_0^1[dy]\,V(x_i,y_i)\,\phi(y,Q) \tag{4.12}$$

where

$$V(x_i,y_i) = 2C_F\left\{x_1y_2\,\theta(y_1-x_1)\left(\delta_{h_1\bar{h}_2} + \frac{\Delta}{y_1-x_1}\right) + (1\leftrightarrow 2)\right\} \tag{4.13}$$

($\delta_{h_1\bar{h}_2} = 1$ when the q and $\bar{q}$ helicities are opposite) and

$$\Delta\tilde{\phi}(y_i,Q) = \tilde{\phi}(y_i,Q) - \tilde{\phi}(x_i,Q) \quad . \tag{4.14}$$

The $\tilde{\phi}(x_1,Q)$ subtraction is due to the $\gamma_F\phi$ term — i.e., the infrared dependence at $y_i = x_i$ is cancelled for color-singlet hadrons. Thus given the initial condition $\phi(x_i,Q_0)$, perturbation theory determines the evolution of $\phi(x,Q)$ for $Q > Q_0$. The solution to the evolution equation is [19]

$$\phi(x_i,Q) = x_1x_2\sum_{n=0}^{\infty} a_n(Q_0^2)\, C_n^{3/2}(x_1-x_2)(\log Q^2/\Lambda^2)^{-\gamma_n} \tag{4.15}$$

where the Gegenbauer polynomials $C_n^{3/2}$ (orthogonal on $\int[dx]x_1x_2$) are eigenfunctions of $V(x_i,y_i)$. The corresponding eigenvalues are the "non-singlet" anomalous dimensions:

$$\gamma_n = \frac{C_F}{\beta_0}\left[1 + 4\sum_2^{n+1}\frac{1}{k} - \frac{2\delta_{h_1\bar{h}_2}}{(n+1)(n+2)}\right] \geq 0 \quad . \tag{4.16}$$

These results can also be derived by using the operator product expansion for the distribution amplitude. [49] By definition

$$\phi(x,Q) = \Lambda^+\int\frac{dz^-}{2\pi}\,e^{ixz^-/2}\, i\langle 0|\bar{\psi}(z)\psi(0)|\pi\rangle^Q\Big|_{z^+=0,\; z^2=-z_\perp^2 = \mathcal{O}(-1/Q^2)} \tag{4.17}$$

($\Lambda^+$ is the positive-energy spinor projection operator). The relative separation of the q and $\bar{q}$ thus approaches the light cone $z^2 = 0$ as $Q^2\to\infty$. Equation (4.16) then follows, by expanding $\psi(z)\psi(0)$ in local operators.

The coefficients $a_n$ are determined from $\phi(x_i,Q_0)$:

$$a_n\left(\log\frac{Q^2}{\Lambda^2}\right)^{-\gamma_n} = \frac{2(2n+3)}{(2+n)(1+n)}\int_{-1}^{1} d(x_1-x_2)\, C_n^{3/2}(x_1-x_2)\,\phi(x_i,Q_0) \quad . \tag{4.18}$$

For $Q^2\to\infty$, only the leading $\gamma_0 = 0$ term survives

$$\lim_{Q^2\to\infty}\phi(x,Q) = a_0x_1x_2 \tag{4.19}$$

where

$$\frac{a_0}{6} = \int_0^1 dx\,\phi(x,Q) = \int_0^1 dx\int^Q\frac{d^2k_\perp}{16\pi^3}\,\psi^Q(x,k_\perp) \tag{4.20}$$

is the meson wavefunction at the origin as measured in the decay $\pi\to\mu\nu$:

$$\frac{a_0}{6} = \frac{1}{2\sqrt{n_c}}\,f_\pi \quad . \tag{4.21}$$

More generally, the leptonic decay ($\rho^0 \to e^+e^-$, etc.) of each meson normalizes its distribution amplitude by the "sum rule"

$$\int_0^1 dx\, \phi_M(x,Q) = \frac{f_M}{2\sqrt{n_c}} \quad , \tag{4.22}$$

independent of Q. The fact that $f_\pi \neq 0$ implies that the probability of finding the $|q\bar{q}\rangle$ Fock state in the pion is non-zero. In fact all the Fock-state wavefunctions $\psi_n^\kappa(x_i,\vec{k}_{\perp i})(|\mathcal{E}| < \kappa^2)$ are well defined, even in the infrared limit $x_i \to 0$ (since $|\mathcal{E}| \sim \langle k_\perp^2\rangle/x_i$ and $\langle k_\perp^2\rangle$ is non-zero for a state of finite radius).

The pion form factor at high $Q^2$ can thus be written [11,19,50]

$$F_\pi(Q^2) = \int_0^1 dx\, \phi^*(x,Q) T_H(x,y;\, Q)\, \phi(y,Q) \tag{4.23}$$

$$T_H = \frac{16}{3\pi} \frac{\alpha_s\big((1-x)(1-y)Q^2\big)}{(1-x)(1-y)Q^2} \quad .$$

Thus

$$F_\pi(Q^2) = \left|\sum_{n=0} a_n \log^{-\gamma_n} Q^2/\Lambda^2\right|^2 \frac{16\pi}{3} \frac{\alpha_s(\bar{Q}^2)}{Q^2} \tag{4.24}$$

$$\times \left[1 + \mathcal{O}\left(\frac{\alpha_s(Q^2)}{\pi}\right) + \mathcal{O}\left(\frac{m^2}{Q^2}\right)\right]$$

where $\bar{Q}^2 \cong \langle(1-x)(1-y)\rangle Q^2$. Finally, for the asymptotic limit where only the leading anomalous dimension contributes: [51]

$$\lim_{Q^2 \to \infty} F_\pi(Q^2) = 16\pi\, f_\pi^2 \frac{\alpha_s(Q^2)}{Q^2} \quad . \tag{4.25}$$

The analysis of the $F_{\pi\gamma}(Q^2)$ form factor, measurable in $ee \to ee\pi^0$ reactions, proceeds in a similar manner (see Fig.1a). An interesting result is [19]

$$\alpha_s(Q^2) = \frac{F_\pi(Q^2)}{4\pi Q^2 |F_{\pi\gamma}(Q^2)|^2} \left[1 + \mathcal{O}\left(\frac{\alpha_s(Q^2)}{\pi}\right)\right] , \tag{4.26}$$

which provides a definition of $\alpha_s$ independent of the form of the distribution function $\phi_\pi$. Higher-order corrections to $F_\pi(Q^2)$ and $F_{\pi\gamma}(Q^2)$ are discussed in Ref. 50.

### 3.4.4 Large Momentum Transfer Exclusive Processes [19]

The meson form factor calculation which we outlined above is the prototype for the calculation of the QCD hard scattering contribution for the whole range of exclusive processes at large momentum transfer. Away from possible special points in the $x_i$ integrations (see below) a general hadronic amplitude can be written to leading order in $1/Q^2$ as a convolution of a connected hard scattering amplitude $T_H$ convoluted with the meson and baryon distribution amplitudes:

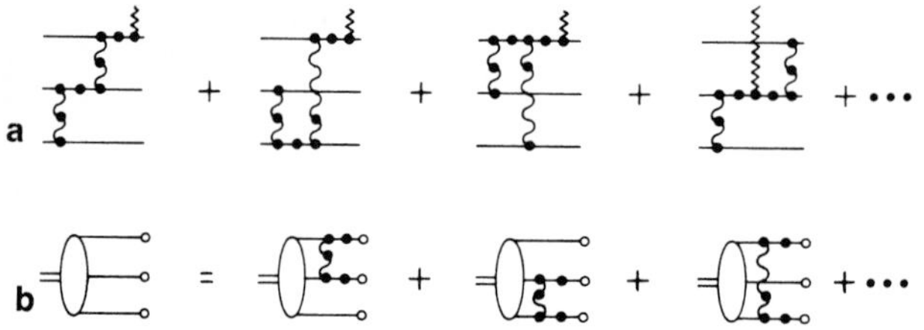

Fig.7. (a) Leading contributions to $T_H$ for the baryon form factors corresponding to the four terms of Eqs. (4.31) and (4.32), respectively. (b) Contributions to the kernel for the evolution of the baryon distribution amplitude

$$\phi_M(x,Q) = \int^{|\mathcal{E}|<Q^2} \frac{d^2k_\perp}{16\pi^2} \psi^Q_{q\bar{q}}(x,\vec{k}_\perp) \quad , \tag{4.27a}$$

and

$$\phi_B(x_i,Q) = \int^{|\mathcal{E}|<Q^2} [d^2k_\perp] \, \psi_{qqq}(x_i,\vec{k}_{\perp i}) \quad . \tag{4.27b}$$

The hard scattering amplitude $T_H$ is computed by replacing each external hadron line by massless valence quarks, each collinear with the hadrons momentum $p_i^\mu \cong x_i \, p_H^\mu$. For example, the baryon form factor at large $Q^2$ has the form [9,19] (see Fig.2a and Fig.7)

$$G_M(Q^2) = \int [dx][dy] \, \phi^*(y_i,\hat{Q}) \, T_H(x,y;\, Q^2) \, \phi(s,\hat{Q}) \tag{4.28}$$

where $T_H$ is the $3q + \gamma \to 3q'$ amplitude. (The optimal choice for $\hat{Q}$ is discussed in Ref. 19.) For the proton and neutron we have to leading order ($C_B = 2/3$)

$$T_p = \frac{128\pi^2 \, C_B^2}{\left(Q^2 + M_0^2\right)^2} \, T_1 \tag{4.29}$$

$$T_n = \frac{128\pi^2 \, C_B^2}{3\left(Q^2 + M_0^2\right)^2} \left[T_1 - T_2\right] \tag{4.30}$$

where

$$T_1 = -\frac{\alpha_s(x_3y_3Q^2)\,\alpha_s\left((1-x_1)(1-y_1)Q^2\right)}{x_3(1-x_1)^2\, y_3(1-y_1)^2} + \frac{\alpha_s(x_2y_2Q^2)\,\alpha_s\left((1-x_1)(1-y_1)Q^2\right)}{x_2(1-x_1)^2\, y_2(1-y_1)^2} \tag{4.31}$$

$$- \frac{\alpha_s(x_2y_2Q^2)\,\alpha_s(x_2y_2Q^2)}{x_2x_3(1-x_3)\, y_2y_3(1-y_1)} \quad ,$$

and

$$T_2 = -\frac{\alpha_s(x_1y_1Q^2)\,\alpha_s(x_3y_3Q^2)}{x_1x_3(1-x_1)\, y_1y_3(1-y_3)} \quad . \tag{4.32}$$

$T_1$ corresponds to the amplitude where the photon interacts with the quarks (1) and (2) which have helicity parallel to the nucleon helicity, and $T_2$ corresponds to the amplitude where the quark with opposite helicity is struck. The running coupling constants have arguments $\hat{Q}^2$ corresponding to the gluon momentum transfer of each diagram. Only large $Q^2$ behavior is predicted by the theory; we utilize the parameter $M_0$ to represent the effect of power-law suppressed terms from mass insertions, higher Fock states, etc.

The $Q^2$ evolution of the baryon distribution amplitude can be derived from the operator product expansion of three-quark fields or from the gluon exchange kernel, in parallel with the derivation of (4.12). The baryon evolution equation to leading order in $\alpha_s$ is [19]

$$x_1x_2x_3\left\{\frac{\partial}{\partial\zeta}\tilde{\phi}(x_i,Q)+\frac{3}{2}\frac{C_F}{\beta_0}\tilde{\phi}(x_i,Q)\right\}=\frac{C_B}{\beta_0}\int_0^1[dy]\,V(x_i,y_i)\,\tilde{\phi}(y_i,Q) \quad . \qquad (4.33)$$

Here $\phi = x_1x_2x_3\tilde{\phi}$, $\zeta = \log(\log Q^2/\Lambda^2)$ and (see Fig.7b)

$$V(x_i,y_i) = 2x_1x_2x_3\sum_{i\neq j}\theta(y_i-x_i)\,\delta(x_k-y_k)\,\frac{y_j}{x_j}\left(\frac{\delta_{h_i\bar{h}_j}}{x_i+x_j}+\frac{\Delta}{y_i-x_i}\right) \qquad (4.34)$$

$$= V(y_i,x_i) \quad .$$

The infrared singularity at $x_i = y_i$ is cancelled because the baryon is a color singlet. The evolution equation has the general solution

$$\phi(x_i,Q) = x_1x_2x_3\sum_{n=0}^{\infty}a_n\,\tilde{\phi}_n(x_i)\left(\log\frac{Q^2}{\Lambda^2}\right)^{-\gamma_n^B} \quad . \qquad (4.35)$$

The leading (polynomial) eigensolution $\tilde{\phi}_n(x_i)$ and corresponding baryon anomalous dimensions are given in Refs.19 and 52. Thus at large $Q^2$, the nucleon magnetic form factors have the form [9,19]

$$G_M(Q^2) \to \frac{\alpha_s^2(Q^2)}{Q^4}\sum_{n,m}b_{nm}\left(\log\frac{Q^2}{\Lambda^2}\right)^{-\gamma_n^B-\gamma_n^B}\left[1+\mathcal{O}\left(\alpha_s(Q^2),\frac{m^2}{Q^2}\right)\right] \quad . \qquad (4.36)$$

We can also use this result to obtain results for ratios of various baryon and isobar form factors assuming isospin or SU(3)-flavor symmetry for the basic wavefunction structure. Results for the neutral weak and charged weak form factors assuming standard SU(2) × U(1) symmetry are given in Ref. 46.

As we see from Eq. (4.28), the integrations over $x_i$ and $y_i$ have potential endpoint singularities. However, it is easily seen that any anomalous contribution [e.g., from the region $x_2,x_3 \sim \mathcal{O}(m/Q)$, $x_1 \sim 1 - \mathcal{O}(m/Q)$] is asymptotically suppressed at large $Q^2$ by a Sudakov form factor arising from the virtual correction to the $\bar{q}\gamma q$ vertex when the quark legs are near-on-shell [$p^2 \sim \mathcal{O}(mQ)$]. [19,54] This Sudakov suppression of the endpoint region requires an all orders resummation of perturbative contributions, [57] and thus the derivation of the baryon form factors is not as rigorous as for the meson form factor, which has no such endpoint singularity.

The most striking feature of the QCD prediction (4.36) is the $1/Q^4$ power-law behavior of $G_M^p$ as $G_M^n$. The power-law dependence [8] reflects:

(1) The essential scale invariance of the qq scattering subprocesses within $T_H$.

(2) The fact that the minimal Fock state of a baryon is the 3-quark state.

We will discuss the phenomenology of the baryon form factors and the resulting constraints on the baryon wavefunction in Sect. VI.

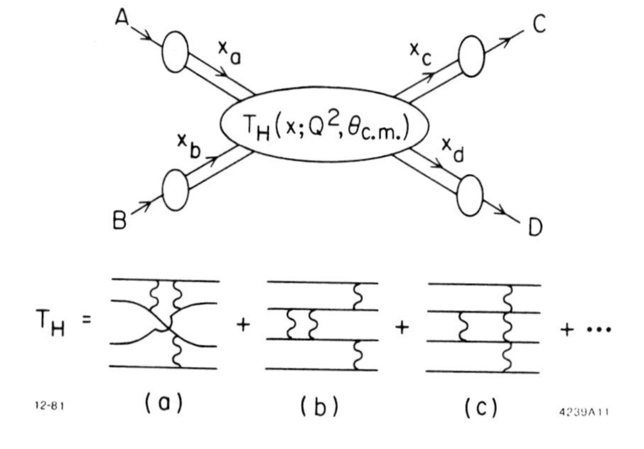


Fig. 8. QCD contributions to meson-meson scattering at large momentum transfer. Diagram (c) corresponds to the Landshoff pinch singularity which is suppressed by quark form factor effects

In the case of hadron scattering amplitudes $A+B \to C+D$, photoproduction, Compton scattering, etc., the leading hard scattering QCD contribution at large momentum transfer $Q^2 = tu/s$ has the form [19] (helicity labels and suppressed)(see Fig.8)

$$\mathcal{M}_{A+B\to C+D}(Q^2,\theta_{c.m.}) = \int [dx]\phi_C(x_c,\tilde{Q})\ \phi_D(x_d,\tilde{Q})\ T_H(x_i;\ Q^2,\theta_{c.m.}) \qquad (4.37)$$
$$\times\ \phi_A(x_a,\tilde{Q})\ \phi_B(x_b,\tilde{Q}) \quad .$$

The essential behavior of the amplitude is determined by $T_H$, computed where each hadron is replaced by its (collinear) quark constituents. We note again that $T_H$ is "collinear irreducible," i.e., the transverse momentum integrations of all reducible loop integration are restricted to $k_\perp^2 > \mathcal{O}(Q^2)$, since the small $k_\perp$ region is already contained in $\phi$. If the internal propagators in $T_H$ are all far-off-shell $\mathcal{O}(Q^2)$ (as in Fig.8a), then a perturbative expansion in $\alpha_s(Q^2)$ can be carried out. However, this is not true for all hadron-hadron scattering amplitudes, since one can have multiple quark-quark scattering processes which allow near-on-shell propagation in intermediate states at finite values of the $x_i$. [17] The classic example is meson-meson scattering, where two pairs of quarks scatter through the same angle (see Fig.7c). However, the near-on-shell region of integration is again suppressed by Sudakov factors. (Physically this suppression occurs because the near-on-shell quarks must scatter without radiating gluons.) A model calculation by MUELLER [10] for $\pi-\pi$ scattering in QCD (using an exponentiated form of the Sudakov form factor) shows that the leading contribution comes in fact from the off-shell region $|k^2| \sim \mathcal{O}(Q^2)^{1-\varepsilon}$ where $\varepsilon = (2c+1)^{-1}$, $c = 8C_F/(11 - 2/3\ n_f)$ (for four flavors $\varepsilon \cong 0.281$). This region gives the contribution [10]

$$\mathcal{M}_{\pi\pi\to\pi\pi} \sim \mathcal{O}(Q^2)^{-3/2\ -\ c\ell n\,(2c+1/2c)} \qquad (4.38)$$
$$\cong (Q^2)^{-1.922}$$

compared to $(Q^2)^{-2}$ from the hard scattering $|k^2| \sim \mathcal{O}(Q^2)$ region.

Thus, even when pinch singularities are present, the far-off-shell hard scattering quark and gluon processes dominate large-momentum-transfer hadron scattering amplitudes. Given this result, we can abstract some general QCD features common to all exclusive processes at large momentum transfer:

(1) All of the non-perturbative bound-state physics is isolated in the process-independent distribution amplitudes.

(2) The nominal power-law behavior of an exchange amplitude is $(1/Q)^{n-4}$ where n is the number of external elementary particles (quarks, gluons, leptons, photons in $T_H$). This immediately implies the dimensional counting rules: [8]

$$\frac{d\sigma}{dt}\,(A+B\to C+D)\sim\left(\frac{1}{Q^2}\right)^{n-2} f(\theta_{c.m.}) \tag{4.39}$$

where $n = n_A + n_B + n_C + n_D$, and

$$F_H(Q^2)\sim\left(\frac{1}{Q^2}\right)^{n_H-1} \tag{4.40}$$

where $F_H$ is the helicity-conserving [18,19] form factor. These power-law predictions are modified by (a) the $Q^2$ dependence of the factors of $\alpha_s$ in $T_H$, (b) the $Q^2$ evolution of the distribution amplitudes and (c) a possible small power associated with the almost complete Sudakov suppression of pinch singularities in hadron-hadron scattering. The dimensional counting rules appear to be experimentally well established for a wide variety of processes (see Ref. 19 and Fig.9):

$$G_M(Q^2)\sim(Q^2)^{-2}\quad,\qquad F_\pi(Q^2)\sim(Q^2)^{-1} \tag{4.41}$$

and

$$\frac{d\sigma}{dt}\,(\gamma p\to\pi p)\sim(Q^2)^{-7} \tag{4.42}$$

$$\frac{d\sigma}{dt}\,(\pi p\to\pi p)\sim(Q^2)^{-8}$$

$$\frac{d\sigma}{dt}\,(pp\to pp)\sim(Q^2)^{-10}$$

$$\frac{d\sigma}{dt}\,(\gamma p\to\gamma p)\Big/\frac{d\sigma}{dt}\,(\gamma p\to\pi p)\sim Q^2$$

at fixed $\theta_{c.m.}$. The application to $\gamma\gamma\to M\bar{M}$ processes is discussed in Sect. IV-E.

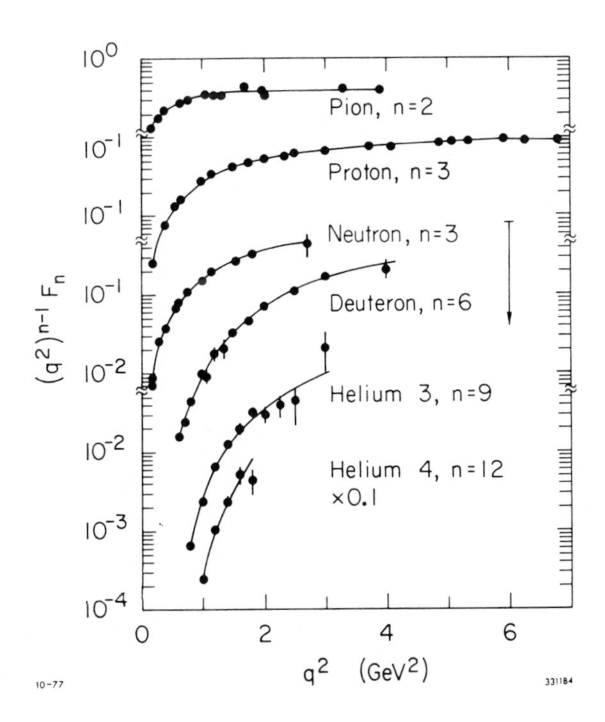


Fig. 9. Hadronic form factors multiplied by $(Q^2)^{n-1}$. (From Ref. 1.)

(3) Since the distribution amplitudes $\phi_M$ and $\phi_B$ are $L_z = 0$ angular momentum projections of the hadronic wavefunctions, the sum of the quark spin along the hadron's momentum equals the hadron spin: [18]

$$\sum_{i \in H} S_i^z = S_H^z \; . \tag{4.43}$$

(In contrast, in inclusive reactions there are any number of non-interacting quark and gluon spectators, so that the spin of the interacting constituents is only statistically related to the hadron spin — except possibly at the edge of phase-space $x \sim 1$.) Furthermore, since all propagators in $T_H$ are hard, the quark and hadron masses can be neglected at large $Q^2$ up to corrections of order $\sim m/Q$. The vector-gluon interactions conserve quark helicity when all masses are neglected. Thus total quark helicity is conserved in $T_H$ at large $Q^2$. Combining this with (4.43), we have the QCD selection rule:

$$\sum_{\text{initial}} \lambda_H = \sum_{\text{final}} \lambda_H \tag{4.44}$$

i.e., total hadron helicity is conserved up to corrections of order $\mathcal{O}(m/Q)$.

Hadron helicity conservation thus applies for all large-momentum-transfer exclusive amplitudes involving light meson and baryons. Notice that the photon spin is not important: QCD predicts that $\gamma p \to \pi p$ is proton helicity conserving at fixed $\theta_{c.m.}$, $s \to \infty$, independent of the photon polarization. Exclusive amplitudes which involve hadrons with quarks or gluons in higher orbital angular momentum states are also suppressed by powers of the momentum transfer. An important corollary of this rule is that helicity-flip form factors are suppressed, e.g.:

$$F_{2p}(Q^2) \Big/ F_1(Q^2) \sim \mathcal{O}(m^2/Q^2) \quad . \tag{4.45}$$

The helicity rule, Eq. (4.44), is one of the most characteristic features of QCD, being a direct consequence of the gluon's spin. A scalar or tensor gluon-quark coupling flips the quark's helicity. Thus, for such theories, helicity may or may not be conserved in any given diagram contributing to $T_H$, depending upon the number of interactions involved. Only for a vector theory, like QCD, can we have a helicity selection rule valid to all orders in perturbation theory.

The study of timelike hadronic form factors using $e^+e^-$ colliding beams can provide very sensitive tests of this rule, since the virtual photon in $e^+e^- \to \gamma^* \to h_A\bar{h}_B$ always has spin $\pm 1$ along the beam axis at high energies. Angular momentum conservation implies that the virtual photon can "decay" with one of only two possible angular distributions in the center of momentum frame: $(1 + \cos^2\theta)$ for $|\lambda_A - \lambda_B| = 1$, and $\sin^2\theta$ for $|\lambda_A - \lambda_B| = 0$ where $\lambda_{A,B}$ are the helicities of hadron $h_{A,B}$. Hadronic helicity conservation, Eq. (4.44), as required by QCD greatly restricts the possibilities. It implies that $\lambda_A + \lambda_B = 0$ (since the photon carries no "quark helicity"), or equivalently that $\lambda_A - \lambda_B = 2\lambda_A = -2\lambda_B$. Consequently, angular momentum conservation requires $|\lambda_A| = |\lambda_B| = 1/2$ for baryons, and $|\lambda_A| = |\lambda_B| = 0$ for mesons; furthermore, the angular distributions are now completely determined:

$$\frac{d\sigma}{d\cos\theta}(e^+e^- \to B\bar{B}) \propto 1 + \cos^2\theta \quad \text{(baryons)} \tag{4.46}$$

$$\frac{d\sigma}{d\cos\theta}(e^+e^- \to M\bar{M}) \propto \sin^2\theta \quad \text{(mesons)} \; . \tag{4.47}$$

We emphasize that these predictions are far from trivial for vector mesons and for all baryons. For example, one expects distributions like $1 + \alpha\cos^2\theta$, $-1 < \alpha < 1$, in theories with a scalar or tensor gluon. So simply verifying these angular distributions would give strong evidence in favor of a vector gluon.

Table 3. Exclusive channels in $e^+e^-$ annihilation. The $h_A\bar{h}_B\gamma^*$ couplings in allowed processes are $-ie(p_A - p_B)^\mu F(s)$ for mesons, $-ie\bar{v}(p_B)\gamma^\mu G(s)u(p_A)$ for baryons, and $-ie^2\varepsilon_{\mu\nu\rho\sigma}p_M^\nu\varepsilon^\rho p_\gamma^\sigma F_{M\gamma}(s)$ for meson-photon final states. Similar predictions apply to decays of heavy-quark vector states, like the $\psi,\psi',\ldots$, produced in $e^+e^-$ collisions

| | $e^+e^- \to h_A(\lambda_A)\ \bar{h}_B(\lambda_B)$ | Angular Distribution | $\frac{\sigma(e^+e^- \to h_A\bar{h}_B)}{\sigma(e^+e^- \to \mu^+\mu^-)}$ |
|---|---|---|---|
| Allowed in QCD | $e^+e^- \to \pi^+\pi^-, K^+K^-$ | $\sin^2\theta$ | $\frac{1}{4}\vert F(s)\vert^2 \sim c/s^2$ |
| | $\rho^+(0)\rho^-(0), K^{*+}K^{*-}$ | $\sin^2\theta$ | $\frac{1}{4}\vert F(s)\vert^2 \sim c/s^2$ |
| | $\pi^o\gamma(\pm 1), \eta\gamma, \eta'\gamma$ | $1 + \cos^2\theta$ | $(\pi\alpha/2)s\vert F_{M\gamma}(s)\vert^2 \sim c/s$ |
| | $e^+e^- \to p(\pm\frac{1}{2})\bar{p}(\mp\frac{1}{2}), n\bar{n},\ldots$ | $1 + \cos^2\theta$ | $\vert G(s)\vert^2 \sim c/s^4$ |
| | $p(\pm\frac{1}{2})\bar{\Delta}(\mp\frac{1}{2}), \bar{n}\Delta,\ldots$ | $1 + \cos^2\theta$ | $\vert G(s)\vert^2 \sim c/s^4$ |
| | $\Delta(\pm\frac{1}{2})\bar{\Delta}(\mp\frac{1}{2}), y^*\bar{y}^*,\ldots$ | $1 + \cos^2\theta$ | $\vert G(s)\vert^2 \sim c/s^4$ |
| Suppressed in QCD | $e^+e^- \to \rho^+(0)\rho^-(\pm 1), \pi^+\rho^-, K^+K^{*-},\ldots$ | $1 + \cos^2\theta$ | $< c/s^3$ |
| | $\rho^+(\pm 1)\rho^-(\pm 1),\ldots$ | $\sin^2\theta$ | $< c/s^3$ |
| | $e^+e^- \to p(\pm\frac{1}{2})\bar{p}(\pm\frac{1}{2}), p\bar{\Delta}, \Delta\bar{\Delta},\ldots$ | $\sin^2\theta$ | $< c/s^5$ |
| | $p(\pm\frac{1}{2})\bar{\Delta}(\pm\frac{3}{2}), \Delta\bar{\Delta},\ldots$ | $1 + \cos^2\theta$ | $< c/s^5$ |
| | $\Delta(\pm\frac{3}{2})\bar{\Delta}(\pm\frac{3}{2}),\ldots$ | $\sin^2\theta$ | $< c/s^5$ |

The power-law dependence in s of these cross sections is also predicted in QCD, using the dimensional counting rule. Such "all orders" predictions for QCD-allowed processes are summarized in Table 3.

Processes suppressed in QCD are also listed there; these all violate hadronic helicity conservation, and are suppressed by powers of $m^2/s$ in QCD. This would not necessarily be the case in scalar or tensor theories.

The exclusive decays of heavy-quark atoms $(\psi,\psi',\ldots)$ into light hadrons can also be analyzed in QCD. [18] The decay $\psi \to p\bar{p}$, for example, proceeds via diagrams such as those in Fig.2b. Since $\psi$'s produced in $e^+e^-$ collisions must also have spin $\pm 1$ along the beam direction, and since they can only couple to light quarks via gluons, all the properties listed in Table 3 apply to $\psi$, $\psi'$, $\Upsilon$, $\Upsilon'$,... decays as well. There are considerable experimental data for the $\psi$ and $\psi'$ decays. [55]

Perhaps the most significant tests are the decays $\psi,\psi' \to p\bar{p}, n\bar{n},\ldots$ . The predicted angular distribution $1 + \beta^2\cos^2\theta$ is consistent with published data. [35] This is important evidence favoring a vector gluon, since scalar or tensor gluon theories would predict a distribution of $\sin^2\theta + \mathcal{O}(\alpha_s)$. Dimensional counting rules can be checked by comparing the $\psi$ and $\psi'$ rates into $p\bar{p}$, normalized by the total rates into light-quark hadrons so as to remove dependence upon the heavy-quark wavefunctions. Theory predicts

$$\frac{BR(\psi \to p\bar{p})}{BR(\psi' \to p\bar{p})} \sim \left(\frac{M_{\psi'}}{M_\psi}\right)^8 \tag{4.48}$$

where

$$BR(\psi \to p\bar{p}) \equiv \frac{\Gamma(\psi \to p\bar{p})}{\Gamma(\psi \to \text{light-quark hadrons})} \quad . \tag{4.49}$$

Existing data suggest a ratio $(M_{\psi'}/M_\psi)^n$ with $n \sim 6 \pm 3$, in good agreement with QCD.

Many more examples of exclusive reactions which test the basic scaling laws and spin structure of QCD are discussed in Refs. 18 and 19. The essential point is that exclusive reactions have the potential for isolating the QCD hard scattering subprocesses in situations where the helicities of all the interaction constituents are controlled. In contrast, in inclusive reactions the absence of restrictions on the spectator quark and gluons allows only a statistical correlation between the constituent and hadronic helicities.

### 3.4.5 Two-Photon Processes [12]

One of the most important applications of perturbative QCD is to the two-photon processes $d\sigma/dt$ ($\gamma\gamma \to M\bar{M}$), $M = \pi, K, \rho, \omega$ at large $s = (k_1 + k_2)^2$ and fixed $\theta_{c.m.}$. These reactions, which can be studied in $e^+e^- \to e^+e^-M\bar{M}$ processes, provide a particularly important laboratory for testing QCD, since these "Compton" processes are, by far, the simplest calculable large-angle exclusive hadronic scattering reactions. As we discuss below, the large-momentum-transfer scaling behavior, the helicity structure, and often even the absolute normalization can be rigorously computed for each two-photon channel.

Conversely, the angular dependence of the $\gamma\gamma \to M\bar{M}$ amplitudes can be used to determine the shape of the process-independent meson "distribution amplitudes," $\phi_M(x,Q)$, the basic short-distance wavefunctions which control the valence quark distributions in high-momentum-transfer exclusive reactions.

A critically important feature of the $\gamma\gamma \to M\bar{M}$ amplitude is that the contributions of LANDSHOFF [17] pinch singularities are power-law suppressed at the Born level — even before taking into account Sudakov form factor suppression. There are also no anomalous contributions from the $x \sim 1$ endpoint integration region. Thus, as in the calculation of the meson form factors, each fixed-angle helicity amplitude can be written to leading order in $1/Q$ in the factorized form [$Q^2 = p_T^2 = tu/s$; $\tilde{Q}_x = \min(xQ, (1-x)Q)$] (see Fig.9):

$$\mathcal{M}_{\gamma\gamma \to M\bar{M}} = \int_0^1 dx \int_0^1 dy\, \phi_{\bar{M}}(y,\tilde{Q}_y)\, T_H(x,y;\, s,\theta_{c.m.})\, \phi_M(x,\tilde{Q}_x) \qquad (4.50)$$

where $T_H$ is the hard scattering amplitude $\gamma\gamma \to (q\bar{q})(q\bar{q})$ for the production of the valence quarks collinear with each meson and $\phi_M(x,Q)$ is the (process-independent) distribution amplitude for finding the valence $q$ and $\bar{q}$ with light-cone fractions of the meson's momentum, integrated over transverse momenta $k_\perp < Q$. The contribution of nonvalence Fock states are power-law suppressed. Further, the spin selection rule (4.44) of QCD predicts that vector mesons $M$ and $\bar{M}$ are produced with opposite helicities to leading order in $1/Q$ and all orders in $\alpha_s(Q^2)$.

Dimensional counting [8] predicts that for large $s$, $s^4\, d\sigma/dt$ scales at fixed $t/s$ or $\theta_{c.m.}$ up to factors of $\ln s/\Lambda^2$.

Some forty diagrams contribute to the hard scattering amplitudes for $\gamma\gamma \to M\bar{M}$ (for nonsinglet mesons). These can be derived from the four independent diagrams in Fig.10b by particle interchange. The resulting amplitudes for helicity-zero mesons are:

$$\left.\begin{matrix} T_{++} \\ T_{--} \end{matrix}\right\} = \frac{16\pi\alpha_s}{3s}\, \frac{32\pi\alpha}{x(1-x)y(1-y)} \left[\frac{(e_1 - e_2)^2 a}{1 - \cos^2\theta_{c.m.}}\right] \qquad (4.51)$$

$$\left.\begin{matrix} T_{+-} \\ T_{-+} \end{matrix}\right\} = \frac{16\pi\alpha_s}{3s}\, \frac{32\pi\alpha}{x(1-x)y(1-y)} \left[\frac{(e_1 - e_2)^2(1-a)}{1 - \cos^2\theta_{c.m.}} + \frac{e_1 e_2 a\big(y(1-y) + x(1-x)\big)}{a^2 - b^2\cos^2\theta_{c.m.}}\right] \qquad (4.52)$$

where $\left.\begin{matrix} a \\ b \end{matrix}\right\} = (1-x)(1-y) \pm xy$, the subscripts $++, --, \ldots$ refer to photon helicities, and $e_1$, $e_2$ are the quark charges [i.e., the mesons have charges $\pm(e_1 - e_2)$].

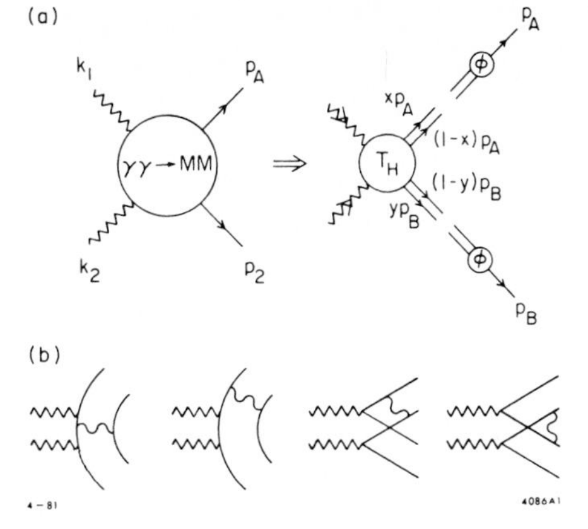


Fig. 10. (a) Factorized structure of the $\gamma\gamma \to M\bar{M}$ amplitude in QCD at large momentum transfer. The $T_H$ amplitude is computed with quarks collinear with the outgoing mesons. (b) Diagram contributing to $T_H(\gamma\gamma \to M\bar{M})$ to lowest order in $\alpha_s$

To compute the $\gamma\gamma \to M\bar{M}$ amplitude $\mathcal{M}_{\lambda\lambda'}$ [Eq.(4.50)], we now need only know the x dependence of the meson's distribution amplitude $\phi_M(x,Q)$; the overall normalization of $\phi_M$ is fixed by the 'sum rule' ($n_c = 3$)

$$\int_0^1 dx\ \phi_M(x,Q) = \frac{f_M}{2\sqrt{3}} \tag{4.53}$$

where $f_M$ is the meson decay constant as determined from leptonic decays. Note that the dependence in x and y of several terms in $T_{\lambda\lambda'}$ is quite similar to that appearing in the meson's electromagnetic form factor (4.23):

$$F_M(s) = \frac{16\pi\alpha_s}{3s} \int_0^1 dx\ dy\ \frac{\phi_M^*(x,\tilde{Q}_x)\ \phi_M^*(y,\tilde{Q}_y)}{x(1-x)\ y(1-y)} \tag{4.54}$$

when $\phi_M(x,Q) = \phi_M(1-x,Q)$ is assumed. Thus much of the dependence on $\phi(x,Q)$ can be removed from $\mathcal{M}_{\lambda\lambda'}$ by expressing it in terms of the meson form factor — i.e.,

$$\left.\begin{matrix}\mathcal{M}_{++}\\ \mathcal{M}_{--}\end{matrix}\right\} = 16\pi\alpha\ F_M(s) \left[\frac{\langle(e_1-e_2)^2\rangle}{1-\cos^2\theta_{c.m.}}\right] \tag{4.55}$$

$$\left.\begin{matrix}\mathcal{M}_{+-}\\ \mathcal{M}_{-+}\end{matrix}\right\} = 16\pi\alpha\ F_M(s) \left[\frac{\langle(e_1-e_2)^2\rangle}{1-\cos^2\theta_{c.m.}} + 2\langle e_1e_2\rangle\ g[\theta_{c.m.};\ \phi_M]\right] \tag{4.56}$$

up to corrections of order $\alpha_s$ and $m^2/s$. Now the only dependence on $\phi_M$, and indeed the only unknown quantity, is in the θ-dependent factor

$$g\left[\theta_{c.m.};\ \phi_M\right] = \frac{\displaystyle\int_0^1 dx\ dy\ \frac{\phi_M^*(x,\tilde{Q})\ \phi_M^*(y,\tilde{Q})}{x(1-x)\ y(1-y)}\ \frac{a[y(1-y)+x(1-x)]}{a^2-b^2\cos^2\theta_{c.m.}}}{\displaystyle\int_0^1 dx\ dy\ \frac{\phi_M^*(x,\tilde{Q})\ \phi_M^*(y,\tilde{Q})}{x(1-x)\ y(1-y)}}\ . \tag{4.57}$$

The spin-averaged cross section follows immediately from these expressions:

$$\frac{d\sigma}{dt} = \frac{2}{s}\frac{d\sigma}{d\cos\theta_{c.m.}} = \frac{1}{16\pi s^2}\frac{1}{4}\sum_{\lambda\lambda}|\mathcal{M}_{\lambda\lambda'}|^2 \tag{4.58}$$

$$= 16\pi\alpha^2\left|\frac{F_M(s)}{s}\right|^2\left\{\frac{\left\langle(e_1-e_2)^2\right\rangle^2}{\left(1-\cos^2\theta_{c.m.}\right)^2}+\frac{2\langle e_1e_2\rangle\left\langle(e_1-e_2)^2\right\rangle}{1-\cos^2\theta_{c.m.}}\right.$$

$$\left.\times\, g\left[\theta_{c.m.};\phi_M\right]+2\langle e_1e_2\rangle^2 g^2\left[\theta_{c.m.};\phi_M\right]\right\} .$$

In Fig.11 the spin-averaged cross sections (for $\gamma\gamma\to\pi\pi$) are plotted for several forms of $\phi_M(x,Q)$. At very large energies, the distribution amplitude evolves to the form

$$\phi_M(x,Q) \underset{Q\to\infty}{\longrightarrow} \sqrt{3}\, f_M\, x(1-x) \quad , \tag{4.59}$$

and the predictions [curve (a)] become exact and parameter free. However, this evolution with increasing $Q^2$ is very slow (logarithmic), and at current energies $\phi_M$ could be quite different in structure, depending upon the details of hadronic binding. Curves (b) and (c) correspond to the extreme examples $\phi_M \propto [x(1-x)]^{1/4}$ and $\phi_M \propto \delta(x - 1/2)$, respectively. Remarkably, the cross section for charged mesons is essentially independent of the choice of $\phi_M$, making this an essentially parameter-free prediction of perturbative QCD. By contrast, the predictions for neutral helicity-zero mesons are quite sensitive to the structure of $\phi_M$. Thus we can study the x dependence of the meson distribution amplitude by measuring the angular dependence of this process.

The cross sections shown in Fig.8 are specifically for $\gamma\gamma\to\pi\pi$, where the pion form factor has been approximated by $F_\pi(s) \sim 0.4\ \mathrm{GeV}^2/s$. The $\pi^+\pi^-$ cross section is quite large at moderate s:

$$\frac{\frac{d\sigma}{dt}(\gamma\gamma\to\pi^+\pi^-)}{\frac{d\sigma}{dt}(\gamma\gamma\to\mu^+\mu^-)} \sim \frac{4|F_\pi(s)|^2}{1-\cos^4\theta_{c.m.}} \sim \frac{0.6\ \mathrm{GeV}^4}{s^2} \quad \text{at} \quad \theta_{c.m.} = \pi/2 \tag{4.60}$$

Similar predictions are possible for other helicity-zero mesons. The normalization of $\gamma\gamma\to M\bar{M}$ relative to the $\gamma\gamma\to\pi\pi$ cross section is completely determined by the ratio of meson decay constants $(f_M/f_\pi)^4$ and by the flavor-symmetry of the wavefunctions, provided only that $\phi_M$ and $\phi_\pi$ are similar in shape. Note that the cross section for charged $\rho$'s with helicity zero is almost an order of magnitude larger than that for charged $\pi$'s.

Finally, notice that the leading order predictions [Eq. (4.58)] have no explicit dependence on $\alpha_s$. Thus they are relatively insensitive to the choice of renormalization scheme or of a normalization scale. This is not the case for either the form factor or the two-photon annihilation amplitude when examined separately. However, by combining the two analyses as in Eq. (4.58) we obtain meaningful results without computing $O(\alpha_s)$ corrections. The corresponding calculations for helicity-one mesons are given in Ref. 12. Hadronic helicity conservation implies that only helicity-zero mesons can couple to a single highly virtual photon. So $F_{M_\perp}$, the transverse form factor, cannot be measured experimentally. For simplicity we will assume that the longitudinal and transverse form factors are equal to obtain a rough estimate of the $\gamma\gamma\to\rho_\perp\rho_\perp$ cross section (Fig.12). Again we see strong dependence on $\phi_{M_\perp}$ for all angles except $\theta_{c.m.} \sim \pi/2$, where the terms involving $g_\perp$ vanish. Consequently, a measurement of the angular distribution would be very sensitive to the x dependence of $\phi_{M_\perp}$, while measurements at $\theta_{c.m.} = \pi/2$ determine

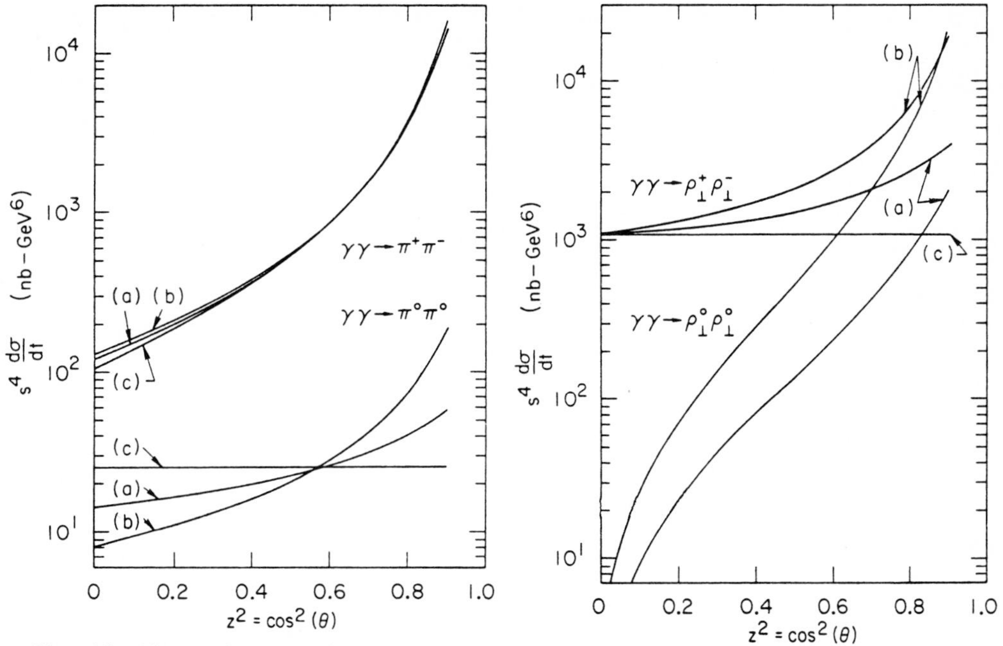


Fig. 11. QCD predictions for $\gamma\gamma \to \pi\pi$ to leading order in QCD. The results assume the pion form factor parameterization $F_\pi(s) \sim 0.4\ \mathrm{GeV}^2/s$. Curves (a), (b) and (c) correspond to the distribution amplitudes $\phi_M = x(1-x)$, $[x(1-x)]^{1/4}$, and $\delta(x-1/2)$, respectively. Predictions for other helicity-zero mesons are obtained by multiplying with the scale constants given in Ref. 15

Fig. 12. QCD predictions for $\gamma\gamma \to \rho_\perp\bar{\rho}_\perp$ with opposite helicity $\pm 1$ to leading order in QCD. The normalization given here assumes that the $\rho$ distribution amplitude is helicity independent

$F_{M_\perp}(s)$. Notice also that the number of charged $\rho$-pairs (with any helicity) is much larger than the number of neutral $\rho$'s, particularly near $\theta_{c.m.} = \pi/2$. The cross sections are again quite large with

$$\left.\frac{d\sigma/dt\left(\gamma\gamma \to \rho_\perp^+\rho_\perp^-\right)}{d\sigma/dt\left(\gamma\gamma \to \mu^+\mu^-\right)}\right|_{\theta_{c.m.}=\frac{\pi}{2}} \sim \frac{5\ \mathrm{GeV}^4}{s^2} \quad . \tag{4.61}$$

Results for other mesons are given in Ref. 12.

The $\gamma\gamma \to M\bar{M}$ and $\gamma^*\gamma \to M$ processes thus provide detailed checks of the basic Born structure of QCD, the scaling behavior of the quark and gluon propagators and interactions, as well as the constituent charges and spins. Conversely, the angular dependence of the $\gamma\gamma \to M\bar{M}$ amplitudes can be used to determine the shape of the process-independent distribution amplitude $\phi_M(x,Q)$ for valence quarks in the meson $q\bar{q}$ Fock state. The $\cos\theta_{c.m.}$ dependence of the $\gamma\gamma \to M\bar{M}$ amplitude determines the light-cone x dependence of the meson distribution amplitude in much the same way that the $x_{Bj}$ dependence of deep-inelastic cross sections determines the light-cone x dependence of the structure functions (quark probability functions) $G_{q/M}(x,Q)$.

The form of the predictions given here is exact to leading order in $\alpha_s(Q^2)$. Power-law $(m/Q)^2$ corrections can arise from mass insertions, higher Fock states, pinch singularities and nonperturbative effects. In particular, the predictions are only valid when s-channel resonance effects can be neglected. It is likely that the background due to resonances can be reduced relative to the leading order QCD contributions if one measures the two-photon processes with at least one of the photons tagged at moderate spacelike momentum $q^2$, since resonance contributions are expected to be strongly damped by form factor effects. In contrast, the leading order QCD $\gamma_1\gamma_2 \to M\bar{M}$ amplitudes are relatively insensitive to the value of $q_1^2$ or $q_2^2$ for $|q_i^2| \ll s$.

Finally, we note that the amplitudes given above have simple crossing properties. In particular, we can immediately analyze the Compton amplitude $\gamma M \to \gamma M$ in the region t large enough with $s \gg |t|$ in order to study the leading Regge behavior in the large-momentum-transfer domain. In the case of helicity $\pm 1$ mesons, the leading contribution to the Compton amplitude has the form ($s \gg |t|$)

$$\mathcal{M}_{\gamma M \to \gamma M} = 16\pi\alpha F_{M_\perp}(t)\left(e_1^2 + e_2^2\right) \qquad (4.62)$$

$$(\lambda_\gamma = \lambda'_\gamma , \qquad \lambda_M = \lambda'_M)$$

which corresponds to a fixed Regge singularity at J = 0. [56] In the case of helicity-zero mesons, this singularity actually decouples, and the leading J-plane singularity is at J = -2.

## 3.5 Deep Inelastic Lepton Scattering

The crucial evidence that the electromagnetic current within hadrons is carried by point-like spin-1/2 quarks comes from deep-inelastic electron, muon and neutrino scattering. At large momentum transfer, $Q^2 \gtrsim 2\ \mathrm{GeV}^2$, the lepton-nucleon inelastic cross section displays a scale-invariant behavior consistent with the simplest type of impulse approximation — where the electron scatters directly against point-like quark constituents of the target. [57] The deviations which are observed at very large $Q^2$ are consistent with the color radiative corrections predicted by QCD. In addition, at low values of $Q^2$, there is evidence for power-law "higher twist" corrections associated with coherent multiquark processes, interference effects, and final-state corrections — quite in analogy to the corrections to impulse approximation expected in nuclear physics inelastic breakup calculations.

The Fock-state representation we discussed in Sect. III provides a particularly simple and elegant basis for calculating the deep-inelastic cross section in QCD. We first consider the forward Compton amplitude $\gamma^* p \to \gamma^* p$ with virtual photon mass $q^2 = -Q^2 < 0$, and then calculate the $ep \to eX$ cross section from the absorptive part. An ideal Lorentz frame is

$$p = \left(p^+, p^-, \vec{p}_\perp\right) = \left(p^+, \frac{M^2}{p^+}, \vec{0}_\perp\right) \qquad (5.1)$$

$$q = \left(q^+, q^-, q_\perp\right) = \left(0, \frac{2p\cdot q}{p^+}, \vec{q}_\perp\right) \qquad (5.2)$$

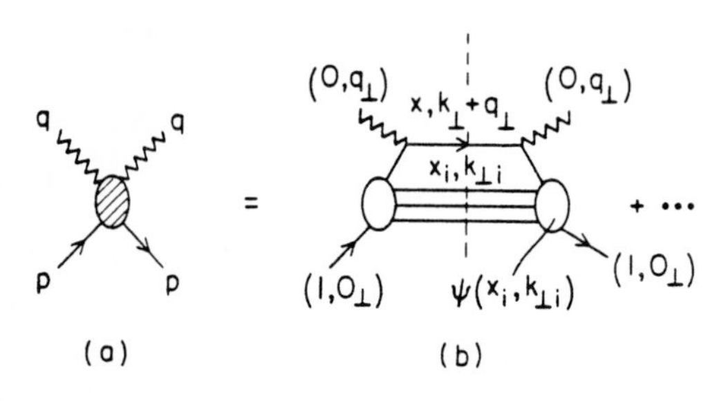


Fig. 13. Calculation of the forward virtual Compton amplitude. Diagram (b) gives the impulse approximation, neglecting final-state and multiquark interactions

with $q^2 = -Q^2$ and $p \cdot q = M\nu$. For the diagram 13b, which has no final-state interactions, the (light-cone) energy denominator between the photon interactions is

$$D = M^2 + 2M\nu - \frac{(\vec{k}_\perp + \vec{q}_\perp)^2 + m^2}{x} - \sum_{i \neq 1} \left( \frac{k_\perp^2 + m^2}{x} \right)_i + i\varepsilon \tag{5.3}$$

where m is the struck quark mass, and the sum over $i \neq 1$ gives the spectator quark and gluon contributions. For states with

$$|\mathscr{E}| = \left| M^2 - \sum_i \left( \frac{k_\perp^2 + m^2}{x} \right)_i \right| << 2M\nu \text{ and } k_\perp^2 << Q^2$$

we can write

$$D \cong 2M\nu - \frac{Q^2}{x} + i\varepsilon \tag{5.4}$$

$$\mathrm{Im} D^{-1} = \frac{x\pi}{2M\nu} \delta\left(x - \frac{Q^2}{2M\nu}\right) \tag{5.5}$$

i.e., the electron scattering on a quark with light-cone momentum fraction

$$x \equiv \frac{k^o + k^3}{p^o + p^3} \cong \frac{Q^2}{2M\nu} = x_{Bj} \ . \tag{5.6}$$

The corresponding impulse approximation cross section is $(x \to x_{Bj})$

$$\frac{d\sigma}{dQ^2\, dx} (\ell p \to \ell' X) = \sum_q G_{q/p}(x,Q) \frac{d\sigma}{dQ^2} (\ell q \to \ell' q) \Bigg|_{p_q = xp} \tag{5.7}$$

where [21]

$$G_{q/p}(x,Q) = \sum_{n \geq 3} \int^Q \left[d^2 k_\perp\right][dx] \left|\psi_n^Q(x,k_\perp)\right|^2 \delta(x - x_q) \tag{5.8}$$

gives the probability distribution for finding the quark with fractional light-cone momentum collinear up to the scale $k_\perp^2 < Q^2$, $|\mathscr{E}| < 2M\nu$. Unlike large-momentum-transfer exclusive amplitudes, all Fock states contribute to the inclusive cross section. The subprocess cross section $d\sigma/dQ^2(\ell q \to \ell' q)$ is evaluated for a quark collinear with the proton momentum $p_q^+ = xp^+$, $\vec{k}_\perp \simeq 0$. Since all the loop corrections to the subprocess cross section are hard $[k_\perp^2 \gtrsim O(Q^2)]$, it can be developed as a power series in $\alpha_s(Q^2)$. Thus the only correction to perfect scale invariance of $d\sigma/dx\, dQ^2$ at large $Q^2$ and fixed $x_{Bj}$ comes from the $Q^2$ dependence of the probability distribution $G(x,Q^2)$. This in turn can only arise from the wavefunction renormalization or from contributions $\psi_n \sim \mathcal{O}(1/k_\perp)$ at large $k_\perp$. In QCD these occur only from the perturbative processes $q \to qg$, and $g \to gg$, $g \to q\bar{q}$, as illustrated in Fig.14.

In parallel to the derivation of the evolution equation for the distribution amplitude, we then can derive evolution equations for the distributions $G_{q/H}(x,Q^2)$ and $G_{g/H}(x,Q^2)$ of the form [58,59]

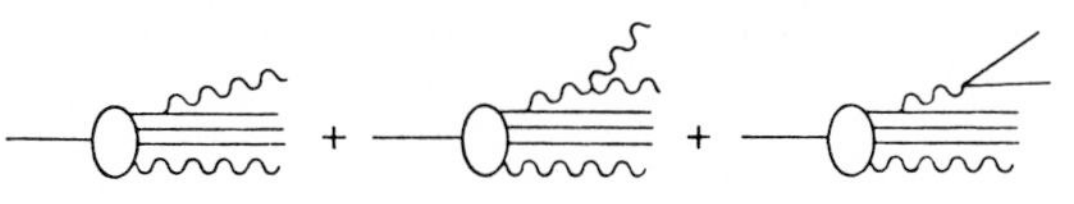

Fig.14. Contributions to the hadron Fock-state wavefunction which give $\psi \sim 1/k_\perp$ at large $k_\perp$ and thus structure function evolution

$$\frac{\partial}{\partial \log Q^2} G(x,Q) = \frac{\alpha_s(Q^2)}{2\pi} \int_x^1 P\left(\frac{x}{y}\right) G(y,Q) \frac{dy}{y} . \tag{5.9}$$

For example, for the "non-singlet" distribution

$$G_{q/H}(x,Q) = G_{q/H}(x,Q) = G_{\bar{q}/H}(x,Q) \tag{5.10}$$

we have to lowest order in $\alpha_s(Q^2)$, ($C_F = 4/3$)

$$P_{q/q}(\bar{z}) = C_F \left(\frac{1+z^2}{1-z}\right)_+ = C_F\left[\frac{1+z^2}{1-z} - \delta(1-z)\int_0^1 dx \frac{1+x^2}{1-x}\right] . \tag{5.11}$$

[The subtraction term, which ensures finite behavior at $x_g = 0$, arises from the wavefunction renormalization, as in Eq. (4.14)]. The $Q^2$ dependence can be displayed most simply by taking moments:

$$M_n(Q^2) = \int_0^1 G(x,Q^2)\, x^n\, dx . \tag{5.12}$$

Then

$$M_n^{NS} = M_n^{NS}(Q_0^2)\left(\frac{\log Q^2/\Lambda^2}{\log Q_0^2/\Lambda^2}\right)^{-\gamma_n} \tag{5.13}$$

where the $\gamma_n$ are defined in Eq. (4.16). The higher-order corrections to the $Q^2$ evolution of $M_n$ are discussed in Refs. 1 and 2. A critical feature [21] is the fact that the higher loop corrections (e.g., from the higher Fock states) are constrained kinematically to $k_\perp^2 < (1-y)Q^2 < (1-x)Q^2$, where y is labelled in the figure; i.e., the evolution is reduced at large x and for large n. A detailed discussion is given in Ref. 41.

Equation (5.7) displays an essential feature of the QCD predictions for inclusive reactions: the factorization of the physical cross section into a hard scattering subprocess cross section, controlled by short-distance perturbative QCD, convoluted with structure functions $G(x,Q^2)$ which contain the long-distance hadronic bound-state dynamics. Notice that the $Q^2$ evolution of $G(x,Q)$ is also completely specified by the perturbative QCD processes and is independent of the nature of the target.

All the corrections to the perturbative QCD impulse approximation from final-state interactions, finite $k_\perp^2$ effects, interference contributions, mass corrections, etc. are of higher order in $1/Q^2$, at least when analyzed using perturbative methods. In the operator product analysis these contributions correspond to matrix elements of "higher twist" operators which have non-minimal dimensions. The most important higher twist terms for deep-inelastic lepton scattering are expected to correspond to processes where the lepton scatters on multiparticle clusters in the target ($qq$, $q\bar{q}$, virtual mesons, $qg$, etc.). We thus obtain a sum of contributions (see Fig.15): [15]

$$\frac{d\sigma}{dQ^2 dx}(\ell H \to \ell' X) = \sum_{a\in H} G_{a/H}(x) \frac{d\sigma}{dQ^2}(ea \to ea)\bigg|_{p_a = xp_H} \tag{5.14}$$

where, in general $d\sigma_a/dQ^2$ falls in $Q^2$ according to the compositeness of a:

$$\frac{d\sigma}{dQ^2}(\ell a \to \ell' a) \sim \frac{4\pi\alpha^2}{Q^4} |F_a(Q^2)|^2 . \tag{5.15}$$

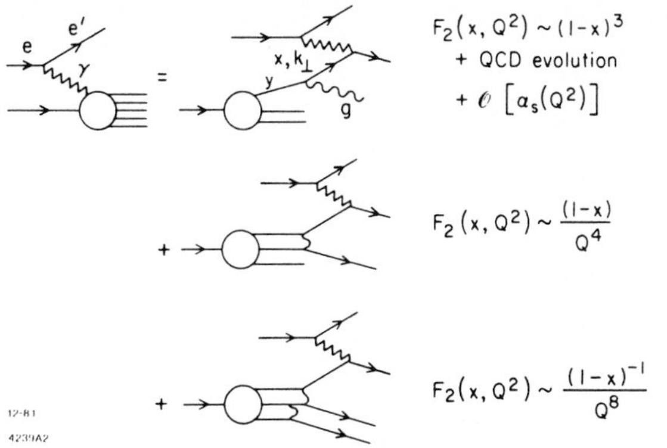

$F_2(x, Q^2) \sim (1-x)^3$
+ QCD evolution
$+ \mathcal{O}\left[\alpha_s(Q^2)\right]$

$F_2(x, Q^2) \sim \frac{(1-x)}{Q^4}$

$F_2(x, Q^2) \sim \frac{(1-x)^{-1}}{Q^8}$

Fig. 15. QCD contributions to inelastic electron-nucleon scattering, including radiative and higher twist (diquark, triquark) corrections

For example, the "diquark" eqq → eqq gives a contribution to ep → eX of relative order $(m^2/Q^2)^2$. Since the qq can carry a large fraction of the proton's momentum, this contribution can be significant at large x. For a guide to this effect one can use the spectator counting rule: [60,8]

$$G_{a/H}(x) \underset{x \to 1}{\sim} (1-x)^{2n_s-1} \tag{5.16}$$

where $n_s$ is the minimum number of spectator quarks (or gluons) in the Fock state required to stop at $x \to 1$. The minimal Fock states containing a give the dominant contribution.

The simplified rule (5.16) can be derived from minimally connected tree-graph diagrams, ignoring spin effects, or from simple phase-space considerations if one ignores the spectator quark masses [61] (see Sect, VI). Using this simple counting we can then classify the contributions to the hadron structure functions, as illustrated in Fig.15. The diquark contribution is expected to give a large contribution to the longitudinal structure function,since it acts coherently as a boson current. The order $\alpha_s(Q^2)$ contribution from the hard gluon radiative corrections with $k_\perp^2 > (1-x)Q^2$ also gives a significant contribution to $\sigma_L$.

A detailed derivation of the behavior of structure functions at $x \sim 1$ from perturbative QCD is given in Ref. 21. At $x \sim 1$ all of the hadron's momentum must be carried by one quark, and each quark and gluon quark and gluon propagator which transfers this momentum becomes far off shell:

$$k^2 \sim \mathcal{O}\left(-\frac{k_\perp^2 + m^2}{1-x}\right) \quad .$$

Perturbative QCD predictions thus become relevant. An important result is that at large x the struck quark tends to have the same helicity as the target nucleon: [21,62]

$$G_{q\uparrow/p\uparrow} \sim (1-x)^3; \qquad G_{q\downarrow/p\uparrow} \sim (1-x)^5 \tag{5.17}$$

This type of spin correlation is consistent with the SLAC-Yale polarized electron/polarized target data. Combined with the SU(6) symmetry of the nucleon wavefunction, this implies that the leading quark in the proton is five times more likely to be an up quark than a down quark, and thus [62] $\left(F_2 = \sum_q e_q^2 x G_{q/n}\right)$

$$F_{2n}(x,Q^2)/F_{2p}(x,Q^2) \underset{x \sim 1}{\longrightarrow} 3/7 \; . \tag{5.18}$$

For the case of mesons, the perturbative QCD gluon exchange prediction is [63]

$$G_{q/m} \sim (1-x)^2 . \tag{5.19}$$

In addition, the same QCD analysis predicts a large $C/Q^2$ contribution to the meson longitudinal structure function (see Fig.3b): [22,64]

$$F_L^{\pi}(x,Q^2) = \frac{2x^2}{Q^2} C_F \int_{\sim m^2/(1-x)}^{Q^2} dk^2 \, \alpha_s(k^2) \, F_{\pi}(k^2) \tag{5.20}$$

which numerically is $F_L \sim x^2/Q^2$ in $GeV^2$ units. This contribution, which can dominate leading twist quark distributions in mesons, is normalized in terms of the meson distribution amplitude, which in turn is normalized by the pion form factor.

The dominance of the longitudinal structure functions in the fixed W limit for mesons is an essential prediction of perturbative QCD. Perhaps the most dramatic consequence is in the Drell-Yan process $\pi p \to \ell^+\ell^- X$; one predicts [22] that for fixed pair mass Q, the angular distribution of the $\ell^+$ (in the pair rest frame) will change from the conventional $(1+\cos^2\theta_+)$ distribution to $\sin^2(\theta_+)$ for pairs produced at large $x_L$. A recent analysis of the Chicago-Illinois-Princeton experiment [65] at FNAL appears to confirm the QCD high twist prediction with about the expected normalization. Striking evidence for the effect has also been seen in a Gargamelle analysis [66] of the quark fragmentation functions in $\nu p \to \pi^+\mu^- X$. The results yield a quark fragmentation distribution into positive charged hadrons which is consistent with the predicted form: $dN^+/dzdy \sim B(1-z)^2 + (C/Q^2)(1-y)$ where the $(1-y)$ behavior corresponds to a longitudinal structure function. It is also crucial to check that the $e^+e^- \to MX$ cross section becomes purely longitudinal $(\sin^2\theta)$ at large z at moderate $Q^2$. [62]

The results (5.17) and (5.19) for $G_{q/B}$ and $G_{q/M}$ give the behavior of the leading QCD contribution to the structure function before QCD evolution is applied; e.g., the results are valid for $F_2(x,Q^2)$ at $Q^2$ of order of $\langle k_\perp^2\rangle_H$. The large $Q^2$ behavior is determined by the evolution equations (5.9), taking account of the phase-space limits of the radiated gluons at $x \sim 1$. [41]

## 3.6 The Phenomenology of Hadronic Wavefunctions

Thus far, most of the phenomenological tests of QCD have focused on the dynamics of quark and gluon subprocesses in inclusive high-momentum-transfer reactions. The Fock-state wavefunctions $\psi_n^{\kappa}(x_i,\vec{k}_{\perp i};\lambda_i)$ which determine the dynamics of hadrons in terms of their quark and gluon degrees of freedom are also of fundamental importance. If these wavefunctions were accurately known, then an extraordinary number of phenomena, including decay amplitudes, exclusive processes, higher twist contributions to inclusive phenomena, structure functions, and low-transverse-momentum phenomena (such as diffractive processes, leading particle production in hadron-hadron collisions and heavy flavor hadron production) could be interrelated. Conversely, these processes can provide phenomenological constraints on the Fock-state wavefunctions which are important for understanding the dynamics of hadrons in QCD. In addition, as we discuss in Sect, VII, the structure of nuclear wavefunctions in QCD is essential for understanding the syntheses of nuclear physics phenomenology with QCD.

### 3.6.1 Measures of Hadron Wavefunctions

As we have shown in Sect. III, the central measures of the hadron wavefunctions are the distribution amplitudes

$$\phi(x_i,Q) = \int^Q \left[d^2k_\perp\right] \psi_V^Q(x_i,\vec{k}_{\perp i}) \tag{6.1}$$

which control high-momentum-transfer form factors and exclusive processes:

$$\mathcal{M} \cong \Pi\, \phi \otimes T_H \tag{6.2}$$

and the quark and gluon structure functions

$$G_{q/H}(x,Q) = \sum_n \int^Q \left[d^2k_\perp\right][dx]\,|\psi_n(x_i,k_{\perp i})|^2\,\delta(x-x_q) \tag{6.3}$$

which control high-momentum-transfer inclusive reactions

$$d\sigma \cong \Pi\, G \otimes d\hat{\sigma}\ . \tag{6.4}$$

Examples are shown in Figs.1 through 3. A summary of the basic properties, logarithmic evolution, and power-law behavior of these quantities is given in Table 4.

Table 4. Comparison of exclusive and inclusive cross sections

| Exclusive Amplitudes | Inclusive Cross Sections |
|---|---|
| $\mathcal{M} \sim \Pi\, \phi(x_i,Q) \otimes T_H(x_i,Q)$ | $d\sigma \sim \Pi\, G(x_a,Q) \otimes d\hat{\sigma}(x_a,Q)$ |
| $\phi(x,Q) = \int^Q \left[d^2k_\perp\right] \psi^Q_{val}(x,k_\perp)$ | $G(x,Q) = \sum_n \int^Q \left[d^2k_\perp\right][dx]'\,\lvert\psi^Q_n(x,k_\perp)\rvert^2$ |
| Measure $\phi$ in $\gamma\gamma \to M\bar{M}$ | Measure G in $\ell p \to \ell X$ |
| $\sum_{i\epsilon H} \lambda_i = \lambda_H$ | $\sum_{i\epsilon H} \lambda_i \neq \lambda_H$ |
| EVOLUTION | |
| $\frac{\partial\phi(x,Q)}{\partial \log Q^2} = \alpha_s \int [dy] V(x,y)\phi(y)$ | $\frac{\partial G(x,Q)}{\partial \log Q^2} = \alpha_s \int dy\, P(x/y)G(y)$ |
| $\lim_{Q\to\infty} \phi(x,Q) = \prod_i x_i \cdot C_{flavor}$ | $\lim_{Q\to\infty} G(x,Q) = \delta(x)\ C$ |
| POWER-LAW BEHAVIOR | |
| $\frac{d\sigma}{dx}(A+B \to C+D) \cong \frac{1}{s^{n-2}} f(\theta_{CM})$ | $\frac{d\sigma}{d^2p/E}(AB \to CX) \cong \sum \frac{(1-x_T)^{2n_s-1}}{(Q^2)^{n_{act}-2}} f(\theta_{CM})$ |
| $n = n_A + n_B + n_C + n_D$ | $n_{act} = n_a + n_b + n_c + n_d$ |
| $T_H$: expansion in $\alpha_s(Q^2)$ | $d\hat{\sigma}$: expansion in $\alpha_s(Q^2)$ |
| COMPLICATIONS | |
| Endpoint singularities<br>Pinch singularities<br>High Fock states | Multiple scales<br>Phase-space limits on evolution<br>Heavy-quark thresholds<br>Heavy twist multiparticle processes<br>Initial- and final-state interactions |

The exclusive formula (6.2) also includes applications to large-momentum-transfer multiparticle production [68,8] $e^+e^- \to H_1 \ldots H_n$ with $p_i \cdot p_j \sim \mathcal{O}(Q^2)$, and the elastic and inelastic weak and electromagnetic form factors. We also note that hard scattering higher twist subprocesses to inclusive reactions such as $\gamma q \to Mq$, $gq \to Mq$, $q\bar{q} \to M\bar{M}$, $qq \to B\bar{q}$., etc. are absolutely normalized in terms of the distribution amplitudes. [69] In particular, some amplitudes such as $\gamma q \to \pi q$, $q\bar{q} \to \pi g$ and $gq \to \pi q$ can be rigorously related to the pion form factor, since the same integral

$$\int_0^1 \frac{dx}{1-x}\, \phi_\pi(x,Q) \tag{6.5}$$

enters in each of the quantities. [70] The $p_T^{-6}$ processes [24] $gq \to Mq$ (see Fig.3a) and $q\bar{q} \to Mq$ are particularly interesting and important in high-$p_T$ meson production processes such as $pp \to MX$, since the meson is produced directly in the subprocess without the necessity for quark or gluon jet fragmentation. In fact, the contributions of standard $p_T^{-4}$ scaling processes such as $qq \to qq$, $gq \to gq$, and $gg \to gg$ to hadron production are strongly suppressed by two to three orders of magnitude because of the suppression of jet fragmentation $D_M/q(z)$ at large momentum fraction z and the fact that the subprocesses must occur at a significantly larger momentum transfer than that of the triggered particle. [71]

Despite much effort there is at this time no systematic understanding of high-$p_T$ hadron production in QCD. A comprehensive attack must take into account not only the leading twist subprocesses and directly coupled higher twist contributions such as those listed above, but also the effects of initial-state multiple scattering effects. One of the most important experiments which could clarify the nature of these effects is the measurement of the ratio of direct photon to meson at high $p_T$: $(x_T + 2p_T/\sqrt{s})$

$$R_{\gamma/\pi}(x_T, s, \theta_{c.m.}) = \frac{d\sigma}{d^3p/E}(pp \to \gamma X) \Big/ \frac{d\sigma}{d^3p/E}(pp \to \pi X)\ . \tag{6.6}$$

For example, if leading twist QCD processes dominate these reactions, then $R_{\gamma/\pi} \sim f(x_T) \sim (1-x_T)^{-2}$ at $\theta_{c.m.} \sim \pi/2$. If directly coupled processes such as $gq \to \pi q$ dominate the meson production, then one predicts $R_{\gamma/\pi} \sim p_T^2$ at fixed $x_T$ and $\theta_{c.m.}$ [72]. Measurements of this ratio in nuclear targets are important for clarifying the contribution of final-state multiple scattering processes.

The photon probe plays a crucial role in high-$p_T$ hadron reactions, since the photon couples directly to the quark and gluon subprocesses at short distances. The most dramatic example of these point-like phenomena are the recent observations at PETRA [6-8] of high-transverse-momentum hadrons in $\gamma\gamma$ collisions. The results at $p_T \geq 3$ GeV appear to be consistent with the scale-invariant QCD prediction [73]

$$\frac{d\sigma(\gamma\gamma \to \text{jet} + \text{jet})}{d\sigma(\gamma\gamma \to \mu^+\mu^-)} = 3\sum e_q^4, \quad q = u,d,s,c \tag{6.7}$$

$$\left[1 + \mathcal{O}\left(\frac{\alpha_s(p_T^2)}{\pi}\right)\right]\ .$$

These results also indicate that, unlike typical meson-induced reactions, an incident photon often produces high-$p_T$ hadronic jets without leaving hadronic energy in the beam fragmentation direction. [74] One also expects analogous results for directly coupled photons in $\gamma p \to HX$ and $\gamma p \to \text{Jet} + X$ reactions. The point-like behavior of on-shell photons is in direct contrast to the predictions of vector-meson dominance models.

A surprising feature of QCD is that even a hadron can produce jets at large $p_T$ without beam fragmentation. [70] For example, the existence of high twist subprocesses such as $Mq \to gq$ and $Mg \to q\bar{q}$ leads to high-$p_T$ jet events in meson-induced collisions $Mp \to$ Jet + Jet + X where there is no hadronic energy left in the meson beam fragmentation direction (see Fig.3c). The inclusive cross section, which scales as $p_T^{-6}$ at fixed $x_T$ and $\theta_{c.m.}$, is absolutely normalized to the meson form factor. As in the case of the photon-induced reactions, the directly coupled meson has no associated color radiation or structure function evolution. An experimental search for these unique and highly kinematically constrained events is very important in order to confirm the presence of these subprocesses, which involve the direct coupling of meson $q\bar{q}$ Fock state to quarks and gluons at short distance.

In general, we can replace any direct photon interaction by a direct-coupled meson interaction in the subprocess cross section by the replacement $\alpha \rightleftarrows F_\pi(p_T^2)$. Furthermore, one can compute direct-coupled processes which isolate the valence Fock state of baryons, e.g., $pp \to pX$ (production of isolated large-$p_T$ protons via the $qq \to pq$ subprocesses), and reactions $pp \to qqX$ (from $\bar{q}p \to qq$) (see Fig.3b), $pp \to qqqX$ (from $gq \to qqq$) etc., each of which produce jets at high $p_T$ without beam spectators or fragmentation.

### 3.6.2 Constraints on the Pion and Proton Valence Wavefunction [27]

The central unknown in the QCD analysis of hadronic matrix elements is the hadron wavefunction in the non-perturbative domain $\kappa^2 \gtrsim 1\ \mathrm{GeV}^2$. For illustration we shall assume that in this region the $\psi_n$ fall off exponentially in the off-shell energy:

$$\psi_n^\kappa(x_i, k_{\perp i}) = A_n\, e^{b_n^2 \mathscr{E}_n} \tag{6.8}$$

$$\mathscr{E}_n = M^2 - \sum_{i=1}^{n} \left(\frac{k_\perp^2 + m^2}{x}\right)_i < 0\ . \tag{6.9}$$

The parameterization is taken to be independent of spin; the full wavefunction is then obtained by multiplying by free spinors $u/\sqrt{k^+}$ . The form (6.8) has the advantage of analytic simplicity: for example, the resulting baryon distribution amplitude at small $\kappa$ is

$$\phi(x_i,\kappa) = A_\phi\, x_1 x_2 x_3\, e^{-b_3^2 \sum_{i=1}^{3} \frac{m_i^2}{x_i}}\ . \tag{6.10}$$

At large $\kappa$, $\phi$ is determined from the evolution equation (4.33). At very large $k_\perp$ the $\psi_n$ for non-valence Fock states should match onto the power-law fall-off $k_\perp^{-1}$ predicted by perturbative QCD. It should be emphasized that the form (6.8) is chosen just for simplicity. An equally plausible parameterization is $\psi_n \sim A_n \mathscr{E}_n^{-p}$ with $p = 3$, which is suggested by the Schroedinger equation assuming a linear potential and the correspondence given in Eq. (3.41).

In the case of the pion we can derive two important constraints on the valence wavefunction from the $\pi \to \mu\nu$ and $\pi^0 \to \gamma\gamma$ decay amplitudes:

$$\int \frac{d^2k_\perp}{16\pi^3} \int_0^1 dx\, \psi^\kappa(x,k_\perp) = \frac{f_\pi}{2\sqrt{n_c}} \left[1 + \mathcal{O}\left(\frac{m_\pi^2}{\kappa^2}\right)\right] \tag{6.11}$$

and [27]

$$\psi^\kappa(x,k_\perp = 0)\, \frac{Z_2(m_\pi^2)}{Z_2(\kappa^2)} = \frac{\sqrt{n_c}}{f_\pi}\ . \tag{6.12}$$

The derivation of the second constraint assumes that the radius of the pion is much smaller than its Compton length:

$$m_q^2, m_\pi^2 << \frac{6}{R_\pi^2} \quad . \tag{6.13}$$

Let us now assume the form

$$\psi_{q\bar{q}}^{\kappa} \propto e^{-b_V^2\left(\frac{k_\perp^2 + m^2}{x(1-x)}\right)}, \quad (\kappa^2 < 1\ \mathrm{GeV}^2) \tag{6.14}$$

where

$$-\frac{d}{dQ^2} F_\pi^V(Q^2)\bigg|_{Q^2=0} = \frac{1}{6}\left(R_\pi^{q\bar{q}}\right)^2 = b_V^2 \tag{6.15}$$

is the contribution to the slope of the meson form factor from the valence Fock state [see Eq. (4.2)]. The two conditions (6.11) and (6.12) then determine $R_\pi^{q\bar{q}} = 0.42$ fm, and [27]

$$P_{q\bar{q}/\pi}^{\kappa} = \int^{\kappa} \frac{d^2k_\perp}{16\pi^3} \int_0^1 dx \left|\psi_{q\bar{q}/\pi}^{\kappa}(x, \vec{k}_\perp)\right|^2 = \frac{1}{4}\left(\frac{Z_2(\kappa^2)}{Z_2(m_\pi^2)}\right)^2 \leq \frac{1}{4} \quad . \tag{6.16}$$

Thus the probability that the pion contains only the valence Fock state at small $\kappa^2$ is less than 1/4. Furthermore, the radius of the valence state turns out to be smaller than that of the total state: $R_\pi^{expt} \cong 0.7$ fm. One can also verify that the bound $P_{q\bar{q}/\pi} \leq 1/4$ is also true for power-law wavefunctions $\psi \sim \mathcal{E}^{-p}$, $p > 2$.

The existence of other Fock states at equal $\tau$ in the pion is to be expected considering the fact that its quark and gluon constituents are relativistic. The existence of large $m_\rho/m_\pi$ and $m_\Delta/m_N$ spin splittings (due to transverse-polarized gluon exchange) also implies that there is a non-zero gluon component intrinsic to both meson and nucleon bound states.

In the case of the baryon wavefunction, one can obtain non-trivial constraints on the form of the 3-quark valence wavefunction by making a simultaneous analysis of the proton and neutron form factors and the $\psi \to pp$ decay amplitude, assuming the $\psi$ decays via a 3-gluon intermediate state (see Fig.6). The observed angular distribution [53] for $\psi \to p\bar{p}$ is in fact consistent with the predicted form $1 + \beta^2 \cos^2\theta$ (where $\beta$ is the nucleon velocity) and is a non-trivial check of hadron helicity conservation for exclusive processes in QCD.

The $\psi \to p\bar{p}$ ratio is given to leading order in $\alpha_s$ by (Fig.1b) [18]

$$\frac{\Gamma(\psi \to 3g \to p\bar{p})}{\Gamma(\psi \to 3g \to \mathrm{all})} + 3.2 \times 10^6\ \alpha_s^3(s)\ \frac{|\vec{p}_{CM}|}{\sqrt{s}}\ \frac{\langle T\rangle^2}{s^4} \tag{6.17}$$

where $|p_{CM}|/\sqrt{s} \simeq 0.4$, $s = 9.6\ \mathrm{GEV}^2$, and

$$\langle T\rangle \equiv \int_0^1 [dx][dy]\ \frac{\phi^*(y_i,s)}{y_1y_2y_3}\ \frac{x_1y_3 + x_3y_1}{[x_1(1-y_1) + y_1(1-x_1)]\ [x_3(1-y_3) + y_3(1-x_3)]} \times \frac{\phi(x_i,s)}{x_1x_2x_3} \tag{6.18}$$

is a well-defined function of the baryon distribution amplitude. In the case of the nuclear form factors [see Eqs. (4.31, 4.32)] it is important to use the correct argument for each $\alpha_s$ in the hard scattering amplitude $T_H$ corresponding to the actual momentum transfer which flows through each exchanged gluon in Fig.7b. This effect is expected to yield the most important contribution to next to leading order in $\alpha_s$ and is an integral part of the QCD predictions. It is interesting to note that if $\phi_B = A_\phi x_1 x_2 x_3$ and if all the $\alpha_s$ have the same argument (which is, in fact, the situation in the asymptotic $Q^2 \to \infty$ limit [9,19]) then Eqs. (4.28-4.32) give $\lim_{Q^2\to\infty} G_M^P(Q^2)/G_M^n(Q^2) = 0$. However, the fact that $\alpha_s$ is not a constant and has different arguments for each diagram in $T_1$ allows one to obtain empirically consistent results for the normalization [75] of $G_M^P(Q^2)$, $G_M^n(Q^2)$ and the $\phi \to p\bar{p}$ decay rate. To first approximation one requires [27]

$$\frac{\alpha_s(x_i y_i Q^2)}{\alpha_s\left((1-x_i)(1-y_i)Q^2\right)} \cong \frac{\alpha_s(Q^2/9)}{\alpha_s(4Q^2/9)} \sim 1.5 \text{ to } 2.0 \text{ at } Q^2 \cong 10\ \text{GeV}^2 \quad . \tag{6.19}$$

The QCD predictions (4.28-4.30) for the proton and neutron form factors are only valid at large $Q^2$ where the effects of mass corrections, higher Fock states and finite transverse momentum can be neglected. In order to understand these effects we extend the parameterization of the 3-quark valence Fock-state contribution by using $(Q^2 + M_0^2)^{-2}$ in the denominators of (4.29, 4.30) and replacing $\alpha_s(Q^2) \to \alpha_s(Q^2 + M^2) = 4\pi/\beta_0 \log\left((Q^2 + M^2)/\Lambda^2\right)$ to reflect the fact that at low $Q^2$ the transverse momenta intrinsic to the bound-state wavefunctions flow through all the propagators.

Although we have not tried to optimize the parameterizations, a typical fit which is compatible with the proton and neutron form factors (see Fig.16) and $\psi \to p\bar{p}$ decay data is $M_0 \cong 1.5$ GeV, $\mu \cong 450$ MeV, $m_q \cong 300$ MeV, and $\Lambda = 280$ MeV, so that $\alpha_s(Q^2 = 10\ \text{GeV}^2) \cong 0.29$. (Analyses [50] of higher-order QCD corrections to the meson form factors suggest that one can identify the $\Lambda$ used here with $\Lambda_{mom} = 2.16\ \Lambda_{\overline{MS}}$.) The computed radius of the 3-quark valence state [computed from $G_M^{\prime}$ via

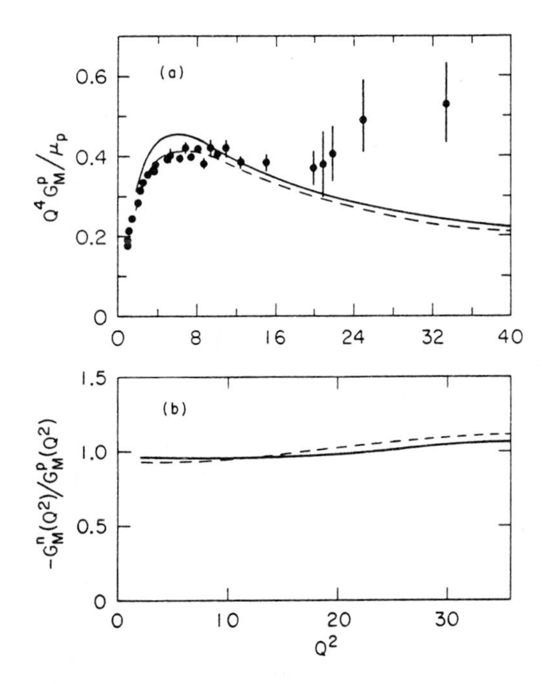


Fig.16. Fit nucleon form factor data described in the text. (From Ref.27)

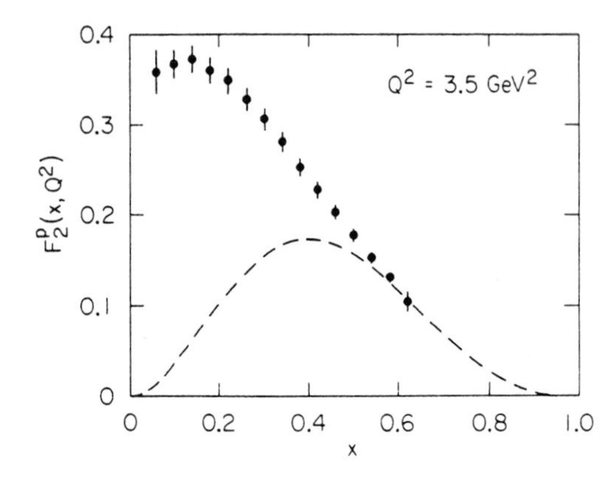


Fig.17. Predicted valence quark contribution to the proton structure function. Evolution and higher Fock-states are not included. (From Ref.27)

Eq. (4.2)] is, however, quite small: $R_V \cong 0.23$ fm, and the valence Fock-state probability is $P_{qqq/p} \gtrsim 1/4$. If this preliminary analysis is correct, then, as in the meson case, the valence state is much smaller in transverse size than the physical hadron (which receives contributions to its charge radius from all Fock states).

The most crucial prediction from this analysis is that $Q^4 G_M^p(Q^2)$ should decrease by a factor of 2 for $Q^2 = 10$ to $Q^2 = 40\ GeV^2$, a trend not at all indicated by the data! Further measurements of $G_M(Q^2)$ are clearly crucial in order to check this essential prediction of asymptotic freedom.

Given the above parameterization of the nucleon valence Fock state we can use Eq. (5.8) to compute the 3-quark non-perturbative contribution to the proton structure function at large x (see Fig.17)

$$G^V_{q/p}(x,Q_0^2) \propto x(1-x)^3\, e^{-2m^2b^2\left(\frac{1}{x}+\frac{2}{1-x}\right)} \quad . \tag{6.20}$$

Since $4\, m^2b^2 \sim 0.05$, the exponential factor is not very important away from the edge of phase space, and so it is difficult to distinguish between the non-perturbative and $(1-x)^3$ perturbative contributions at large x (see Sect. V). Higher Fock states $|qqqg\rangle$, $|qqq\, q\bar{q}\rangle$ are expected to give the dominant contribution at lower x. Despite the freedom in this parameterization it is reassuring that one can simultaneously fit a number of diverse nucleon properties with QCD formulae and parameters which are in the expected range.

At low $Q^2$ the exact formula (4.2) can be used as a further constraint on the baryon Fock states. Eventually one hopes to extend the predictions to other domains of baryon phenomenology such as the baryon decay amplitude in grand unified models and the normalization of higher twist subprocess contributions to inelastic lepton-nucleon scattering.

### 3.6.3 Quark Jet Diffractive Excitation [30]

The fact that the wavefunction of a hadron is a superposition of (infrared and ultraviolet finite) Fock amplitudes of fixed particle number but varying spatial and spin structure leads to the prediction of a novel effect in QCD. [30] We first note that the existence of the decay amplitude $\pi \to \mu\nu$ requires a finite probability amplitude for the pion to exist as a quark and diquark at zero transverse separation:

$$\psi(x,\vec{r}_\perp = 0) = \sqrt{4\pi}\,\sqrt{n_c}\; x(1-x) f_\pi \quad . \tag{6.21}$$

In a QCD-based picture of the total hadron-hadron cross section, the components of a color-singlet wavefunction with small transverse separation interact only weakly with the color field, and thus can pass freely through a hadronic target while the other components interact strongly. A large nuclear target will thus act as a filter removing from the beam all but the short-range components of the projectile wavefunction. The associated cross section for diffractive production of the inelastic states described by the short-range components is then equal to the elastic scattering cross section of the projectile on the target multiplied by the probability that sufficiently small transverse separation configurations are present in the wavefunction. In the case of the pion interacting in a nucleus one computes the cross section

$$\left.\frac{d\sigma}{dx\, d^2 r_\perp}\right|_{r_\perp^2 \sim 0} \cong \sigma_{e\ell}^{\pi A}\, 12\pi\, f_\pi^2\, x^2 (1-x)^2 \qquad (6.22)$$

corresponding to the production of two jets just outside the nuclear volume. The x distribution corresponds to $d\sigma/d\cos\theta \sim \sin^2\theta$ for the jet angular distribution in the $q\bar{q}$ center of mass. By taking into account the absorption of hadrons in the nucleus at $\vec{r}_\perp \neq 0$ one can also compute the $k_\perp$ distribution of the jets and the mass spectrum of the diffractive hadron system. Details are given in Ref. 30.

### 3.6.4 The "Unveiling" of the Hadronic Wavefunction and Intrinsic Charm

The renormalizability of QCD implies that all of the dynamics of the hadron wavefunctions $\psi_n^\kappa(x_i, \vec{k}_{\perp i})$ at scales $\kappa^2$ much larger than mass thresholds is completely contained in the structure of the running coupling constant $\alpha_s(\kappa^2)$ and running mass $m(\kappa^2)$ and the quark and gluon external line renormalization constants. Nevertheless, the fact that there are different hadronic scales and thresholds in QCD does imply non-trivial dynamical structure of the wavefunctions. In the case of Compton scattering, $\gamma p \to \gamma p$, the energy denominators [see Eq. (5.3)] are a function of $2M\nu - \mathscr{E}_n$, so that the cross section is sensitive to wavefunctions up to the scale $\kappa^2 \sim 2M\nu$.

As an example of the change of wavefunction physics with the resolution scale let us consider a deuteron target. For very low $\kappa^2 \ll 2M\varepsilon_{B.E.}$ the deuteron acts as a coherent object. At the scale $\kappa^2 \gg 2M\varepsilon_{B.E.}$, the wavefunction corresponds to an n-p bound state. As the scale increases to $\kappa^2 \gtrsim 1$ GeV$^2$, the quark degrees of freedom become relevant and the deuteron wavefunction in QCD must be described in terms of six-quark (and higher) Fock states: [76]

$$|D\rangle = a|(uud)_1(ddu)_1\rangle + b|(uud_8(ddu)_8\rangle + c|(uuu)_1(ddd)_1\rangle + d|(uuu)_8(ddd)_8\rangle + \ldots \qquad (6.23)$$

The first component corresponds to the usual n-p structure of the deuteron. The second component corresponds to "hidden color" or "color polarized" configurations where the three-quark clusters are in color octets, but the overall state is a color-singlet. The last two components are the corresponding isobar configurations. If we suppose that at low relative momentum the deuteron is dominated by the n-p configuration, then quark-quark scattering via single gluon exchange generates the color-polarized states (b) and (d) at high $k_\perp$; i.e., there must be mixing with color-polarized states in the deuteron wavefunction at short distances. [67]

The deuteron's Fock-state structure is thus much richer in QCD than it is in nuclear physics where the only degrees of freedom are hadrons.

It is interesting to speculate on whether the existence of these new configurations in normal nuclei could be related to the repulsive core of the nucleon-nucleon potential, [76] and the enhancement [77] of parity-violating effects in nuclear capture reactions. One may also expect that there are resonance states with nuclear quantum numbers which are dominantly color polarized. The mass of these states is not known. It has also been speculated [78] that such long-lived states could have an anomalously large interaction cross section, and thus account for the JUDEK [79] anomaly in cosmic-ray and heavy-ion experiments. [80] Independent of these speculations, it is clearly important that detailed high-resolution searches for these states be conducted, particularly in inelastic electron scattering and tagged photon nuclear target experiments, such as $\gamma d \to \gamma d$ scatter at large angles.

The structure of the photon's Fock states in QCD is evidently richer than that expected in the vector-meson dominance model. [81] For example, consider the one-gluon exchange correction to the $\gamma \to q\bar{q}$ vertex. For $\ell_\perp^2 > \mathcal{O}(\kappa^2)$ the vertex correction renormalizes the point vertex. For the soft domain $\ell_\perp^2 < \mathcal{O}(\kappa^2)$ one expects large corrections which eventually by dispersion theory correspond to the usual $\rho$, $\omega$, $\phi$, ... interpolating fields. The soft corrections thus give the usual hadron-like component of real photon interactions. Nevertheless, the point-like component survives at any momentum scale, [81] producing point-like corrections to photon shadowing, $J = 0$ fixed pole phenomena in the Compton amplitude, and the "anti-scaling" QCD structure function of the photon. [13] As the resolution scale $\kappa^2$ increases past the heavy-quark thresholds, one adds the $\gamma \to c\bar{c}$, $b\bar{b}$, etc. components to the photon's wavefunctions.

It is also interesting to consider the dynamical changes to the nucleon wavefunction as one passes heavy-quark thresholds. For $\kappa^2 > 4m_c^2$ the proton Fock-state structure contains charm quarks, e.g., states $|p\rangle \sim |uud\,c\bar{c}\rangle$. We can distinguish two types of contributions to this Fock state. [31] (1) The "extrinsic" or interaction-dependent component generated from quark self-energy diagrams as shown in Fig.18b — a component which evolves by the usual QCD equations with the photon mass scale $Q^2$; and (2) the "intrinsic" or interaction-independent component which is generated by the QCD potential and equations of motion for the proton, as in Fig.18a — a component which contributes to the proton Fock state without regard to QCD evolution. Since the intrinsic component is maximal for minimum off-shell energy

$$\mathcal{E} = M^2 - \sum_i \left[(k_\perp^2 + m^2)/x\right]_i \ ,$$

the charm quarks tend to have the largest momentum fraction x in the Fock state. (This also agrees with the physical picture that all the constituents of a bound state tend to have the same velocity in the rest frame, i.e., strong correlations in rapidity.) Thus, heavy quarks (though rare) carry most of the momentum in the Fock state in which they are present — in contrast to the usual parton-model assumption that non-valence sea quarks are always found at low x. One can also estimate using the bag model and perturbative QCD that the probability of finding intrinsic charm in the proton is ~1-2%. [82]

The diffractive dissociation of the proton's intrinsic charm state [30,31] provides a simple explanation why charmed baryons and charmed mesons which contain no valence quarks in common with the proton are diffractively produced at large $x_L$ with sizeable cross sections at ISR energies. Further discussion may be found in Ref. 31.

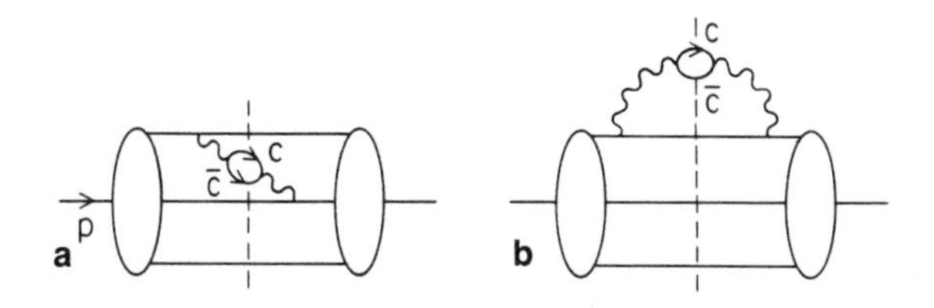


Fig. 18. Intrinsic (a) and extrinsic (b) contributions to the proton $|uudc\bar{c}\rangle$ Fock state

## 3.7 The Synthesis of QCD and Nuclear Physics

In this section we will discuss applications of quantum chromodynamics to nuclear physics where the basic quark and gluon substructure of hadrons plays an essential role at the nuclear level. [83] Because of asymptotic freedom we can make detailed predictions for nuclear form factors and nuclear scattering processes at large momentum transfer, as well as predict the asymptotic short-distance features of the nucleon-nucleon interaction and nuclear wavefunctions. [84,85] We shall also discuss areas where QCD places constraints on or actually conflicts with standard nuclear physics models. In particular, the fact that the nuclear wavefunction has "hidden color" Fock components [86] implies that the conventional meson and nucleon degrees of freedom of nuclear physics are not sufficient to fully describe nuclei in QCD.

### 3.7.1 The Deuteron Form Factor and Nuclear States at Short Distances

The most direct application of perturbative quantum chromodynamics to nuclei is the structure of the Fock-state wavefunctions and the form factors of nuclei at large momentum transfer. In analogy with the meson and nucleon form factor calculations discussed in Secs. III and VI we can write the deuteron form factor at large momentum transfer in the factorized form (see Fig.19): [85]

$$F_D(Q^2) = \int_0^1 [dx] \int_0^1 [dy]\ \phi_D^*(x_i,Q)\ T_H(x_i,y_i;\ Q)\ \phi_D(y,Q) \tag{7.1}$$

where $T_H \sim [\alpha_s(Q^2)/Q^2]^5$ is computed from the sum of hard scattering diagrams $6q + \gamma^* \to 6q$ where the initial and final quarks are collinear with the initial and final deuteron momentum $p$ and $p+q$, respectively. The distribution amplitude

$$\phi_D(x_i;\ Q) = \int^\phi \left[d^2k_\perp\right] \psi_{6q}(x_i,k_{\perp i}) \tag{7.2}$$

is defined in terms of the deuteron's six-quark valence wavefunction evaluated at equal time on the light cone. As in the case of the meson and baryon distribution amplitudes, the log $Q^2$ dependence of $\phi_D$ is determined from an evolution equation of the form (4.33) where to leading order in $\alpha_s(Q^2)$, the interaction kernel is determined from the sum of single gluon exchange amplitudes.

Because of the helicity selection rules, the leading form factor of the deuteron corresponds to the helicity zero — helicity zero electron-deuteron scattering amplitude:

$$F_D(Q^2) = \sqrt{A_D(Q^2)} \quad .$$

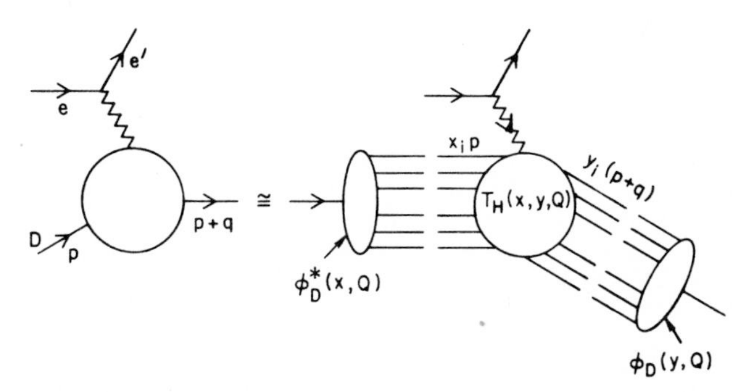


Fig.19. QCD factorization of the deuteron form factor at large momentum transfer. $T_H$ is computed for six quarks collinear with the incident and final directions

The other deuteron form factors are suppressed by at least one extra power of $Q^2$. As in the case of the meson form factors, the leading deuteron form factor is not affected by endpoint singularities in the $x_i$ and $y_i$ integration. Thus asymptotically, to leading order in $m^2/Q^2$ and $\alpha_s(Q^2)$ we have

$$F_D(Q^2) = \left[\frac{\alpha_s(Q^2)}{Q^2}\right]^5 \sum_{n,m=0}^{\infty} d_{nm} \left[\log \frac{Q^2}{\Lambda^2}\right]^{-\gamma_n^D - \gamma_m^D} \tag{7.3}$$

where the deuteron anomalous dimensions $\gamma_n^D$ can be computed from the eigenvalues of the evolution equation for $\phi_D(x_i,Q)$ or the operator product expansion for six fermion fields near the light cone.

The nominal QCD power-law prediction $F_D(Q^2) \sim (Q^2)^{-5}$ at large $Q^2$ is consistent with the dimensional counting rule [8] $F(Q^2) \sim (Q^2)^{n-1}$ where n is the minimum number of elementary constituents in the Fock state. The prediction thus reflects the QCD substructure of the nucleus and the essential scale invariance of the renormalizable quark interactions in the tree graphs for $T_H$. A comparison with data [87] for $\pi$, p, n, D, $H_e^3$ and $H_e^4$ is shown in Fig.9.

As we have indicated in Fig.20, the deuteron form factor receives contributions from six quark wavefunction components which are in both the standard color $|(uud)_1(udd)_1\rangle$ and "hidden color" $|(uud)_8(udd)_8\rangle$ configurations (see Sect. VI). It should be emphasized that the QCD equation of state for $\psi_{6q}$ automatically leads to mixed color components, at least at short distances. For example, if we impose the boundary condition that the deuteron is effectively an n - p bound state at large distances, then the one gluon exchange kernel in the evolution equation for $\phi_D(x,Q)$ automatically leads to hidden color components at large $Q^2$.

The perturbative structure of the QCD equation of state for $\psi_D$ at large $k_\perp$ also determines the power law and anomalous dimension structure of the valence wavefunction. [47] For example, if one quark has large $k_\perp$ relative to the deuteron, then $\psi_D(x_i,k_\perp) \sim (k_\perp^2)^{-1}$. On the other hand, if we consider the deuteron as two nucleon clusters, then at large transverse separation we have

$$\psi_D(x_i,k_{\perp i}) \sim \left(\frac{1}{k_{\perp N}^2}\right)^5 . \tag{7.4}$$

This power law reflects the fact that the effective nucleon-nucleon interaction large momentum transfer is $T_{np\to np} \sim (1/Q^2)^4$, which is again consistent with dimensional counting.

The specific connection of the asymptotic deuteron form factor to the nucleon-nucleon interaction is as follows: [84] the deuteron form factor is the probability amplitude for the deuteron to remain intact after absorbing a large momentum trans-

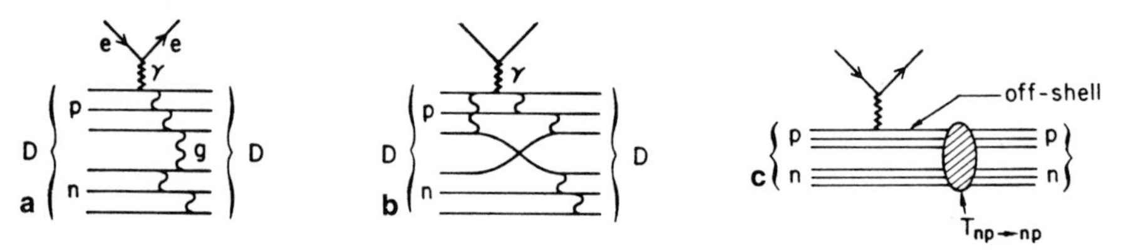


Fig. 20. Hard scattering contributions to the deuteron form factor. The contributions of diagram (a) require an internal hidden color state. Diagram (b) corresponds to quark interchange. Diagram (c) shows the relationship of the deuteron form factor to the N-N off-shell scattering amplitude

fer $p \to p+q$. If we consider the deuteron to be a loosely bound n-p system, with each constituent sharing almost equally the deuteron-four momentum, then each nucleon scatters from $\sim p/2$ to $\sim(p+q)/2$. The coupling of the electromagnetic current to the struck nucleon is effectively point-like as in the case of deep-inelastic scattering at large $q^2$, since the intermediate nucleon state $(p/2 + q)^2 \sim q^2/2$ is far off shell. The required n-p scattering amplitude (evaluated at $t = q^2/4 = u$, with one leg space-like at $p_N^2 = q^2/2$) scales at $T_{np\to np} \sim (1/Q^2)^4$. This scaling, combined with the off-shell propagator, then gives the results $F_D(Q^2) \sim (Q^2)^{-5}$. The normalization of $F_D(Q^2)$ can then be related to the non-relativistic deuteron wavefunction at the origin (see Ref. 84). It should be emphasized that the relativistic calculation of the deuteron form factor is incompatible with the conventional nuclear physics parameterization [88]

$$F_D(Q^2) = F_N(Q^2)\, F_{Body}(Q^2) \quad . \tag{7.5}$$

In the case of (static) non-relativistic models this form removes the structure of the struck nucleon. Equation (7.5) is, however, incorrect in the large $Q^2$ domain, since the struck nucleon cannot be on shell both before and after the interaction with the electromagnetic current.

### 3.7.2 Reduced Form Factors [84]

For a general nucleus, the asymptotic power behavior for the minimal helicity-conserving form factor is $F_A(Q^2) \sim (Q^2)^{1-3A}$, reflecting the fact that one must pay a penalty of $\alpha_s(Q^2)/(Q^2)$ to move each quark constituent from p to p+q. The fact that the momentum transfer must be partitioned among the constituents implies that the asymptotic domain increases with the nuclear number A.

However, as we shall now show, the introduction of the reduced form factor $F_A(Q^2)$ will allow interesting QCD predictions to be made even at relatively low momentum transfers. The basic idea is as follows: the deuteron form factor $F_D(Q^2)$ is the probability amplitude for the nucleus to remain intact after absorbing momentum transfer Q. Clearly $F_D(Q^2)$ must fall at least as fast as $G_M^p(Q^2/4) \cdot G_M^n(Q^2/4)$, since each nucleon must change momentum from p/2 to (p+q)/2 and stay intact. Thus we should define the "reduced form factor" $f_D(Q^2)$ via

$$F_D(Q^2) \equiv F_N^2\left(\frac{Q^2}{4}\right) f_D(Q^2) \quad . \tag{7.6}$$

Note that $f_D(Q^2)$ must itself decrease at large $Q^2$, since it can be identified as the probability amplitude for the n-p system to remain a ground-state deuteron. In fact, the dimensional counting rules $F_D(Q^2) \sim (Q^2)^{-5}$, $F_N(Q^2) \sim (Q^2)^{-2}$ imply

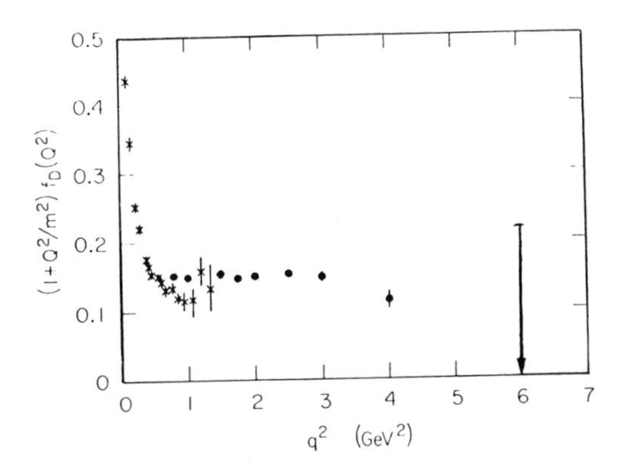


Fig.21. Comparison of deuteron from factor data with the QCD prediction $(1+Q^2/m^2)f_D(Q^2) \to$ const. at large $Q^2$. The data are from Ref.87

the asymptotic behavior $f_D(Q^2) \sim (Q^2)^{-1}$. This is precisely what one expects for a composite of two elementary systems once the nucleon structure has been removed.

We can also understand the origin of the simple result for $f_D(Q^2)$ from $T_H$ diagrams such as Fig.20c where a gluon immediately transfers momentum $1/2\ q^\mu$ to the other nucleon. Such diagrams give contributions of the form

$$F_D(Q^2) \cong F_N^2\left(\frac{Q^2}{4}\right) \frac{\alpha_s(Q^2/4)}{1+Q^2/m^2} \quad . \tag{7.7}$$

The mass parameter can be estimated from the corresponding parameters in the meson and nucleon form factors and is expected to be small, $m^2 \cong 0.3\ \text{GeV}^2$. The comparison of the data for $f_D(Q^2)$ with the prediction $(Q^2 + 0.3\ \text{GeV}^2)\ f_D(Q^2) \to$ const. is given in Fig.21. Remarkably, the predicted flat behavior for $Q^2 f_D(Q^2)$ appears to be accurate from $Q^2$ below 1 $\text{GeV}^2$ out to the limits of the data. The prediction is also verified at larger $Q^2$ when one uses inelastic deuteron form factor data at fixed mass $(p+q)^2$.

In general, we can define reduced nuclear form factors [84]

$$f_A(Q^2) \equiv \frac{F_A(Q^2)}{\left[F_N(Q^2/A^2)\right]^A} \quad . \tag{7.8}$$

QCD then predicts the power behavior $f_D(Q^2) \sim (Q^2)^{1-A}$ (as if the nucleons were elementary). Comparisons with data for $H_e^3$ and $H_e^4$ are given in Ref. 87. The definition of the reduced form factor takes into account the correct partitioning of the nuclear momenta, and thus to first approximation represents the nuclear form factor in the limit of point-like nucleon constituents. One can also extend the definition to reduced elastic nuclear scattering amplitudes

$$t_A(Q^2) \equiv \frac{T_A(Q^2)}{\left[F_N(Q^2/A^2)\right]^A} \tag{7.9}$$

e.g., in meson-deuteron elastic scattering at large momentum transfer. It should be of interest to see whether a consistent parameterization of nuclear amplitudes can be obtained if in each nuclear scattering process, reduced "point" amplitudes are defined by dividing out all of the constituent nucleon form factors at the correct partitioned momentum. Again, we emphasize that the standard method based on Eq. (7.5) is invalid in a relativistic theory. The measurements of hadron-nucleus elastic scattering are also interesting from the standpoint of testing basic QCD scattering mechanisms. [84] For example, the $K^+ - A$ scattering amplitude should scale as $A+Z$ at large momentum transfers if the scattering is dominated by u-quark interchange.

### 3.7.3 The Nucleon-Nucleon Interaction at Short Distances

The basic measure of the nuclear force is nucleon-nucleon scattering. As we have discussed in Sect. IV, two general features of the $N-N$ amplitude at large momentum transfer can be predicted from perturbative QCD: hadron helicity conservation and power-law scaling at fixed angle. In general there are five independent parity-conserving and time-reversal-invariant helicity amplitudes. The QCD selection rules [18] $h_{\text{initial}} = h_{\text{final}}$ imply that $\mathcal{M}(++ \to +-)$ and $\mathcal{M}(-- \to ++)$ are power-law suppressed relative to $\mathcal{M}(++ \to ++)$, $\mathcal{M}(+- \to +-)$, $\mathcal{M}(-+ \to +-)$. The helicity-conserving amplitudes thus are predicted in first approximation to scale as $\mathcal{M}_{\Delta h=0} \sim (Q^2)^{-4}$, yielding the dimensional counting prediction

$$s^{10} \frac{d\sigma}{dt}(s,\theta_{cm}) = F(\theta_{cm}) \tag{7.10}$$

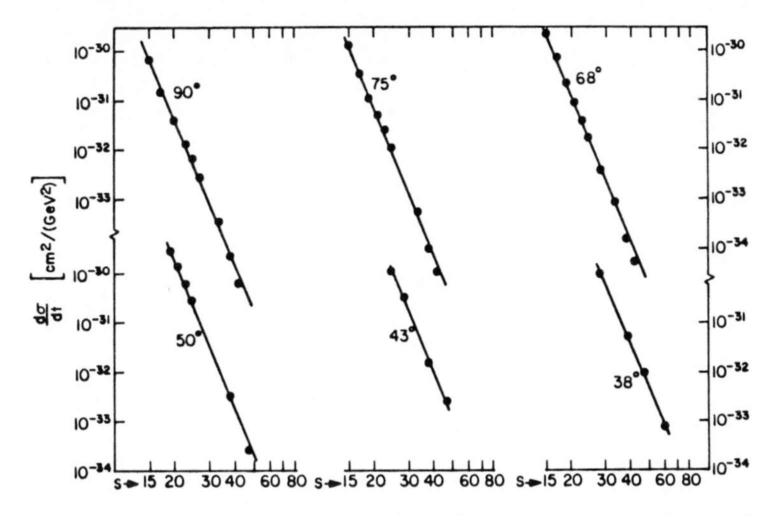


Fig.22. Differential cross sections fro $pp \to pp$ scattering at large center-of-mass angles. The straight lines correspond to the predicted power-law fall-off at $1/s^{10}$. The data compilation is from Ref.89

for nucleon-nucleon scattering at fixed angle and $s \gg M^2$. More precisely, the nominal power law is slightly modified by the Landshoff pinch singularity contributions and the logarithm factors from 10 powers of $\alpha_s(Q^2)$ and the anomalous dimensions of the distribution amplitudes. Remarkably, the $pp \to pp$ data is consistent (within a factor ~2) with the fixed-angle scaling predicted by (7.10) as the cross section falls more than 4 decades in the range $4 < p_T^2 < 12\ \mathrm{GeV}^2$, $38^o < \theta_{cm} < 90^o$. (See Fig.22.) The simplest interpretation of the results is that the variation of $\alpha_s(Q2)$ is very slow in this domain, as in the case of the $Q^4G_M(Q^2)$ scaling of the nucleon form factors. The presence of the Landshoff pinch singularities, however, could act to compensate for the fall-off of $\alpha_s$. In addition, there is some evidence [90] that the data is systematically oscillating about the $s^{10}d\sigma/dt$ const prediction, possibly suggesting the presence of an interfering subasymptotic amplitude.

The computation of the angular dependence and normalization of each of the helicity-conserving N - N amplitudes in QCD is a formidable task, since even to lowest order in $\alpha_s$, there are of the order of $3 \times 10^6$ connected Feynman diagrams in which five gluons interact with six quarks; [91] in addition,a detailed representation of the Sudakov suppression is needed in order to integrate over the Landshoff singularities. [92] Considerable phenomenological progress has, however, been made simply by assuming that the dominant diagrams involve quark interchange; [15] i.e., exchange of the common valence quarks. This ansatz seems to yield a good approximation to the observed large-angle meson-baryon and baryon-baryon scattering amplitude angular distributions, as well as the correct crossing behavior between the hadronic amplitudes, including $pp \to pp$ to $p\bar{p} \to \bar{p}p$. A useful analytic form for the interchange amplitude in terms of light-cone Fock-state wavefunctions is given in Ref. 93. A simple model for the quark interchange amplitude for $pp \to pp$ which has such properties is $\mathcal{M} \propto G_M^p(t)G_M^p(u)$.

The most sensitive tests of the hard scattering QCD prediction involve the polarization effects. The spin asymmetry $A_{NN}$ is defined as

$$A_{NN} = \frac{\frac{d\sigma}{dt}(\uparrow\uparrow) + \frac{d\sigma}{dt}(\downarrow\downarrow) - \frac{d\sigma}{dt}(\uparrow\downarrow) - \frac{d\sigma}{dt}(\downarrow\uparrow)}{\frac{d\sigma}{dt}(\uparrow\uparrow) + \frac{d\sigma}{dt}(\downarrow\downarrow) + \frac{d\sigma}{dt}(\uparrow\downarrow) + \frac{d\sigma}{dt}(\downarrow\uparrow)} , \tag{7.11}$$

which measures the difference of cross sections when both nucleons are polarized parallel to the normal ($\hat{x}$) of the scattering plane or are anti-parallel. Similarly $A_{LL}$ refers to the polarization asymmetry where the initial spins are polarized along the laboratory beam direction ($\hat{z}$) versus anti-parallel spins, and $A_{SS}$ refers to initial spins polarized (sideways) along the third direction ($\hat{y}$).

For the scattering of identical particles at $90^o$, all amplitudes involving a single helicity flip vanish, e.g., $(++ \to +-)$. This implies the sum rule [97,98]

$$A_{NN} - A_{LL} - A_{SS} = 1 \quad (\theta_{c.m.} = 90^o) \quad . \tag{7.12}$$

If in addition the double-flip amplitude $(++ \to --)$ vanishes, as in the case of the perturbative QCD predictions, then we have $A_{NN} = -A_{SS}$ (all angles) and the above sum rule becomes

$$2A_{NN} - A_{LL} = 1 \quad (\theta_{c.m.} = 90^o) \quad . \tag{7.13}$$

The striking CRABB et al., Argonne measurements for $A_{NN}$ (see Fig.23) can now be combined with preliminary results [96] for $A_{LL}$ at $90^o$ and $p_{lab} = 11.75$ GeV ($p_T \cong 2.4$ GeV): $2A_{NN} - A_{LL} \cong 2(0.58 \pm 0.04) - (0.18 \pm 0.09) = 0.98 \pm 0.17$, which is consistent with helicity conservation. On the other hand, it should be noted that the change of $A_{NN}$ is very rapid: $A_{NN} \cong 0.05$ at $\theta_{c.m.} \le 60^o$ to $A_{NN} \cong 0.60$ at $\theta_{c.m.} \ge 70^o$, which is in marked contrast to the generally smooth behavior predicted from calculations of $T_H$ for proton-proton scattering. For example, hard scattering diagrams with only quark interchange (see, e.g., Fig.20b) between the nucleons (which gives a good representation of the pp $\to$ pp angular distribution and crossing to $p\bar{p} \to p\bar{p}$) leads to the simple prediction [97, 98]

$$A_{NN} = -A_{LL} = -A_{SS} = 1/3 \quad (\theta_{c.m.} = 90^o) \quad , \tag{7.14}$$

with a very slow variation ($\lesssim 2\%$) over all $\theta_{c.m.}$. Diagrams with quark interchange plus gluon exchange between nucleons give a smaller value for $A_{NN}$. [99] The angular distribution predicted for diagrams with only gluon exchange is incompatible with the large-angle data; furthermore, if these amplitudes are normalized to the small-angle regime, then they are negligible at $90^o$. [19]

At this stage, there does not seem to be a convincing explanation of the nucleon-nucleon polarization effects at large angle. [100] It seems possible that whatever interference of amplitudes causes the oscillation of $d\sigma/dt$ around the smooth $s^{-10}$ behavior can also lead to striking interference effects in the polarization correlations. [90,97] One possibility is that the quark interchange amplitude is asymptotically dominant, but that in the present experimental range there is significant interference with multi-Regge exchange contributions. [97] An important

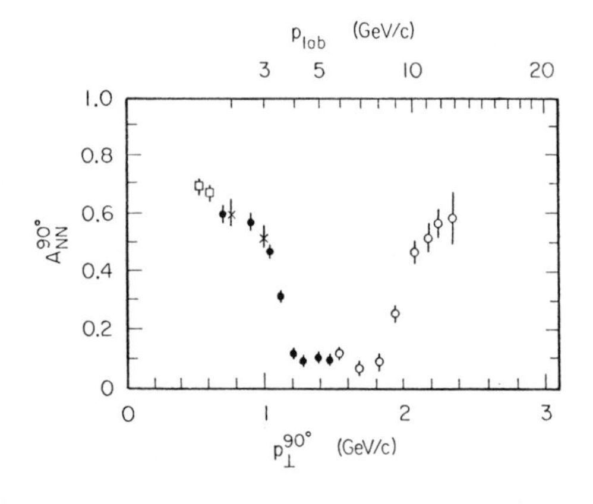


Fig.23. Data for the spin asymmetry $A_{NN}$ (normal to the scattering plane) for pp scattering at $90^o$ as a function of $p_{lab}$ and $p_T$. (From Ref.95)

point is that the Landshoff pinch contribution for pp → pp scattering includes three sequential qq → qq scatterings, each at approximately the same momentum transfer $\hat{t} \sim 1/9$. Since $|\hat{t}| < 1.1\ \text{GeV}^2$ is not very large, ordinary Reggeon exchange could still be playing a role in the quark-quark scattering amplitude. Unfortunately, the introduction of such contributions necessarily includes extra parameters and considerable model dependence. Nevertheless, a simple estimate of the rotating phase associated with triple Regge exchange is consistent with the interference pattern indicated by the pp → pp large-angle data. [91]

### 3.7.4 Continuity of Nuclear Physics and Quantum Chromodynamics

The syntheses of nuclear dynamics with QCD is clearly an important and fascinating fundamental problem in hadron dynamics. The short-distance structure of the nucleon-nucleon interaction as determined by perturbative QCD must join smoothly and analytically with the large-distance constraints (meson-exchange dynamics) of the N - N potential. The length scale of QCD is comparable with the inverse nucleon radius, so it is difficult to find a specific domain where nuclear physics can be studied in isolation from QCD.

The grand goal of QCD would be to actually derive the nuclear force from fundamental QCD interactions. The difficulty is that the nucleon-nucleon interaction in QCD is a remnant of the color forces and is analogous in complexity to calculating the molecular force between neutral atoms, e.g., positronium. The basic ingredients are quark interchange which is evidently related at long distances to pion and other meson exchange, and multiple gluon exchange, which despite the zero mass of the gluon must have an inverse range shorter than the mass of the lowest-lying gluonium state. It is possible that numerical results for the N - N potential will eventually be obtained from lattice gauge theory calculations. Model calculations of these exchange forces have also been given in the context of bag [101] and potential models. [102]

The constraints of asymptotic QCD behavior, especially its power-law scaling and helicity selection rules, have only begun to be exploited. For example, dispersion relations and superconvergent relations for the nuclear-nuclear helicity amplitudes should yield sum rules and constraints on hadronic couplings and their spectra. One could try to enforce a form of duality which equates the q-$\bar{q}$-g exchange amplitudes with the sum over meson exchange degrees of freedom. However, this cannot be strictly correct, since the existence of hidden color configurations — whether mixed with ordinary nuclear states or appearing as resonance excitations — implies that duality in terms of the low-lying hadrons cannot be a true identity.

One missing ingredient in nuclear physics model calculations of meson exchange amplitudes and currents is the form of the effective off-shell meson-nucleon-nucleon vertices. In principle, the effective form factors of these couplings are determined by QCD. Let us return to the form of the ultraviolet regularized QCD Lagrangian density discussed in Sect. II. If the cutoff $\kappa^2$ is comparable to hadronic scales, then extra contributions will be generated in the effective Lagrangian:

$$\begin{aligned}\mathscr{L}^{\kappa}_{QCD} = \mathscr{L}^{\kappa}_{0} &+ \frac{em(\kappa)}{\kappa^2}\,\bar{\psi}\sigma_{\mu\nu}\,\partial^{\mu}\,\psi A^{em}_{\nu} \\ &+ e\,\frac{f_\pi^2}{\kappa^2}\,\phi_\pi\,\overleftrightarrow{\partial}^{\mu}\,\phi_\pi\,A^{em}_{\mu} \\ &+ e\,\frac{f_p^2}{\kappa^2}\,\bar{\psi}_N\,\gamma^{\mu}\,\psi_N\,A^{em}_{\mu} \\ &+ \frac{f_p^2\,f_\pi}{\kappa^6}\,\partial_{\mu}\,\bar{\psi}_N\,\gamma_5\,\gamma^{\mu}\,\psi_N\,\phi_\pi \\ &+ \ldots\end{aligned} \qquad (7.15)$$

where $\mathscr{L}_0^\kappa$ is the standard contribution, and the higher twist terms of order $\kappa^{-2}$, $\kappa^{-4}$, ... are schematic representations of the quark Pauli form factor, the pion and nucleon Dirac form factors, and the $\pi$-N-N coupling. The pion and nucleon fields represent composite operators constructed and normalized from the valence Fock amplitudes and the leading interpolating quark operators. Our main point for writing down Eq. (7.15) is just to estimate the effective asymptotic power-law behaviors of the couplings, e.g., $F^{quark}_{Pauli} \sim 1/Q^2$, $F_\pi \sim f_\pi^2/Q^2$, $G_M \sim f_p^2/Q^4$ and the effective $\pi\,\bar{N}\gamma_5 N\, F_{\pi N\bar{N}}$ coupling: $F_{\pi N\bar{N}}(Q^2) \sim M_N f_p^2 f_\pi/Q^6$. The net pion exchange amplitude thus falls off very rapidly at large momentum transfer $M^\pi_{NN\to NN} \sim (Q^2)^{-7}$, much faster than the leading quark interchange amplitude $M^{q\bar{q}}_{NN\to NN} \sim (Q^2)^{-4}$. Similarly, the vector exchange contributions give contributions $M^\rho_{NN\to NN} \sim (Q^2)^{-6}$.

Thus, meson exchange amplitudes and currents, even summed over their excited spectra, do not contribute to the leading asymptotic behavior of the N - N scattering amplitudes or deuteron form factors, once proper account is taken of the off-shell form factors which control the meson-nucleon-nucleon vertices.

There is a further difficulty extending nuclear physics models based on an effective nucleon-nucleon-meson field theory. If one uses pointlike $\bar{N}\not{\rho}N$ isospin-invariant couplings of the nucleons to the rho meson, then the theory is not renormalizable without the full apparatus of non-Abelian gauge theories, including triple $\rho$ and four-point $\rho$ meson couplings, and a spontaneous symmetry-breaking mechanism to generate the $\rho$ mass. We emphasize that a non-renormalizable field theoretic model requires a new cutoff in each order of perturbation theory and thus is not predictive.

In addition to the above problems, it is difficult to understand within the context of QCD the role of $N\bar{N}$ pair production contributions as conventionally used in nuclear physics model calculations of electromagnetic exchange currents, etc. Nucleon pair (i.e., $qqq\bar{q}\bar{q}\bar{q}$) terms are far off shell and highly suppressed by off-shell form factors in QCD. On the other hand, anomalous "contact" terms are automatically generated in QCD time-ordered perturbation theory for the Z-graph term in the quark electromagnetic current. In the case of light-cone perturbation theory these are the instantaneous quark propagator terms described in Sect. III.

### 3.7.5 Structure Functions of Nuclei

If the nucleus were simply a loosely bound collection of nucleons, then the nuclear structure functions should reflect simple additivity:

$$G_{q/A}(x,Q) = Z\, G_{q/p}(x,Q) + (A-Z)\, G_{q/n}(x,Q) \tag{7.16}$$

$$G_{g/A}(x,Q) = A\, G_{g/N}(x,Q)$$

where $x = A\left(k^o + k^3/p^o_A + p^3_A\right)$ is the quark light-cone momentum fraction scaled to the nucleon momentum. The interesting physics is the derivation from simple additivity, which arises from the following sources:

(1) The nuclear structure functions $G_{q/A}$ and $G_{g/A}$ do not vanish at x = 1 but extend kinematically all the way out to x = A where one quark or gluon has the entire available light-cone momentum of the nucleus. For $x \gtrsim 1$ this is related to ordinary fermi motion. At larger x the structure functions are sensitive to far-off-shell QCD dynamics. [84,103,104] Modulo logarithms, the power behavior of perturbative QCD contributions to the inclusive distributions, are given by the spectator counting rule [60] (see Fig.24)

$$G_{a/A} = \frac{dN_{a/A}}{dx_a} \underset{x_a\to A}{=} C_{a/A}\,(A-x_a)^{2n_s-1} \tag{7.17}$$

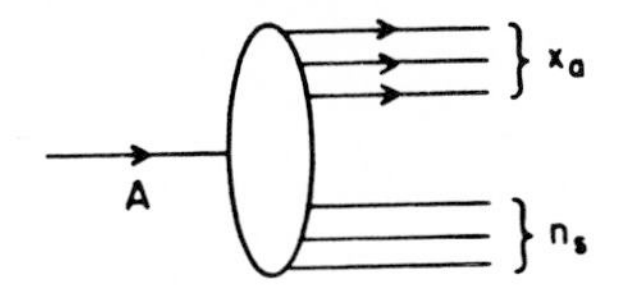


Fig. 24. Application of spectator counting rule to general composite or nuclear systems. The subsystem a has light-cone momentum fraction $x_a = k_a^+/p_a^+$. There are $n_s$ quark spectators

where $n_s$ is the number of spectator (quark) constituents in the bound system A forced to carry small light-cone momentum fraction: $x_s \to 0$. The power law is derived by simply counting the minimum number of off-shell propagators ($\mathcal{E} \to -\infty$ as $x \to x_{max}$) which are required to transfer all the momentum of A to a. Since the end result only depends on the number of spectators, it is easily shown that the system a can be a quark, gluon, or multiparticle cluster of constituents. However, the rule (7.17) holds only for the case where the helicities of a and A are identical; otherwise there are additional power-law suppressions. Examples of the spectator counting rule are $dN/dx \sim (1-x)^3$ for q/p, $(3-x)^{15}$ for $q/H_e^3$ and $(3-x)^{11}$ for $p/H_e^3$. These rules can be tested not only in deep-inelastic lepton-nucleus scattering, but also in forward inclusive nuclear scattering reactions where hadrons are produced with large longitudinal momentum fractions; e.g., $dN/dx\ (A_1 + A_2 \to p + X) \sim dN/dx\ (P/A_1)$. [103,105] The data for large x for these reactions does appear to be generally consistent with the power-law fall-off predicted by QCD spectator counting. Further discussions and tests can be found in Refs. 83, 103, and 105. In the case of the deuteron (and other even-spin nuclei) the mismatch between the quark and nuclear helicity implies that the deuteron structure function vanishes at the kinematic limit as [106] $F_{2D} \sim G_{q/D}(x) \sim (2-x)^{10}$ rather than $(2-x)^9$. (In each case, the power is logarithmically increased by QCD evolution.) One also expects an anomalous contribution to $F_{LD}$ at $x \sim 2$ analogous to the pion longitudinal structure function. Such contributions cannot be obtained from simple convolutions of the nucleon structure functions with nuclear distributions. The testing of these predictions is, of course, difficult because of the rapid fall-off of the structure functions, and the necessity for high $Q^2$ in order to avoid higher twist contributions. As we have discussed in Sect. V, we expect, in general, a sum of impulse approximation contributions [84,107]

$$\frac{d\sigma}{dQ^2\,dx}(\ell A \to \ell' X) = \sum_a \frac{d\sigma}{dQ^2}(\ell a \to \ell' a)\,\frac{dN_{a/A}}{dx} \tag{7.18}$$

representing incoherent contributions, each of which corresponds to lepton scattering on one quark or clusters of quarks in the nuclear target. We also note that the transverse momentum distributions $dN_{a/A}/d^2k_\perp$ can also be predicted from the perturbative QCD processes which control the high momentum tail of the bound-state wavefunctions.

(2) The deviations from simple additivity of $G_{a/A}$ at $x \sim 0$ are related to the important question of whether the leading twist nucleon structure functions are shadowed; i.e., $F_{2A}(x,Q^2) \sim A^{\alpha(x,Q^2)} F_{2N}(x,Q^2)$ at large $Q^2$, with $\alpha(x,Q^2) < 1$ (see Fig.25). A simple duality argument [109] based on the assumption of continuity of the structure function at $x = x_{Bj} = Q^2/2M\nu \to 0$ with the photoabsorption cross section $\sigma_{\gamma A}(\nu)$ (which is shadowed because of coherent vector-meson photoproduction processes) obviously implies shadowing of $F_{2A}(x,Q^2)$. However, as emphasized in Ref. 110, the QCD momentum sum rule then implies that a region of x must exist (probably at $x \sim m_\pi/M_N$) where the structure function obeys "anti-shadowing," i.e., $\alpha(x) > 1$. The existing data on lepton-nucleon scattering [108] clearly show shadowing at low x and low $Q^2$, but the data are not sufficient to demonstrate whether the shadowing occurs in the leading twist Bjorken-scaling contributions to the structure function, rather than in higher twist contributions associated with vector-meson electroproduction.

There are several arguments which indicate that QCD actually predicts the <u>absence</u> of shadowing for the leading twist structure functions, i.e., $\alpha(x,Q^2) \cong 1$

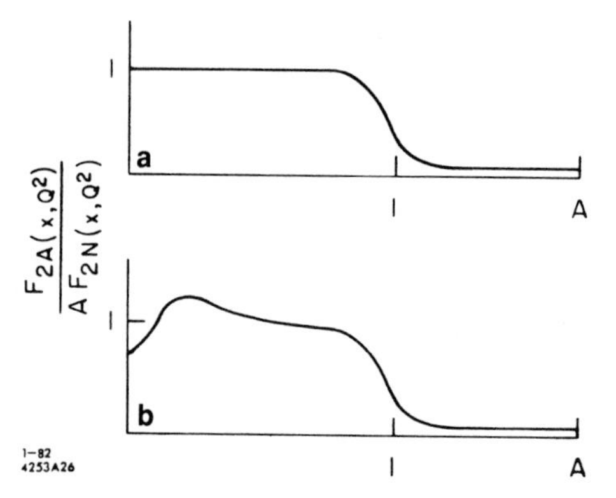


Fig. 25. Schematic representation of the deep-inelastic nuclear structure function normalized to its nucleon components.
(a) The case of zero shadowing.
(b) Shadowing and anti-shadowing

at $Q^2 \to \infty$ and fixed $x < 1$. Since shadowing is associated with initial-state (Glauber) interactions, [20] let us consider the representative initial-state contributions to the virtual photo-absorption cross section $\sigma_{\gamma^*}(x,Q^2)$ shown in Fig.26. At low $Q^2$, soft vector gluon exchange (finite transverse momentum $\ell_\perp$, and small light-cone momentum fraction $\ell^+ \sim \mathcal{O}(1/\sqrt{s})$ between the incident quark and the nuclear quark spectators gives an energy-independent initial-state correction to the photon-nucleus cross section as in meson-nucleus reactions. However, at high $Q^2 >> \ell_\perp^2$, the contributions of Figs.26a and 26b exactly cancel — corresponding to the vanishing of the hadronic radius of the photon. A complimentary argument for the absence of shadowing corrections based on explicit consideration of coherent shadowing contributions and their damping at large $Q^2$ is given in Ref. 105.

(3) In addition to the above considerations, simple additivity of the nuclear structure functions will be violated by the fact that the nuclear Fock-state spectrum is more complex than that of the individual nucleon. For example, the nuclear binding associated with meson exchange contributions leads to a modification to the sea quark and antiquark distributions in the nuclear structure functions. The number of strange quarks in the α-nucleus structure function may be different than the extrapolation from a nucleon target. We also emphasize that the existence of hidden color components in the Fock-state expansion of the nuclear state also implies new contributions to the nuclear structure functions, particularly in the $x > 1$ far-off-shell domain.

The definitive experimental identification of additivity-violating effects in the nucleus will also require a careful study of the nuclear target dependence of lepto-production channels, e.g., the reaction $eA \to eK^+X$ which is sensitive to the intrinsic strange quark composition of the nucleus, i.e., contributions not due to QCD evolution (see Sect. VI). The identification of specific $ed \to eN^*N^*$ channels in electron-deuteron scattering may be an important clue to the ΔΔ and hidden color Fock states of the deuteron as in Eq. (6.23).

### 3.7.6 Nuclei as Probes of Particle Physics Dynamics

Thus far in this section we have discussed applications of QCD specific to the dynamics and structure of nuclei. Conversely, there are numerous examples where a nuclear target can be used as a tool to probe particular aspects of particle physics. We will only mention a few applications here.

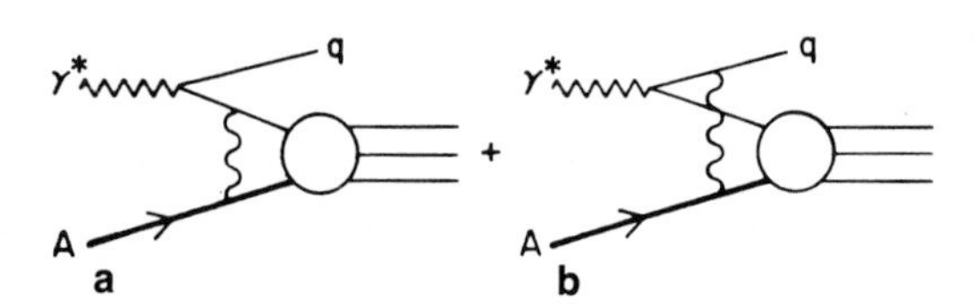


Fig. 26. Example of an initial-state scattering correction to the nuclear photo-absorption cross section leading to Glauber corrections and shadowing of the nuclear structure functions. The contributions of (a) and (b) cancel for $Q^2$ large compared to the momentum transfer of the exchanged gluon

(1) Parity violation in hadronic or nuclear processes. The exchange of a weak W or Z boson between the quarks of a hadron or nucleus leads to a high momentum component in the Fock-state wavefunction

$$\Delta\psi(x,\vec{k}_\perp) \sim \frac{\alpha}{k_\perp^2 + M_W^2} \phi\left(x,M_W^2\right) \tag{7.19}$$

as in the derivation of the distribution amplitude evolution equation. [19] The interference of these amplitudes with normal QCD contributions leads to parity violation in processes such as photodisintegration $\gamma d \to np$ and total hadronic cross sections. [111]

(2) The nucleus as a color filter. As we have discussed in Sect. VI, one can study a new class of diffractive dissociative jet production processes in nuclei which isolate the valence component of meson wavefunctions. [30] One can also use the A dependence of the nuclear cross section to separate central and diffractive mechanisms for heavy flavor production (open charm, etc.). [30,31]

(3) Nuclear corrections to inclusive QCD reactions. When a hadron traverses a nucleus, its Fock-state structure would be expected to be modified by elastic and inelastic collisions. An analysis based on perturbative QCD is given in Ref. 20. We show that multiple scattering in the nucleus increases the transverse momentum fluctuations of the quark and gluon constituents in the hadron, implying a nuclear enhancement for the rate of hadron and photon production at large transverse momentum. At very large $p_T$ the direct photon production cross section in nuclei should have the form

$$\frac{d\sigma}{d^3p/E}(pA \to \gamma X) = A \frac{d\sigma}{d^3p/E}(pN \to \gamma X)\left[1 + \mathcal{O}\left(\frac{A^{1/3}}{p_T^2}\right)\right] \quad . \tag{7.20}$$

In the case of the Drell-Yan cross section $d\sigma/dQ^2\, d^2Q_\perp\ (pA \to \mu^+\mu^- X)$ the transverse momentum $Q_\perp$ distribution of the produced lepton pair is predicted to broaden due to multiple scattering in the nucleus of the quarks in the initial state. Nevertheless, the integrated cross section $d\sigma/dQ^2\ (pA \to \mu^+\mu^- X)$ is proportional to A. Furthermore, as shown in Ref. 20, the light-cone x distribution of a fast quark is not effected by inelastic processes induced by multiple scattering in the nucleus as long as the quark momentum is large compared to a scale proportional to the length of the target. This effect is related to the formation zone analysis of LANDAU and POMERANCHUK [112] which shows that radiation from a classical current propagation between fixed target centers is limited at high energies.

(4) Propagation of quark and gluon jets in nuclear targets. In the conventional parton-model picture based on the impulse approximation, the multiplicity of hadrons produced in deep-inelastic lepton scattering or a nuclear target is expected to be identical to that on a single nucleon, since only one nucleon is "wounded" at large momentum transfer. In fact, the soft gluons radiated by the scattered quark jet in the deep-inelastic process can interact in the nuclear target and produce extra associated multiplicity in the target-fragmentation and central rapidity regions. [113] As shown in Ref. 20 only fast quanta are prevented in QCD from interacting inelastically in a nuclear target. The study of the initial and final interactions of the hadrons and jets in nuclear target, specifically the modification of longitudinal and transverse momentum distributions, can provide important insights into the nature of QCD dynamics.

## 3.8 Conclusions

In these lectures we have discussed the application of QCD to hadron and nuclear dynamics at short distances where asymptotic freedom allows a systematic perturbative approach. We have shown that it is possible to define the perturbative expansion in $\alpha_s(Q^2)$ in such a way as to avoid ambiguities due to choice of renormalization scheme or scale, at least in the first non-trivial orders. [14] Our main emphasis in these lectures, however, has been on how to systematically incorporate the effects of the hadronic wavefunction in large-momentum-transfer exclusive and inclusive reactions — thus leading to a broader testing ground for QCD. We have particularly emphasized the Fock-state wavefunctions $\psi_n(x_i,\vec{k}_{\perp i};\lambda_i)$ which define the hadron or nuclear state in terms of its quark and gluon degrees of freedom at equal time on the light cone. It is clear that a central problem of QCD is to determine not only the spectrum of the theory but also the basic bound-state wavefunctions of the color-singlet sector. Such solutions may be found in the near future using lattice numerical methods, particularly by quantizing at equal time on the light cone, or by more direct attacks on the QCD equations of motion for the $\psi_n$, as discussed in Sect. III.

Even without explicit solutions for the $\psi_n$, we can make a number of basic and phenomenological statements concerning the form of the wavefunctions: [27]

(1) Given the $\psi_n$ we can compute the single and multiple quark and gluon distribution amplitudes and structure functions which appear as the coefficient functions in the QCD predictions for high-momentum-transfer exclusive and inclusive reactions, including dynamical higher twist contributions. We have also emphasized general features of these distributions, including helicity selection rules, Lorentz properties, connections with the Bethe-Salpeter amplitudes, renormalization properties, and correspondence limits in the non-relativistic weak binding approximation.

(2) The perturbative structure of QCD leads to predictions for the high $k_\perp$, $x \to 1$ and far-off-shell behavior of the wavefunction. In particular, the large $k_\perp$ power-law behavior $\psi_V \sim k_\perp^{-2}$ of the valence wavefunctions and the $|\psi|^2 \sim k_\perp^{-2}$ behavior of the higher Fock-state contributions leads to QCD evolution equations and light-cone operator product expansion for the essential measures of the wavefunctions, the distribution amplitudes $\phi_M(x,Q)$, $\phi_B(x_i,Q)$, $\phi_D(x_i,Q)$ and the structure functions. We have also emphasized the fact that the valence wavefunction behavior $\psi_V \sim k_\perp^{-2}$ implies that the high $k_\perp^2$ behavior of quark and gluon jet distributions $dN/dk_\perp^2$ is $\sim 1/k_\perp^4$, not exponential or gaussian.

(3) Important boundary values and constraints on hadronic wavefunctions are obtained from the weak and electromagnetic decay amplitudes, including $\psi \to B\bar{B}$. The meson and baryon distribution amplitudes are measurable in detail from the angular behavior of the $\gamma\gamma \to M\bar{M}$ and [114] $\gamma\gamma \to B\bar{B}$ amplitudes.

(4) By assuming simple analytic forms for the valence wavefunctions in the non-perturbative domain, we have found consistent parameterizations which are compatible with the data for hadron form factors, decay amplitudes, etc. An important feature which emerges from these studies is that the valence state is more compact in transverse dimensions than the physical hadron. Even at low-momentum-transfer scale, higher Fock states play an important role, i.e., there is no scale where the proton can be identified as a 3-quark valence state. This observation may be compatible with the traditional nuclear physics picture of the nucleon as a central core, surrounded by a light-meson cloud. [115]

(5) The fact that there is a finite probability for a hadron to exist as its valence state alone, implies the existence of a new class of "directly coupled" semi-inclusive processes where a meson or baryon is produced singly at large transverse momentum, or interacts in a high-momentum-transfer reactions without accompanying radiation or structure function evolution. [29] As in the case of directly coupled photon reactions, the hadron can interact directly with quark and gluons in the short-distance subprocess, with a normalization specified rigorously in terms of the distribution amplitudes or form factors. Examples of these subprocesses are $qq \to B\bar{q}$, $gq \to Mq$, $Mg \to q\bar{q}$, $B\bar{q} \to qq$. We have also discussed an important contribution to the longitudinal meson structure function $F_L^M \sim C/Q^2$, involving direct coupling of the meson, somewhat analogous to the photon structure function. The finite probability for a meson to exist as a $q\bar{q}$ Fock state at small separation also implies a new class of diffractive dissociation processes. [30]

(6) The Fock-state description of hadrons in QCD also has interesting implications for nuclear states, especially aspects involving hidden color configurations. More generally, we have emphasized the idea that the far-off-shell components of hadron wavefunctions can be "unveiled" as the energy resolution scale is increased. For example, the existence of heavy-quark vacuum polarization processes within the hadronic bound state implies finite probabilities for hidden charm Fock states even in light mesons and baryons. The diffractive dissociations of these rare states appear to provide a natural explanation of the remarkable features of the charm production cross sections measured at the ISR. [31]

(7) We have also emphasized the importance of initial-state interactions in all inclusive reactions involving hadron-hadron collisions. The initial-state interactions disturb the color coherence, $k_\perp$ distributions, and at low energies the x dependence on the incoming hadronic distributions. Despite these profound effects on the hadronic Fock states, many of the essential features of the QCD predictions still are retained. [20] We have also discussed many examples where a nuclear target can be used to analyze the propagation of quarks and gluons through a hadronic medium.

(8) In Sect. VII of these lectures we focussed on the role of QCD at nuclear dimensions and its implications for fundamental nuclear interactions. The existence of hidden color Fock-state components in the nucleon wavefunction implies that the standard nucleon and meson degrees of freedom are not sufficient to describe nuclei. The mixing of the ground state of a nucleus with the extra hidden color states will evidently lower its energy and thus influence the nuclear magnetic moment, charge radius, and other properties. We expect that the hidden color components will be most significant in large-momentum-transfer nuclear processes and reactions such as the parity-violating terms in the photon disintegration of the deuteron, which are sensitive to the structure of the nuclear wavefunction at short distances. Conversely, the new QCD degrees of freedom should also imply the existence of excited nuclear states which are predominantly of hidden color. These states may have narrow width if they are below the pion decay threshold. The six-quark excitation of the deuteron could possibly be found by a careful search for anomalous resonant structure in $\gamma d \to \gamma d$ scattering at large angles. Other speculations [86] concerning the phenomenology of these states are discussed in Sect. VI.

The fact that QCD is a viable theory for hadronic interactions implies that a fundamental description of the nuclear force is now possible. Although detailed work in the synthesis of QCD and nuclear physics is just beginning, it is clear from the structure of QCD as a relativistic field theory that several traditional concepts of nuclear physics will have to be modified. These include conventional treatments of meson and baryon-pair contributions to the electromagnetic current and analyses of the nuclear form factor in terms of factorized on-shell nucleon form factors. On the other hand, the reduced nuclear form factors and scattering matrix elements discussed in Sect. VII give a viable prescription for the extrapolation of nuclear amplitudes to zero nucleon radius. There is the possibility that the present phenomenology of nuclear parameters will be significantly modified.

Independent of the specific dynamical theory, we have emphasized the utility of light-cone perturbation theory as an elegant but calculationally simple extension of non-relativistic quantum mechanics to the relativistic domain. The number of possible applications of this tool to nuclear physics [116] is extensive, since quantization at equal time on the light cone allows a consistent definition of relativistic Fock-state wavefunctions, their equations of state, and a completely relativistic treatment of the dynamics of elementary and composite systems.

Thus, in summary, we have found that the testing ground of perturbative QCD, where rigorous, definitive tests of the theory can be made, can now be extended throughout a large domain of large-momentum-transfer exclusive and inclusive lepton, photon, hadron and nuclear reactions. With the possible exception of inclusive hadron production at large transverse momentum, a consistent picture of these reactions is now emerging. By taking into account the structure of hadronic wavefunctions, we have the opportunity of greatly extending tests of QCD, unifying the short-and long-distance physics of the theory, and making an eventual synthesis with the realm of hadronic spectroscopy, low-momentum-transfer reactions and nuclear physics.

*Acknowledgements*

These lectures are based on recent work done in collaboration with G. P. Lepage (exclusive processes in QCD and light-cone perturbation theory); Tao Huang and G. P. Lepage (the structure of hadronic wavefunctions and their constraints); G. P. Lepage and P. B. Mackenzie (scheme and scale dependence in perturbative QCD); E. L. Berger and G. P. Lepage (calculable higher twist and direct subprocesses); and G. Bodwin and G. P. Lepage (initial-and final-state corrections to QCD processes). The lectures on the applications of QCD to nuclear physics were inspired by my collaborations with Benson T. Chertok, whose experimental and theoretical work greatly motivated this field.

I am also grateful to the organizers of this very productive summer school, Professors D. C. Fries and B. Zeitnitz.

## References

[1]Reviews of QCD are given in: A. J. Buras, Rev. Mod. Phys. 52, 199 (1980); A. H. Mueller, Phys. Rep. 73C, 237 (1981); E. Reya, Phys. Rep. 69, 195 (1981); G. Altarelli, Univ. of Rome preprint 701 (1978); W. Marciano and H. Pagels, Phys. Rep. 36C, 137 (1978); S. J. Brodsky and G. P. Lepage, Proceedings of the SLAC Summer Inst. on Particle Physics, 1979. See also "Perturbative Quantum Chromodynamics" (Tallahassee, 1981). AIP Proceedings No. 74, New York, 1981.

[2]For recent reviews of the phenomenology of QCD see A. J. Buras, FERMILAB-CONF-81/69-THY and A. H. Mueller, CU-TP-219 (1981), to be published in the Proc. of the 1981 International Symposium on Lepton and Photon Interactions at High Energies, Bonn, August 1981.

[3]S. Gupta and H. R. Quinn, SLAC-PUB-2763 (1981).

[4]H. D. Politzer, Phys. Rev. Lett. 30, 1346 (1973); D. J. Gross and F. Wilczek, Phys. Rev. Lett. 30, 123 (1973).

[5]See, e.g., Sau Lan Wu, DESY-81-071 (1981) and Proc. of the 1981 SLAC Summer Inst. on Particle Physics. See also the 1981 Proc. of the IVth Int. Colloq. on Photon-Photon Interactions, Paris (ed. G. London).

[6]TASSO Collaboration: R. Brandelik, et al., DESY Report 81/053 (1981). JADE Collaboration: W. Bartel, et al., DESY Report 81/048 (1981).

[7]PLUTO Collaboration: see W. Wagner, Proc. of the XXth Int. Conf. on High Energy Physics, Madison, 1980; Ch. Berger, et al., DESY preprint 81/051 (1981).

[8]S. J. Brodsky and G. R. Farrar, Phys. Rev. Lett. 31, 1153 (1973), and Phys. Rev. D11, 1309 (1975); V. A. Matveev, R. M. Muradyan and A. V. Tavkheldize, Lett. Nuovo Cimento 7, 719 (1973).

[9]S. J. Brodsky and G. P. Lepage, Phys. Rev. Lett. 43, 545, 1625(E) (1979). S. J. Brodsky, G. P. Lepage, S.A.A. Zaidi, Phys. Rev. D23, 1152 (1981).

[10]For a review of high-momentum-transfer exclusive processes and references, see A. H. Mueller, Ref. 1.

[11]S. J. Brodsky and G. P. Lepage, SLAC-PUB-2294, published in "Quantum Chromodynamics," Wm. Frazer and F. Henyey (eds.) (AIP, 1979), Phys. Lett. 87B, 359 (1979), S. J. Brodsky, Y. Frishman, G. P. Lepage, C. Sachrajda, Phys. Lett. 91B, 239 (1980). See also A. V. Efremov and A. V. Radyushkin, Rev. Nuovo Cimento 3, 1 (1980); Phys. Lett. 94B, 245 (1980). A. Duncan and A. Mueller, Phys. Rev. D21, 1636 (1980); Phys. Lett. 90B, 159 (1980). G. R. Farrar and D. R. Jackson, Phys. Rev. Lett. 43, 246 (1979); V. L. Chernyak and A. R. Vhitnishii, JETP Lett. 25, 11 (1977); G. Parisi, Phys. Lett. 43, 246 (1979); M. K. Chase, Nucl. Phys. B167, 125 (1980).

[12]S. J. Brodsky and G. P. Lepage, Phys. Rev. D24, 1808 (1981).

[13]E. Witten, Nucl. Phys. B120, 189 (1977). V. A. Bardeen and A. J. Buras, Phys. Rev. D20, 166 (1979).

[14]S. J. Brodsky, G. P. Lepage, P. B. Mackenzie (to be published).

[15]For early work on higher twist contributions see: R. Blankenbecler, S. J. Brodsky, J. F. Gunion, Phys. Rev. D18, 900 (1978), and D. Sivers, S. J. Brodsky, and R. Blankenbecler, Phys. Rep. 23C, 1 (1976) and references therein. Higher twist contributions to deep-inelastic scattering were first discussed in detail by: R. Blankenbecler and I. A. Schmidt, Phys. Rev. D16, 1318 (1977); L. F. Abbott and R. M. Barnett, Annals Phys. 125, 276 (1980); G. R. Farrar and D. R. Jackson, Phys. Rev. Lett. 35 (1975). See also L. F. Abbott, E. L. Berger, R. Blankenbecler, and G. Kane, Phys. Lett. 88B, 157 (1979); L. F. Abbott, W. B. Atwood, and R. M. Barnett, Phys. Rev. D22, 582 (1980), and references therein. More general frameworks for dynamical higher twist contributions are given in W. E. Caswell, R. R. Horgan, S. J. Brodsky, Phys. Rev. D18, 2415 (1978); and H. D. Politzer, Nucl. Phys. B172, 349 (1980); and S. J. Brodsky and G. P. Lepage, Ref. 1. Specific QCD calculations are given in G. R. Farrar and G. C. Fox, Nucl. Phys. B167, 205 (1980); E. L. Berger, T. Gottschalk, and D. Sivers, Phys. Rev. D23, 99 (1981); E. L. Berger and S. J. Brodsky, Phys. Rev. Lett. 42, 940 (1979), Phys. Rev. D24, 2428 (1981); E. L. Berger, Phys. Lett. 89B, 241 (1980); R. L. Jaffe and M. Soldate, Phys. Lett. 105B, 467 (1981).

[16]For a review of experimental evidence on higher twist terms in deep-inelastic scattering see C. Matteuzzi, SLAC-PUB-2827 (1981).

[17]P. V. Landshoff, Phys. Rev. D10, 1024 (1974); P. Cvitanovic, Phys. Rev. D10, 338 (1974); S. J. Brodsky and G. Farrar, Phys. Rev. D11, 1309 (1975).

[18]S. J. Brodsky and G. P. Lepage, Phys. Rev. D24, 2848 (1981).

[19]G. P. Lepage and S. J. Brodsky, Phys. Rev. D22, 2157 (1980).

[20]G. T. Bodwin, S. J. Brodsky and G. P. Lepage, Phys. Rev. Lett. 47, 1499 (1981).

[21]S. J. Brodsky and G. P. Lepage, Ref. 1; and S. J. Brodsky and G. P. Lepage, SLAC-PUB-2601 (1980), presented at the XXth Int. Conf. on High Energy Physics, Madison, Wisc. (1980).

[22]G. R. Farrar and D. R. Jackson, Ref. 15; E. L. Berger and S. J. Brodsky, Ref. 15.

[23]For an early discussion see T. A. DeGrand, Y. J. Ng, S.H.H. Tye, Phys. Rev. D16, 3251 (1977). The large $k_\perp$ behavior of hadronic wavefunctions in QCD is discussed by S. J. Brodsky, Y. Frishman, G. P. Lepage, C. Sachrajda, Ref. 11. See also S. J. Brodsky and G. P. Lepage, Ref. 1. These contributions are in principle normalizable in QCD using the methods of Ref. 19.

[24]E. L. Berger, T. Gottschalk, and D. Sivers, Ref. 15. G. Farrar and G. Fox, Ref. 15.

[25]G. Parisi, Phys. Lett. 43, 246 (1978). A. Duncan and A. H. Mueller, Ref. 11.

[26]See, e.g., S. J. Brodsky, F. E. Close, J. F. Gunion, Phys. Rev. D6, 177 (1972). G. Grammer, J. Sullivan, in Electrons in the Interactions of Hadrons (eds. A. Donaghue, G. Shaw), Plenum Press (1977). T. H. Bauer, R. D. Spital, D. R. Yennie, F. M. Pipkin, Rev. Mod. Phys. 50, 261 (1978).

[27]S. J. Brodsky, T. Huang, G. P. Lepage, SLAC-PUB-2540 (1980), and T. Huang, SLAC-PUB-2580 (1980), published in the Proceedings of the XXth Int. Conf. on High Energy Physics, Madison, Wisc. (1980). The parameterization of the hadronic wavefunctions presented here is preliminary; a complete discussion and final values will be given by S. J. Brodsky, T. Huang, and G. P. Lepage to be published.

[28]A more detailed discussion of the regularization procedure is given by G. P. Lepage et al., presented at the Banff Summer Institute on Particle Physics (1981).

[29]E. L. Berger and S. J. Brodsky, Phys. Rev. D24, 2428 (1981).

[30]G. Bertsch, S. J. Brodsky, A. S. Goldhaber and J. F. Gunion, Phys. Rev. Lett. 47, 297 (1981).

[31]S. J. Brodsky, P. Hoyer, C. Peterson, N. Sakai, Phys. Lett. 93B, 451 (1980). S. J. Brodsky, C. Peterson, N. Sakai, Phys. Rev. D23, 2745 (1981).

[32]For review and references see P. Langacker, Phys. Rep. 72, 185 (1981).

[33]P. Cvitanovic, Phys. Rev. D14, 1536 (1976).

[34]See, e.g., K. Callan, R. Dashen and D. Gross, "Quantum Chromodynamics," Proceedings of the 1978 La Jolla Summer Institute, Wm. Frazer and F. Henyey (eds.), AIP (1979).

[35]For reviews and references for inclusive process factorization see A. J. Buras, Ref. 1. For exclusive processes see Ref. 19 and A. H. Mueller, Ref. 1.

[36]G. Parisi and R. Petronzio, Nucl. Phys. B154, 427 (1979).

[37]M. Dine and J. Sapirstein, Phys. Rev. Lett. 43, 668 (1979). K. G. Chetyrkin, A. L. Kataev, F. V. Trachov, Phys. Lett. 85B, 277 (1979).

[38]Alternatively we can choose a standard scheme such as $\overline{MS}$ to define $\alpha_s(Q^2)$.
[39]R. Barbieri, M. Caffo, R. Gatto, E. Remiddi, Phys. Lett. 95B, 93 (1980) and CERN preprint 3071 (1981).
[40]G. P. Lepage and P. B. Mackenzie, Cornell preprint CLNS/81-498 (1981).
[41]S. J. Brodsky and G. P. Lepage, Ref. 1; C. Curci and M. Greco, Phys. Lett. 92B, 175 (1980); D. Amati, A. Bassetto, M. Ciafaloni, G. Marchesini and G. Veneziano, Nucl. Phys. B137, 429 (1980); M. Ciafaloni, Phys. Lett. 95B, 113 (1980).
[42]P.A.M. Dirac, Rev. Mod. Phys. 21, 392 (1949).
[43]S. Weinberg, Phys. Rev. 150, 1313 (1966); L. Susskind and G. Frye, Phys. Rev. 165, 1535 (1968); J. B. Kogut and D. E. Soper, Phys. Rev. D1, 2901 (1970); J. D. Bjorken, J. B. Kogut, D. E. Soper, Phys. Rev. D3, 1382 (1971); S. J. Brodsky, R. Roskies and R. Suaya, Phys. Rev. D8, 4574 (1973).
[44]J. D. Bjorken, et al., Ref. 43.
[45]The regularization can be performed using Pauli-Villars regularization. See Ref. 28.
[46]S. D. Drell and T. M. Yan, Phys. Rev. Lett. 24, 181 (1970).
[47]S. J. Brodsky and S. D. Drell, Phys. Rev. D22, 2236 (1981).
[48]S. J. Brodsky and T. Huang (in preparation).
[49]S. J. Brodsky, Y. Frishman, G. P. Lepage, C. Sachrajda, Ref. 11.
[50]R. D. Field, R. Gupta, S. Otto, L. Chang, Nucl. Phys. B186, 429 (1981); F. M. Dittes and A. V. Radyushkin, Dubna preprint JINR-E2-80-688 (1980); M. Chase, Ref. 11; E. Braaten (private communication).
[51]G. R. Farrar and D. R. Jackson, Phys. Rev. Lett. 43, 246 (1979).
[52]M. Peskin, Phys. Lett. 88B, 128 (1979).
[53]S. J. Brodsky, G. P. Lepage and S.A.A. Zaidi, Phys. Rev. D23, 1152 (1981).
[54]A. Duncan and A. Mueller, Phys. Rev. D21, 636 (1980); Phys. Lett. 98B, 159 (1980); A. Mueller, Ref. 1.
[55]I. Peruzzi, et al., Phys. Rev. D17, 2901 (1978).
[56]See, e.g., S. J. Brodsky, F. E. Close, J. F. Gunion, Phys. Rev. D5, 1384 (1972); D8, 3678 (1973).
[57]J. D. Bjorken, Phys. Rev. 163, 1767 (1967); 179, 1547 (1969).
[58]G. Altarelli and G. Parisi, Nucl. Phys. B126, 298 (1977). See also V. N. Gribov and L. N. Lipatov, Sov. J. Nucl. Phys. 15, 483, 675 (1972); J. Kogut and L. Susskind, Phys. Rev. D9, 697, 706, 3391 (1974); L. N. Lipatov, Sov. J. Nucl. Phys. 20, 94 (1975).
[59]R. D. Field, published in "Quantum Chromodynamics," Wm. Frazer and F. Henyey (eds.) AIP (1979).
[60]R. Blankenbecler, S. J. Brodsky, Phys. Rev. D10, 2973 (1974).
[61]F. Martin and A. DeRujula, Phys. Rev. D22, 1787 (1980).
[62]G. R. Farrar and D. R. Jackson, Phys. Rev. Lett. 35, 1416 (1975); A. I. Vainshtain and V. I. Zakharov, Phys. Lett. 72B, 368 (1978).
[63]Z. F. Ezawa, Nuovo Cimento 23A, 271 (1974).
[64]See also A. Duncan and A. Mueller, Ref. 54.
[65]K. J. Anderson et al., Phys. Rev. Lett. 43, 1219 (1979).
[66]M. Haguenauer et al., Phys. Lett. 100B, 185 (1981).
[67]S. J. Brodsky and G. P. Lepage, Proceedings of the Eugene Few Body Conference 1980, 247C (Nucl. Phys. A353, 1981).
[68]S. Gupta, Phys. Rev. D24, 1169 (1981).
[69]See references 21, 24, 15, and S. J. Brodsky, J. F. Gunion, R. Rückl, Phys. Rev. D18, 2469 (1978). J. Bagger and J. F. Gunion, U.C. Davis preprint 81/3 (1981).
[70]E. L. Berger and S. J. Brodsky, Phys. Rev. D24, 2428 (1981).
[71]S. D. Ellis, P. V. Landshoff, M. Jacob, Nucl. Phys. B108, 93 (1978); P. V. Landshoff and M. Jacob, Nucl. Phys. B113, 395 (1976).
[72]S. J. Brodsky, J. F. Gunion, R. Rückl, Ref. 69.
[73]S. J. Brodsky, T. DeGrand, J. F. Gunion, J. Weis, Phys. Rev. D19, 1418 (1979), Phys. Rev. Lett. 41, 672 (1978); Ch. Llewellyn-Smith, Phys. Lett. 79B, 83 (1978); S. M. Berman, J. D. Bjorken and J. B. Kogut, Phys. Rev. D4, 3378 (1971).
[74]W. Ochs and L. Stodolsky (private communication).
[75]A review of the form factor data is given by B. T. Chertok, CERN preprint PRINT-81-0318, to be published in the Proceedings of the 16th Rencontre de Moriond (1981).
[76]See Ref. 67 and references therein.
[77]A. P. Kobushkin, Yad. Fiz. 28, 495 (1978).
[78]Y. M. Dubovik and A. P. Kobushkin, Kiev preprint ITF-78-85E (1978).

[79]B. Judek, Can. J. Phys. 46, 343 (1968), 50, 2082 (1972), Proceedings of the 14th Int. Cosmic Ray Conf., Vol. 79, p. 2343. F. M. Friedlander et al., LBL preprint 11136 (1980). Y. J. Karant, LBL preprint 9171 (1979). W. J. Romo and P.J.S. Watson, Phys. Lett. 88B, 354 (1979).

[80]S. J. Brodsky, SLAC-PUB-1497 (1974), published in the Proceedings of the Int. Conf. on Few Body Problems in Nuclear and Particle Physics, University of Laval, Quebec (1974).

[81]See, e.g., S. J. Brodsky, SLAC-PUB-2747 (1981) and W. R. Frazer, UCSD 10P10-222 (1981), published in the Proceedings of the 4th Int. Colloq. on Photon-Photon Interactions, University of Paris, 1981; and W. A. Bardeen, to be published in the Proceedings of the 1981 Int. Symp. on Lepton and Photon Interactions at High Energies, Bonn, August 1981.

[82]J. F. Donoghue and E. Golowich, Phys. Rev. D15, 3421 (1977).

[83]For related discussions see also S. J. Brodsky and G. P. Lepage, Ref. 67. S. J. Brodsky, SLAC-PUB-2395 (1979) published in the Proceedings of the Berkeley Workshop on Ultrarelativistic Nuclear Collisions 1979: 419. R. Blankenbecler, SLAC-PUB-2077 (1978) and SLAC-PUB-2667 (1981). M. Chemtob, Nucl. Phys. A339, 229 (1980); A314, 387 (1979), and Saclay preprint (1981).

[84]S. J. Brodsky and B. T. Chertok, Phys. Rev. Lett. 37, 269 (1976); Phys. Rev. D14, 3003 (1976). S. J. Brodsky, Ref. 80.

[85]S. J. Brodsky and G. P. Lepage (to be published).

[86]See also, A. P. Kobushkin, Ref. 77; Y. M. Dobovik and A. P. Kobushkin, Ref. 78; Y. J. Kurant, Ref. 79; W. J. Romo and P.J.S. Watson, Ref. 79; S. Fredrikson and M. Järdel, Stockholm preprint TRITA-TFY-81-11 (1981).

[87]R. G. Arnold et al., Phys. Rev. Lett. 35, 776 (1979), 40, 1429 (1978). B. T. Chertok, in Progress in Particle and Nuclear Physics, Erice (1981); and Ref. 75. R. G. Arnold, SLAC-PUB-2334, published in the Proceedings of the Mainz Conference on Nuclear Physics with Electromagnetic Interactions (1979).

[88]For recent calculations which assume factorization of on-shell nucleon form factors see S. A. Gurvitz, Weizmann preprints WIS-81/47 PH (1981); Phys. Rev. Lett.

[89]P. V. Landshoff and J. C. Polkinghorne, Phys. Lett. 44B, 293 (1973).

[90]A. W. Hendry, Phys. Rev. D10, 2300 (1974); Indiana University preprint IUK-IT-60 (1980). There is, however, some question whether the oscillation is systematically observed when all experiments are combined (B. Chertok, private communication).

[91]S. J. Brodsky and G. P. Lepage, SLAC-PUB-2656, published in the Proceedings of the Lausanne International Symposium on High Energy Physics with Polarized Beams and Polarized Targets 1980: 169.

[92]A. H. Mueller, Ref. 1.

[93]J. F. Gunion, S. J. Brodsky, R. Blankenbecler, Phys. Rev. D8, 287 (1973).

[94]See G. H. Thomas, Proceedings of the Lausanne International Symposium 1980, and references therein.

[95]D. G. Crabb et al., Phys. Rev. Lett. 41, 1257 (1978); E. A. Crosbie, et al., Univ. of Michigan preprint UM HE 80-2 (1980). J. R. O'Fallon, et al., ibid. 39, 733 (1977).

[96]A. Yokosawa, Proceedings of the Lausanne Symposium 1980.

[97]S. J. Brodsky, C. E. Carlson and H. Lipkin, Phys. Rev. D20, 2278 (1978).

[98]G. R. Farrar, S. Gottlieb, D. Sivers and G. H. Thomas, Phys. Rev. D20, 202 (1979).

[99]J. Szwed, Max Planck Inst. preprints MPI-PAE/PTH 33/80 and 34/80 (1980), and Phys. Lett. 93B, 485 (1980).

[100]For additional discussions of the theory of pp polarization effects see C. Avilez, G. Coche and M. Moreno, preprint IFUNAM 80-17 (1980); G. Preparata and J. Soffer, Phys. Lett. 86B, 304 (1979), G. A. Walters, Phys. Rev. Lett. 45, 776 (1980), M. C. Daniel, Barcelona preprint UAB-FT-65 (1980), and the contributions of M. Doncel, G. F. Walters, G. Preparata, J. Soffer, J. Szwed, and M. Moreno to the Lausanne Symposium 1980.

[101]See L. Heller, Proceedings of the KfK Summer School on Quarks and Nuclear Physics.

[102]See, e.g., D. P. Stanley and D. Robsen, Phys. Rev. Lett. 45, 235 (1980), Florida State University preprint (1981), and references therein.

[103]I. A. Schmidt and R. Blankenbecler, Phys. Rev. D15, 3321 (1977). I. A. Schmidt, SLAC Report 203 (1979), R. Blankenbecler, Ref. 83.

[104]L. L. Frankfurt and M. I. Strickman, Phys. Lett. 69B, 93 (1977); 94B, 216 (1980).

[105]S. J. Brodsky, Berkeley Workshop, Ref. 83.

[106]A. I. Vainshtein and V. I. Zakharov, Phys. Lett. 72B, 368 (1978). R. M. Woloshyn, TRIUMF preprint TRI-PP-81-12 (1981).

[107]For an interesting application, see H. J. Pirner and J. P. Vary, Heidelberg preprints (1980).

[108]For a recent measurement, see M. S. Goodman et al., Phys. Rev. Lett. 47, 203 (1981). For a general review, see T. H. Bauer, R. D. Spital, D. R. Yennie, and F. M. Pipkin, Rev. Mod. Phys. 50, 261 (1978).

[109]J. D. Bjorken, Acta Phys. Pol. B2, 5 (1975), and lectures at the DESY Summer Institute (1975). H. Harari, Proc. of the 1971 International Symposium on Electron and Photon Interactions at High Energies, Cornell.

[110]V. I. Zakharov, N. N. Nikolaev, Sov. J. Nucl. Phys. 21, 227 (1975); V. I. Zakharov, Proc. of the 18th International Conference on High Energy Physics, Tbilisi (1976).

[111]See, e.g., W.-Y.P. Huang and E. M. Henley, Annals of Phys. 129, 47 (1980). Y. M. Dubovik and A. P. Kobushkin, Ref. 78.

[112]L. Landau and I. Pomeranchuk, Dok. Acad. Nauk USSR 92, 535 (1953).

[113]For a phenomenological discussion, see S. J. Brodsky, Ref. 83. See also A. Białas, Fermilab-Conf-79/35-THY; A. Białas, W. Czyż and W. Furmanksi, Acta Phys. Pol. 88, 585 (1977); A. Białas, Fermilab-Pub-78/75-THY (1978); A. Białas and E. Białas, Fermilab-Pub-79/48 THY (1979); A. Białas, M. Blesyznski and W. Czyż, Nucl. Phys. B111, 461 (1976).

[114]P. Damgaard, Cornell preprint CLNS 81/519 (1981).

[115]I wish to thank G. E. Brown for discussions of this point.

[116]See also M. Chemtob, Ref. 83. J. M. Nanyslowski and H. J. Weber, Zeit. fur Phys. B158, 497 (1979). V. P. Karnomonov, ITEP-8 (1980).

# 4. Bag Models and Nuclear Forces

L. Heller

With 8 Figures

## 4.1 Introduction to the Lectures

Some of the other lecturers at this school will try to convince you that quarks and gluons are indeed the building blocks of matter, and therefore we should try to understand nuclear forces and nuclear structure in terms of the interactions between these constituents. But this goal is still off in the distance. I shall discuss some phenomenological attempts to make a start on the problem using bag models. This will be based almost entirely on the MIT model, with very little discussion of variants. This model is reviewed in Jaffe's 1979 Erice lectures [1] and also his 1980 Schladming lectures [2]; De Tar has also given some recent lectures on the bag model at Boulder [3]. Those lectures provide many details about the application of the model to hadrons composed of light quarks. I shall emphasize the different physical approximations which are needed for light and heavy quarks, and discuss heavy quark systems in detail.

Another approach which is receiving a good deal of attention in the literature is based on two-body quark-quark potentials. Since the two-nucleon potential has contributed a great deal to the many-nucleon problem, why not try to repeat the process at the quark level? By the time we have discussed heavy quarks in the bag model, I will be in a better position to present arguments--both theoretical and phenomenological--against the idea that confinement can be understood in terms of two-body potentials.

Most of the work on the bag model has dealt with single hadrons. But we want to study nuclear physics, starting with the two-nucleon problem. I will describe some attempts to do this with the bag model and discuss the difficulties involved.

## 4.2 The MIT Bag Model

### 4.2.1 Introduction and Equations

The basic idea of the model is to assume there is a positive energy density B in any region of space which contains quarks and gluons. In order to guarantee that hadrons have a finite size, and therefore finite energy, it is necessary to introduce into the Lagrangian something which makes a distinction between the region of space which constitutes the inside of the hadron and the region of space which is outside the hadron. There are different formulations which achieve this end [1], and I will present Johnson's version [4]. The bag model Lagrangian density is taken to be

$$L_{BAG}(x) = [L_{QCD}(x) - B]\theta(\bar{q}(x)\Gamma q(x)) \quad , \tag{1}$$

where $L_{QCD}(x)$ is the Lagrangian density of Quantum Chromodynamics, which is the subject of Brodsky's lectures at this school.

$$L_{QCD}(x) = \frac{1}{2}\bar{q}(x)\gamma^{\mu}\left[i\partial_{\mu}q(x) - g\frac{\lambda^a}{2}A^a_{\mu}(x)q(x)\right]$$
$$- \frac{1}{2}\left[i\partial_{\mu}\bar{q}(x) + g\bar{q}(x)\frac{\lambda^a}{2}A^a_{\mu}(x)\right]\gamma^{\mu}q(x) - m\bar{q}(x)q(x) + L_G(x), \tag{2}$$

and $L_G$ is the Lagrangian density for the pure gauge theory

$$L_G(x) = -\frac{1}{4}F^a_{\mu\nu}(x)F^{\mu\nu a}(x) \quad . \tag{3}$$

In these expressions the color index a runs from 1 to 8, the $\lambda^a$ are the Gell-Mann matrices, and

$$F^a_{\mu\nu}(x) = \partial_{\mu}A^a_{\nu}(x) - \partial_{\nu}A^a_{\mu}(x) - gf_{abc}A^b_{\mu}(x)A^c_{\nu}(x) \tag{4}$$

with $f_{abc}$ the SU(3) structure constants,

$$\left[\frac{\lambda^a}{2}, \frac{\lambda^b}{2}\right] = if_{abc}\frac{\lambda^c}{2} \quad . \tag{5}$$

The quark field carries color and flavor indices as well as a Dirac index, and these indices are summed over in the above expressions.

The metric and Dirac matrices are the same as in Bjorken and Drell [5].

The two features of the Lagrangian density in (1) which make it the bag model are the constant energy density B, and the theta function, which has the definition

$$\theta(y) = 1 \quad \text{for} \quad y > 0$$
$$\theta(y) = 0 \quad \text{for} \quad y \leq 0 \quad .$$

Only that region of space in which $\bar{q}(x)\Gamma q(x) > 0$ contributes to the action; this defines the inside of the bag. It follows that the surface of the bag is that set of points for which $\bar{q}(x)\Gamma q(x) = 0$.

$\Gamma$ is a Dirac matrix which is restricted by Lorentz invariance and Hermiticity to be a linear combination of I and $i\gamma_5$ , and the general combination can be written

$$\Gamma = e^{i\alpha\gamma_5} \tag{6}$$

apart from a (positive) normalization constant which is of no relevance since it does not affect the value of the theta function. In order that the theory be invariant under the parity transformation, it is necessary to choose $\alpha = 0$ or $\pi$, corresponding to $\Gamma = I$ or $- I$. But we shall leave $\alpha$ arbitrary for now. The further generalization to $\Gamma = \exp(i\vec{\beta}\cdot\vec{\tau}\gamma_5)$ is needed in studying the relationship of the pion to the bag.

Using the relation $d\theta(y)/dy = \delta(y)$, the Euler-Lagrange equations which follow from (1) are, first for the quark field

$$\left[\frac{\partial L_{QCD}}{\partial\bar{q}(x)} - \partial_\mu \frac{\partial L_{QCD}}{\partial(\partial_\mu\bar{q}(x))}\right]\theta(\bar{q}(x)\Gamma q(x))$$

$$+\left[(L_{QCD} - B)\Gamma q(x) - \frac{\partial L_{QCD}}{\partial(\partial_\mu\bar{q}(x))}\partial_\mu(\bar{q}(x)\Gamma q(x))\right]\delta(\bar{q}(x)\Gamma q(x)) = 0 \quad . \tag{7}$$

The corresponding equation for the glue field is

$$\left[\frac{\partial L_{QCD}}{\partial A_\nu^a(x)} - \partial_\mu \frac{\partial L_{QCD}}{\partial(\partial_\mu A_\nu^a(x))}\right]\theta(\bar{q}(x)\Gamma q(x))$$

$$-\left[\frac{\partial L_{QCD}}{\partial(\partial_\mu A_\nu^a(x))}\partial_\mu(\bar{q}(x)\Gamma q(x))\right]\delta(\bar{q}(x)\Gamma q(x)) = 0 \quad . \tag{8}$$

Due to the singular nature of the theta function, each of the equations (7) and (8) is actually two separate equations. Inside the bag where the theta function is unity and the delta function is zero, the vanishing of the brackets which multiply the theta functions leads to the standard Yang-Mills equations. The equation for the quark field follows from (2) and (7)

$$\gamma^\mu(i\partial_\mu - g\frac{\lambda^a}{2}A_\mu^a(x))q(x) = mq(x) \tag{9}$$

and that for the glue field from (2), (3), (4), and (8)

$$\partial_\mu F^{\mu\nu a}(x) = g\left[\bar{q}(x)\gamma^\nu\frac{\lambda^a}{2}q(x) - f_{abc}F^{\mu\nu b}(x)A_\mu^c(x)\right] \equiv J^{\nu a}(x) \quad . \tag{10}$$

As the surface is approached from the interior, (9) and (10) remain valid, but in addition the coefficients of the $\delta$ functions must vanish. This gives two boundary conditions, one from (7) and one from (8). Before writing them down, note that the vanishing of $\bar{q}(x)\Gamma q(x)$ on the boundary means that the gradient of this quantity must be normal to the surface, so write

$$\partial_\mu(\bar{q}(x)\Gamma q(x)) = D(x)n_\mu \tag{11}$$

where $D(x)$ stands for the magnitude of the derivative and $n_\mu$ is a unit vector normal to the surface. Since the surface moves more slowly than the speed of light, there exists a local rest frame in which $n_\mu$ is a purely spatial vector, and therefore $n^2 = -1$.

From (2), (7), and (11), it follows that the boundary condition on the quark field takes the form

$$\frac{i}{2} D(x)\gamma^\mu n_\mu q(x) = (B - L_G(x))\Gamma q(x) \quad . \tag{12}$$

Eq. (12) actually represents two separate conditions, one on the magnitude of $D(x)$ and the other on the quark field. First, it is straightforward to show that $D(x)/2$ must equal $B - L_G(x)$. Act on both sides of (12) with $i\gamma^\mu n_\mu$ to obtain $[(i\gamma^\mu n_\mu)^2 = -n^2 = 1]$

$$\frac{D(x)}{2} q(x) = (B - L_G(x))i\gamma^\mu n_\mu \Gamma q(x) \quad , \tag{13}$$

and then act on both sides with $(B - L_G(x))\Gamma$, and use (12) to obtain

$$\left(\frac{D(x)}{2}\right)^2 i\gamma^\mu n_\mu q(x) = (B - L_G(x))^2 \Gamma i\gamma^\mu n_\mu \Gamma q(x) \quad . \tag{14}$$

For any choice of $\alpha$, $\Gamma = \exp(i\alpha\gamma_5)$ satisfies $\Gamma\gamma^\mu\Gamma = \gamma^\mu$, and therefore either $i\gamma^\mu n_\mu q(x) = 0$ or the stated condition on $D(x)$ must hold. (A minus sign multiplying $D$ can be absorbed into the definition of $n_\mu$.) But $i\gamma^\mu n_\mu q(x) = 0$ would lead to $q(x) = 0$ (by operating with $i\gamma^\mu n_\mu$ again) and since the equation for $q(x)$ is first order, this would lead to $q(x)$ vanishing everywhere inside the bag.

Gathering together the two consequences of (12) we have the boundary conditions

$$i\gamma^\mu n_\mu q_a(x) = \Gamma q_a(x) \tag{15 (BC1)}$$

and

$$\frac{1}{2}\partial_\mu(\bar{q}_a(x)\Gamma q_a(x)) = (B - L_G(x))n_\mu \tag{16 (BC2)}$$

The color and flavor index a is explicitly shown in these equations, and the repeated index is summed over. These two boundary conditions insure that energy and momentum are conserved at the surface of the bag, as can be seen by examining the energy-momentum tensor.

In the instantaneous rest frame, where $n_\mu = (0, -\hat{n})$, the normal component of (BC2) becomes

$$-\frac{1}{2}\,\hat{n}\cdot\nabla(\bar{q}(x)\Gamma q(x)) + L_G(x) = B \quad , \tag{17}$$

which can be interpreted as a pressure balance equation: the sum of the pressures of the quark and glue fields is equal to the pressure of the surrounding vacuum on the bag. Assuming that $L_G < B$, (17) shows that $\hat{n}$ points outward because $\bar{q}(x)\Gamma q(x) > 0$ in the interior.

There is a further conclusion to be drawn from (BC1). Taking its adjoint gives

$$\bar{q}_b(x)\Gamma = -\,\bar{q}_b(x)i\gamma^\mu n_\mu \quad ,$$

and when this is combined with (BC1) it leads to

$$\bar{q}_b(x)\Gamma q_a(x) = 0 \tag{18a}$$

and

$$\bar{q}_b(x)i\gamma^\mu n_\mu q_a(x) = 0 \tag{18b}$$

on the surface. In the original Lagrangian (1) the surface of the bag was defined by the single equation

$$\sum_a \bar{q}_a(x)\Gamma q_a(x) = 0 \quad , \tag{19}$$

but the boundary condition which results from that Lagrangian implies a separate condition for every pair of (color and flavor) indices. Equation (18b) says that there is no quark current normal to the boundary.

The boundary condition on the glue field follows directly from (2), (8) and (11) and says that

$$n_\mu F^{\mu\nu a}(x) = 0 \quad . \tag{20}$$

(BC3)

The time component of (BC3) is

$$\vec{n}\cdot\vec{E}^a(x) = 0 \quad , \tag{21a}$$

and in the local rest frame, the space part is

$$\hat{n}\times\vec{B}^a(x) = 0 \quad . \tag{21b}$$

When combined with Gauss's Law

$$\nabla\cdot\vec{E}^a(x) = \rho^a(x) \quad , \tag{22}$$

(21a) requires that

$$Q^a = \int d^3x \ \rho^a(x) = 0 \ , \tag{23}$$

so that the bag model allows only color singlet objects![1]

It is interesting to ask what would happen if the $\theta$ function in the original Lagrangian(1) were replaced by a non-singular function which varies continuously from 0 to 1 over a small interval. Instead of obtaining the equations of QCD in the interior plus boundary conditions, there would be slightly modified equations in the interior, followed by a region of finite thickness in which modified versions of the boundary conditions would have to be satisfied.

Before proceeding to examine some of the consequences of the bag model, we should ask what sense it makes to have written it down in the first place. After all, many people hope that the strong interaction properties of hadrons can be understood from QCD all by itself. If one believes that the nonlinear nature of QCD will explain confinement, for example, then it is doing too much to modify the QCD Lagrangian in a way which explicitly produces confinement. In most of the applications of the bag model, the nonlinear terms in ${}_G$ have been dropped, so it is conceivable that some ingredients of the theory are not being counted twice. The ultimate tests of the bag model are how successful it is as phenomenology, and whether it emerges as an approximation to a more fundamental theory.

### 4.2.2 Light Quarks vs. Heavy Quarks

The role played by the three boundary conditions is quite different for light quarks and for heavy quarks. If R represents a linear dimension of a bag and m the quark mass, then we shall call the quark 'light' if $mR \ll 1$, and 'heavy' if $mR \gg 1$.

For any quark, the magnitude of the momentum in the nth mode defined by (BC1) is $p \sim n/R$. One expects, therefore, that in the ground state a light quark will occupy the lowest mode because it requires an energy $E \sim 1/R$ to change n by one unit. Such a quark is very relativistic, and its wave function is spread over the entire bag.

For a heavy quark, on the other hand, the energy is primarily rest mass; and the kinetic energy, $n^2/2mR^2$, is much smaller than $1/R$ (and m) even for large values of n. It costs very little energy to promote it to higher modes, and consequently (BC1) is not important for heavy quarks; they can be rather localized and still remain nonrelativistic. This is the basis for applying the Born-Oppenheimer approximation to hadrons which are composed exclusively of heavy quarks.

---

[1] In Reference [2] Jaffe discusses a generalization in which the gluon is given a finite mass $\mu$. Color nonsinglet states are then possible, and their energies become infinite as $\mu \to 0$.

What about mixed bags which contain both light and heavy quarks? Consider a meson composed of one heavy quark and one light antiquark, and suppose that the only interaction between them is the static Coulomb potential. In order to keep the kinetic energy small, the light quark occupies the lowest mode as discussed above. In the absence of any interaction the heavy quark would do the same. But the attractive Coulomb potential energy is lowered if the wave function for the heavy quark gets concentrated in the center of the charge density associated with the light quark's wave function. The corresponding increase in the heavy quark's kinetic energy is small compared to $1/R$ provided the size of its wave function remains large compared to $(R/m)^{1/2}$. The limiting case of this picture, in which the heavy quark was taken to be a point at the center of a sphere, was used by Shuryak[6] to describe the D meson composed of a c quark and a u antiquark.

### 4.2.3 The Fixed Bag Approximation for Hadrons Composed of Light Quarks (u,d,s)

The highly nonlinear $\theta$ function in the Lagrangian (1) makes it very difficult to quantize the bag model. In other formulations [1] it is the lack of any kinetic energy term associated with the motion of the boundary which requires that the boundary degrees of freedom be eliminated. This problem has been studied by Rebbi [7,8] and DeGrand and Rebbi [9] in the approximation of small boundary oscillations, and it is a very difficult task. But most of the bag model phenomenology is based upon a restricted class of solutions of the classical equations involving a static boundary.

In the second MIT paper [10] the requirements of the boundary conditions (15) and (16) were examined for the free quark field, i.e., $L_G = 0$, in a fixed, spherical bag. Following an argument from the first MIT paper [11] which I do not regard as compelling, only the choice $\Gamma = 1$ was considered, and only massless quarks were discussed. But the discussion can be kept general with regard to these two quantities.

One first takes the known solutions of the free Dirac equation in spherical coordinates, and calculates the eigenfunctions $\psi_s(x)$ and the eigenvalues defined by (BC1), which for this problem becomes

$$-i\vec{\gamma} \cdot \hat{r}\, \psi_s(x) = \Gamma\psi_s(x) \quad . \tag{24}$$

s stands for a complete set of quantum numbers, which includes total angular momentum j; z component m; wave number k; and, for $\Gamma = \pm 1$, parity. For given $\Gamma$, $\mu \equiv mR$, and j, an infinite set of wave numbers $k_n$ is determined by the solutions of a transcendental equation. But the unique solution for $\mu$ as a function of kR can be given explicitly in terms of spherical Bessel functions, by

$$\mu = \frac{kR}{2\cos\alpha}\left[\frac{j_{j+\frac{1}{2}}(kR)}{j_{j-\frac{1}{2}}(kR)} - \frac{j_{j-\frac{1}{2}}(kR)}{j_{j+\frac{1}{2}}(kR)}\right] \tag{25}$$

where $\Gamma = \exp(i\alpha\gamma_5) = \cos\alpha + i\gamma_5 \sin\alpha$ . The total quark energy is related to k by $E = (m^2 + k^2)^{1/2}$. For the extremely important case of $j = \frac{1}{2}$, a graph of the solution of this equation is shown in Fig. 1, for the two lowest modes.

The MIT calculations of the properties of light hadrons [12] were all based on the $\alpha = 0(+)$ curve, which has the conventional (even) parity assignment for a ground state. The $\alpha = \pi(-)$ curve is unusual in two ways. First of all, it has odd parity, so a baryon consisting of three quarks in that mode would also have odd parity. Stranger still, as the quark mass increases, not only does the wave number k decrease, but so does the total energy E. When $mR = 3/2$, the wave number has gone to zero and $E = m = 3/2R$. As the quark mass increases still further, the wave number becomes pure imaginary and the wave function becomes a combination of increasing and decreasing exponentials rather than oscillating functions. In this region the total energy E is less than m, and it continues to decrease as m increases until ultimately $E = 1/R$. In this limit the wave function is concentrated entirely on the surface of the sphere.

This solution is not permitted in empty space because the wave function becomes infinite as $r \to \infty$. But in the bag model it is acceptable, and it is interesting to speculate whether nature makes use of it. All the higher $\alpha = \pi$ modes have the conventional property that $E \to m$ as $m \to \infty$.

Next, one examines the second boundary condition (16), which becomes

$$-\frac{1}{2}\nabla(\bar{q}(x)\Gamma q(x)) = B\hat{r} \tag{26}$$

in a spherical bag, and has the consequences that the magnitude of the left side must be independent of both angle and time. Expanding each color and flavor component $q_a(x)$ in terms of the set of eigenfunctions $\psi_s(x)$ defined by (BC1)

$$q_a(x) = \sum_s A_{a,s}\psi_s(x) \tag{27}$$

leads to

$$\sum_a \bar{q}_a(x)\Gamma q_a(x) = \sum_{s,s'}\left[\sum_a A^*_{a,s'} A_{a,s}\right]\bar{\psi}_{s'}(x)\Gamma\psi_s(x) \quad . \tag{28}$$

There are two cases in which $\bar{\psi}_{s'}(x)\Gamma\psi_s(x)$ itself has no time dependence. If the energies of the modes s and s′ are equal, there is certainly no time depen-

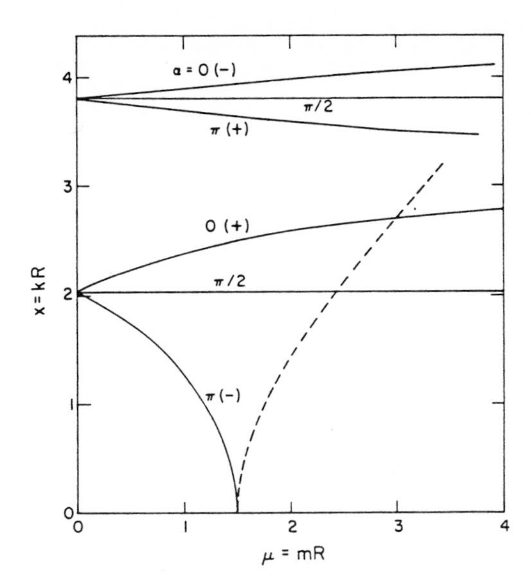


Fig.1 The two lowest $j = \frac{1}{2}$ modes for a quark in a spherical bag. m is the quark mass, R the bag radius, and k the frequency, with $E = (m^2+k^2)^{1/2}$. The three curves for each mode are labelled by the values of $\alpha(0,\pi/2,\pi)$ which occur in the boundary condition on the quark wave function: $i\gamma^\mu n_\mu q(x) = \exp(i\alpha\gamma_5)q(x)$. The parity of each solution is shown in parenthesis. For the $\pi(-)$ case, kR is imaginary beyond $\mu = 3/2$, and this portion of the curve is shown as a dashed line.

dence. In addition, for the special case in which both the energy and parity of s′ are the negatives of their values in s, $\bar{\psi}_s,\psi_s \equiv 0$. In the unquantized theory, one could choose the coefficients so that

$$\sum_a A^*_{a,s'} A_{a,s} = 0 \tag{29}$$

for all other combinations of s and s′, provided the number of modes with non-vanishing coefficients is not too great. But when one performs an approximate quantization of the theory, the numbers A and $A^*$ become creation and annihilation operators and (29) cannot be maintained as an operator equation. For the classical equations one can ensure that (28) has no time dependence by requiring that for each component of the field, a, only one energy value is excited.[2] With the additional requirement that the allowed states are color singlets, only one energy value *in toto* can be excited. This is the basis for the MIT calculation of the properties of light hadrons, with all quarks in the lowest $s_{1/2}$ mode [12].

[2] As already mentioned, the corresponding negative energy state with opposite parity can also be excited.

There remains the task of ensuring that the left side of (26) is independent of angle. For any choice of $\Gamma$ this is achieved by restricting the total angular momentum of the modes to have the value $j = \frac{1}{2}$.

The approximate quantization [10] of this limited set of solutions of the classical equations in a spherical bag consists of replacing the coefficients $A_{a,s}$ for positive energy modes by quark destruction operators, those for negative energy by antiquark creation operators, and imposing the standard anticommutation relations. When this is done, the energy of the system can be expressed as

$$E_Q + E_0 + E_V = \sum_{a,s} E_s\left[N_a(s) + \bar{N}_a(s) - 1\right] + \frac{4}{3}\pi R^3 B \quad . \tag{30}$$

$N_a(s)$ is the number of quarks of type a in mode s, $\bar{N}_a(s)$ is the number of antiquarks, and the sum involving these two terms is designated $E_Q$. The (infinite) zero-point energy is denoted $E_0$, and the bag volume energy is written $E_V$. Only positive values of $E_s$ appear in (30).

So far the glue has not been included. In the MIT spherical bag calculations [12] this was done in second-order perturbation theory, and results in an additional contribution to the energy, $E_G$. There is an important contribution to $E_G$ -- the self-energy of the quarks -- which arises from a gluon which is emitted and reabsorbed by the same quark, and which was not adequately treated. There has been a recent attempt to study this question [13].

The zero-point energy also receives a contribution from the glue field. Only the (finite) geometry-dependent part is of interest, and this has not been satisfactorily calculated either [14], but was simply parameterized as

$$E_0 = -\frac{Z_0}{R} \tag{31}$$

with $Z_0$ an adjustable constant.

We have already discussed restrictions on the quark wave functions imposed by (BC2), but have not yet addressed the question of how to choose the radius of the bag so that (BC2) is actually satisfied quantitatively. The fact that it is a pressure balance equation has the practical consequence that one can calculate the total energy

$$E = E_Q + E_0 + E_V + E_G \tag{32}$$

as a function of R and minimize it with respect to R. In addition to $Z_0$, there were three other free parameters available, the bag constant B, the quark-gluon coupling constant $\alpha_s \equiv g^2/4\pi$, and the mass of the strange quark $m_s$. (The masses of the u and d quarks were taken to be zero or very small.) The values of these four parameters were determined by fitting to the masses of the N, $\Delta$, $\Omega$, and $\omega$.

The resulting spectra of the light baryons and mesons are shown in Fig. 2, taken from DeGrand et al. [12], along with the fitted values of the parameters. I have multiplied their value of $\alpha_c$ by 4 to bring their definition of the coupling constant into agreement with general usage [as in Eq. (2)].

Once the radius of a given hadron is determined (it varies from 0.65 fm for the k meson to 1.1 fm for the $\Delta$), then the quark wave functions can be used to determine other properties, such as the magnetic moment and the charge radius. These were also calculated [12].

Although the original Lagrangian (1) for the bag model is Lorentz invariant, this invariance has been lost in the MIT approximation for treating the light hadrons. Even translation invariance is lost by fixing the (spherical) bag in space and referring the coordinate of each quark to the fixed center. This is the same problem that occurs in the nuclear shell model, and results in the bag state not being an eigenstate of the total momentum operator, $\vec{P}$. The loss of this invariance can represent a significant obstacle to further progress for systems composed of a small number of particles, especially when one tries to do scattering problems. There do exist approximate procedures to patch up this difficulty. One can always project out of a many-body wave function a piece which is an eigenstate of $\vec{P}$ [15]. In nonrelativistic quantum mechanics it is straightforward to go further and take the part of the resulting wave function which refers to the internal coordinates (but which depends on the eigenvalue of $\vec{P}$) and form a superposition over all values of $\vec{P}$, thereby obtaining a true in-

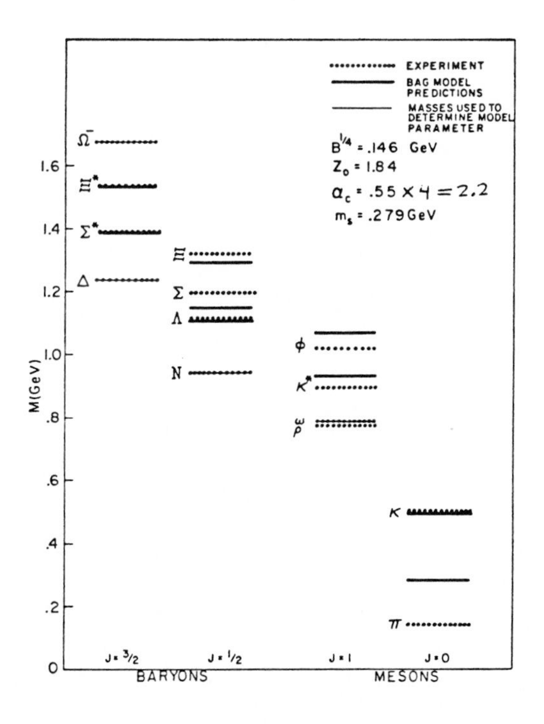


Fig.2 The spectra of the light hadrons in the fixed bag approximation to the MIT bag model, taken from Fig. 4 of De Grand et al. [12].

trinsic state which satisfies the requirements of Galilean invariance. There is even a variational principle for the best way to do this [16].

The corresponding procedure for the bag model for light hadrons would be considerably more complicated. First of all, one is dealing with relativistic quarks rather than nonrelativistic nucleons. Beyond this are the problems which arise because one does not have a true quantum mechanical theory [10]. In the procedures of Peierls and Yoccoz [15] and Peierls and Thouless [16] one has to evaluate matrix elements of various operators, including the Hamiltonian, between states $|\Phi_{\vec{x}}\rangle$ and $|\Phi_{\vec{x}-\vec{r}}\rangle$, where these are (bag model) states centered at $\vec{x}$ and $\vec{x}-\vec{r}$, respectively. We know how to evaluate diagonal matrix elements ($\vec{r}=0$), but what does one do with off-diagonal matrix elements, e.g., for the volume energy[3]?

There has been a recent attempt [17] to formulate bag model boundary conditions in terms of the relative coordinate of a quark and antiquark. The method applies only to the two-body system, and although translation invariance is accounted for, Lorentz invariance is not, because the procedure is explicitly carried out in the center of mass frame.

Before leaving this discussion of hadrons composed of light quarks, note that the fitted mass of the strange quark has the value $m_s = 0.279$ GeV (see Fig. 2). The product of this mass with the bag radius is certainly not small compared to unity, and according to the discussion in Section II.2 the strange quark is not a 'light' quark. It might be advantageous to admix some higher modes into its wave function.

In the next section the strange quark will be treated as a 'heavy' quark, along with the c and b quarks. It will turn out that even though $m_s$ then acquires a much larger value, ~ 0.6 GeV, it is not quite large enough to justify the approximations which will be made. Consequently the strange quark seems to occupy a middle ground which is not treated very accurately by either approximation.

### 4.2.4 The Born-Oppenheimer Approximation for Hadrons Composed of Heavy Quarks (... b,c,s)

*Introduction*

It is fortunate that nature created hadrons like upsilonium ($b\bar{b}$) and charmonium ($c\bar{c}$) composed of quarks which are sufficiently heavy that they move slowly. Familiar techniques from nonrelativistic quantum mechanics and electromagnetism can be carried over fairly directly to the study of these systems, and both

[3] I have mentioned the work of Rebbi and DeGrand, References [7 - 9], in which small oscillations of the bag surface were studied. Here one is talking about shifting the entire bag.

translation and Galilean invariance are built into the description. Still, there are new things to be learned because of the bag setting.

In 1975, a Hungarian group [18] suggested using an adiabatic (Born-Oppenheimer) approximation to study charmonium. They discussed many features of the problem and made a preliminary numerical study of the solution. It will be recalled that this approximation was developed for molecular physics where the electrons move much more rapidly than the nuclei. As applied to the present problem, the approximation consists in assuming that the glue field and bag boundary adjust rapidly compared to the motion of the heavy quarks and antiquarks. Instead of trying to solve for the glue field and quark motion all at once, the problem is broken down into two stages. In the first stage the kinetic energy of the quarks is neglected, and they are fixed at definite positions, $\vec{r}_i$. One solves for the glue field and bag boundary produced by these fixed sources, and obtains the energy as a function of the quark positions, $E(\{\vec{r}_i\})$. [Only $E_0$, $E_V$, and $E_G$ from Eq. (32) contribute here.] Assuming that the glue field adjusts instantaneously as the quarks move, then in the second stage of the problem this energy becomes the potential energy in the Schrödinger equation for the motion of the quarks.

In the MIT treatment of the light quarks, the bag was specified first, and then the quarks were independently put inside it. In the present approximation the bag develops around the quarks wherever they may be, and consequently the energy $E(\{\vec{r}_i\})$ only depends on the relative coordinates $\vec{r}_i - \vec{r}_j$, thereby insuring translation invariance. Note that when this potential energy is added to the quark energy $E_Q = \sum_i (m_i + \vec{p}_i^{\,2}/2m_i)$, the second stage of the Born-Oppenheimer approximation becomes a conventional problem in nonrelativistic quantum mechanics. Galilean invariance is, of course, also satisfied.

### *Approximate Form of the Equations*

Let us return to the first stage of the problem and see what the relevant equations and boundary conditions are. Since the quarks are localized static sources, (9) and (15) play no role, and (16) becomes

$$\frac{1}{2}\sum_a (\vec{E}^a(\vec{x})^2 - \vec{B}^a(\vec{x})^2) = B \tag{33}$$

on the boundary, where the definition of $L_G$ from (3) has been used. We look for a static solution of the remaining equations and boundary conditions. The right-hand side of (10), which is the source of the glue field, contains a quark piece which is first order in g, and a contribution from the non-Abelian nature of the glue which is higher order in g. In the limit that the quark mass becomes infinite, its color magnetic moment contribution to the spatial part of the current vanishes, so we shall omit that. [But even for the charmonium system these terms are not negligible, since the splitting of the singlet ($\eta_c$) and triplet ($\psi$) components of the ground state is 118 MeV[19].] Counting powers of g, $\rho^a$ and $\vec{E}^a$ are O(g), and $\vec{B}^a = O(g^3)$. If terms of relative order $g^4$ are neglected [20], then from (4)

$$\vec{E}^a(\vec{x}) = - \nabla\phi^a(\vec{x}) \quad , \tag{34}$$

and (10) becomes a set of uncoupled Poisson equations

$$\nabla^2\phi^a(\vec{x}) = - \rho^a(\vec{x}) \quad . \tag{35}$$

In this same approximation the boundary condition (33) becomes

$$\frac{1}{2}\sum_a \vec{E}^{a\,2}(\vec{x}) = B, \tag{36}$$

and there is still the boundary condition (21a)

$$\hat{n}\cdot\vec{E}^a(\vec{x}) = 0 \quad . \tag{37}$$

The set of equations (34) - (37) constitutes a problem in electrostatics with a number of complications. First of all, there are two boundary conditions, (36) and (37), rather than one (as in the Dirichlet or Neumann problem). This has the consequence that the boundary must be solved for at the same time that the field is. This is called a 'free boundary' problem.

A second complication arises from the specific form of the boundary conditions, which require that the color electric field be everywhere tangent to the surface, with constant magnitude. There is a theorem which says that the surface for such a problem must contain two singular points [21]. (The cusps are not present if the bag model contains a surface energy term [18].)

Still another departure from conventional electrostatics arises from the nature of the set of charge densities, $\rho^a(x)$, which are matrices in color space rather than simple functions. For N localized quarks this can be written

$$\rho^a(\vec{x}) = g\sum_{j=1}^{N} F_j^a \delta^{(3)}(\vec{x} - \vec{x}_j) \tag{38}$$

where $F_j^a$ is the SU(3) color generator for quark j, and the product $gF_j^a = Q_j^a$ is the color charge of that quark. For quarks $F_Q^a = \lambda^a/2$, and for antiquarks $F_{\bar{Q}}^a = -\lambda^{a*}/2$. (We are working in the direct product space, and it is understood that in the term involving quark j, the color operator for all the other quarks is the identity. Of course $\phi^a(\vec{x})$ and $\vec{E}^a(\vec{x})$ similarly are matrices in color space.) If the N quarks are in a color singlet state, then the sum of the generators vanishes when acting on that state,

$$\sum_{j=1}^{N} F_j^a = 0 \quad . \tag{39}$$

*Energy Variational Principle*

Before discussing what is known about the solutions to Equations (34)-(39), it should again be noted that the problem can be formulated as an energy variational principle. Temporarily omitting (36)--the pressure balance boundary condition--from the set of equations, (35) and (37) constitute a Neumann problem, which can be solved for an arbitrary boundary surface, S. If the sum of the glue energy and the volume energy is then calculated,[4] it can be regarded as a surface functional

$$W[S] \equiv E_G + E_V = \int dV \left[ \frac{1}{2} \sum_a (\vec{E}^a(x))^2 + B \right] \tag{40a}$$

$$= \int dV \left[ \frac{1}{2} \sum_a \rho^a(\vec{x}) \phi^a(\vec{x}) + B \right] . \tag{40b}$$

Provided (35) and (37) are satisfied, the condition that W be an extremum with respect to arbitrary variations in the surface precisely yields (36) [22].

*Tubular Approximation for Large Separations*

For arbitrary positions of the quarks, the solution to these equations must be obtained numerically. But the Hungarian group[18] presented a simple analytic solution to the N = 2 problem in the limit of large separation of the quark and antiquark. First note, though, that the color ingredient is completely trivial in the N = 2 case. Using (39), (38) can be written

$$\rho^a(\vec{x}) = gF_Q^a[\delta^{(3)}(\vec{x} - \vec{x}_1) - \delta^{(3)}(\vec{x} - \vec{x}_2)]$$

$$\equiv F_Q^a \rho(\vec{x}) \tag{41}$$

---

[4] In obtaining (40a) from the Hamiltonian density (see, for example, Ref.[5], volume 2, chapter 15) use has already been made of the boundary condition $\hat{n}\cdot\vec{E}^a(\vec{x}) = 0$. This condition is used again to go from (40a) to (40b).

and the color matrix factors out from the spatial dependence. $\phi^a(\vec{x})$ can also be factored in this way, and the only effect of the color is to make an effective coupling constant

$$g^2F_Q^2 \equiv g^2\sum_a F_Q^{a\ 2} = 4g^2/3 \quad , \quad \text{since} \sum_a(\lambda^a/2)^2 = 4/3 \quad .$$

Now it is straightforward to see that in the limit of large separation of the quark and antiquark, the bag develops into a tube[18] with a limiting radius; and the color electric field lines are parallel to the cylinder with a constant magnitude everywhere (except near the quarks, where there are end effects). See Fig.3a. Equation (37) is satisfied, since the electric field is tangential to the surface, and (36) fixes the magnitude of the field. Equations (35) and (38) then determine the radius of the cylinder via Gauss's theorem, which says that the product of the electric field with the cross-sectional area of the cylinder must equal the quark charge

$$gF_Q^a = E^aA \quad . \tag{42}$$

Squaring this equation and using (6), the limiting radius of the cylinder, $R_\infty$, becomes

$$R_\infty = \left(\frac{2\alpha_sF_Q^2}{\pi B}\right)^{\frac{1}{4}} \tag{43}$$

where $\alpha_s \equiv g^2/4\pi$ .

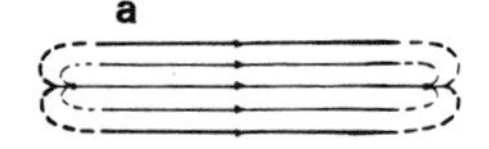


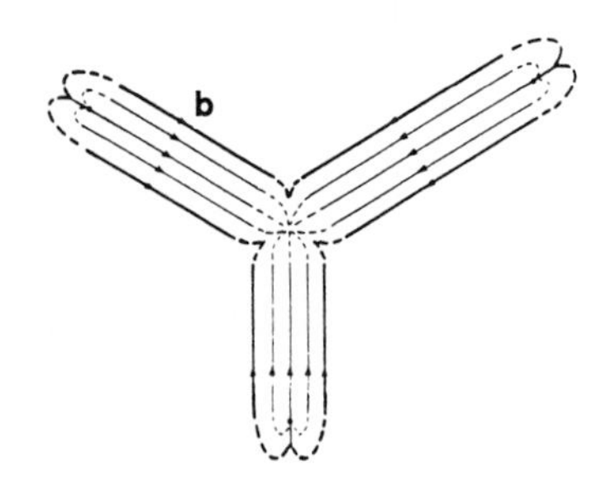


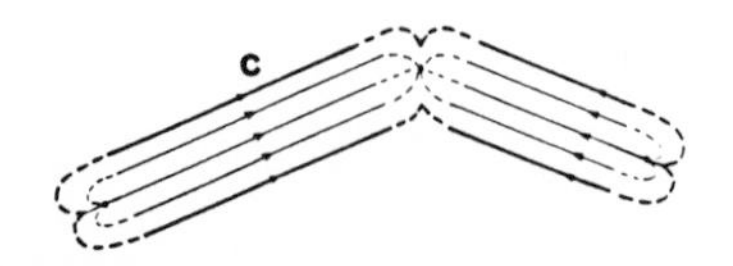


Fig.3 The shape of some bags containing widely separated quarks, showing the color electrostatic field lines. The dashed lines show distortion near the quarks. (a) One quark and one antiquark. (b) Three quarks. (c) Three quarks when one of the angles of the quark triangle exceeds 120°. From Figs. 2 and 3 of [23].

Since the color electric field has a constant magnitude everywhere inside the tube, and that magnitude is fixed by (36), the glue energy and the volume energy contribute equally to (40) in the large separation limit

$$\begin{aligned} E_G + E_V = 2B\int dV &= 2B\pi R_\infty^2 r + V^0_{Q\bar{Q}} \\ &= kr + V^0_{Q\bar{Q}} \end{aligned} \tag{44}$$

where r is the separation of the quark and antiquark and $V^0_{Q\bar{Q}}$ is a constant which accounts for end effects. Remembering that this energy will become the potential energy between the quark and antiquark in the second stage of the Born-Oppenheimer approximation, (44) exhibits the famous linear confining potential, with slope k obtained from (43) and (44)

$$k = (8\pi B\alpha_s F_Q^2)^{1/2} \quad . \tag{45}$$

For orientation, if $\alpha_s$ =1 and $B^{1/4}$ = 0.145 GeV, k = 0.618 GeV/fm and $R_\infty$ = 1.30 fm.

The very same argument leads to tubes in the $Q^3$ system whenever some of the separations are large compared to $R_\infty$[23,24]. They can either connect the quarks directly, forming a triangle, or else a Y shape can form (see Fig. 3b). In the latter case each quark sits at the end of a tube which connects it to a common junction. As viewed by each quark, the junction seemingly has the properties of an antiquark, but the three antiquarks add up to zero net color charge at the junction. Since each of the three tubes is identical to the one tube in the $Q\bar{Q}$ system (except near the junction), the confining term in the potential energy of the $Q^3$ system at large separations is $kr_M$, where $r_M$ is the sum of the lengths of the tubes with the position of the junction $\vec{x}$ chosen in such a way that this total length is minimized,

$$\nabla_x \left(\sum_i |\vec{x} - \vec{x}_i|\right) = 0 \quad .$$

This requires that the junction be located at that point which makes the angles between the legs of the tripod equal to 120°. See Fig. 3b.

Since the value of $r_M$ depends upon the positions of all three quarks simultaneously, the confining potential at large separations is a three-body term. An explicit formula for $r_M$ is given in [23]. If one of the angles of the quark triangle is larger than 120°, then the solution consists of two tubes with the 'middle' quark at the junction, Fig. 3c. Each tube is still identical to the $Q\bar{Q}$ tube.

Although the tubular solution for large separations is instructive, it has little practical value, since the physical systems for which the adiabatic approximation is applicable, such as charmonium and upsilonium, are primarily confined to distances which are smaller than $R_\infty$. In addition, when r becomes large, it is energetically possible for the $Q\bar{Q}$ system to decay into two mesons through the creation of a light quark-antiquark pair.

*Spherical Approximation for Small Separations*

When all the quarks occupy a region of space of linear dimension d which is small compared to $R_\infty$, it can be argued that a sphere is a satisfactory approximation to the shape of the bag. Numerical solutions to be presented later will bolster the argument and also show where the spherical approximation breaks down. The argument goes as follows [22,23,25]. The dominant contribution to the color electric field at a distance R which is large compared to d, is due to the dipole moment, since the net color electric charge is zero: $E^a \sim gF^a_Q d/R^3$. Now at the surface, the pressure boundary condition (36) must be satisfied, and consequently $R^6 \sim \alpha_s F^2_Q d^2/B \sim R^4_\infty d^2$. For $d \ll R_\infty$, $R \gg d$, and the surface is far away from the quarks. This suggests that the energy of such a system is rather insensitive to the details of the bag shape, and a sphere can be expected to provide a good approximation. This same argument also shows that the volume energy is proportional to the linear size of the system, but not necessarily with the same proportionality factor as occurs at large separations [22].

The starting point for analytic and also some numerical solutions is Green's theorem for the color electrostatic potential at an arbitrary interior point, in terms of volume and surface integrals

$$\phi^a(\vec{x}) = \int dV' G(\vec{x},\vec{x}')\rho^a(\vec{x}') + \int dS' [G(\vec{x},\vec{x}')\hat{n}'\cdot\nabla'\phi^a(x') - \phi^a(\vec{x}')\hat{n}'\cdot\nabla' G(\vec{x},\vec{x}')] \tag{46}$$

where G is a Green's function

$$\nabla'^2 G(\vec{x},\vec{x}') = -\delta^{(3)}(\vec{x}-\vec{x}') \quad . \tag{47}$$

Boundary condition (37) eliminates the first term of the surface integral. If one is able to find the Green's function which obeys the Neumann boundary condition [26] on a surface S

$$\hat{n}'\cdot\nabla' G_N(\vec{x},\vec{x}') = -\frac{1}{A} \tag{48}$$

where A is the area of the surface, then the second term of the surface integral is independent of $\vec{x}$. A constant addition to $\phi^a$ has no effect on the energy, because the total color charge is zero. This leads to

$$\phi^a(\vec{x}) = \int dV' G_N(\vec{x},\vec{x}')\rho^a(\vec{x}') \quad , \tag{49}$$

and using (38) and (40b),

$$W[S] = \frac{g^2}{2} \sum_{i,j} F_i \cdot F_j G_N(\vec{x}_i, \vec{x}_j) + BV \tag{50}$$

where $F_i \cdot F_j \equiv \sum_a F_i^a F_j^a$ .

$G_N$ is known analytically for a sphere

$$G_N(\vec{x}, \vec{x}'; R) = \frac{1}{4\pi|\vec{x} - \vec{x}'|} + \frac{1}{4\pi R} \sum_{\ell=1}^{\infty} \frac{\ell + 1}{\ell} \left(\frac{xx'}{R^2}\right)^{\ell} P_\ell(\hat{x} \cdot \hat{x}')$$

$$= \frac{1}{4\pi|\vec{x} - \vec{x}'|} + \frac{1}{4\pi R} \left[ \frac{R^2}{|R^2\hat{x} - x\vec{x}'|} - 1 - \ln\left(\frac{R^2 - \vec{x}\cdot\vec{x}' + |R^2\hat{x} - x\vec{x}'|}{2R^2}\right)\right] \tag{51}$$

where the vectors are measured from the center of the sphere. If (51) is used in (50), the sphere which minimizes the energy can be found by differentiating with respect to R and also with respect to the location of the center of the sphere. This requires the solution of transcendental equations.

*Dipole Approximation*

For sufficiently small separations of the quarks it is adequate to keep only the dipole term from the homogeneous piece of (51), thereby obtaining [23]

$$W_d(R) = \alpha_s \sum_{i>j} \frac{F_i \cdot F_j}{|\vec{x}_i - \vec{x}_j|} + \frac{\alpha_s}{R^3} \sum_{i,j} F_i \cdot F_j \vec{x}_i \cdot \vec{x}_j + \frac{4}{3}\pi R^3 B \quad . \tag{52}$$

In this dipole approximation there is no dependence of the energy on the location of the center of the sphere; because of the color singlet condition, adding a constant vector to all the $\vec{x}_i$'s does not alter (52). The infinite self-energy terms have been dropped from (52), but finite (R-dependent) pieces remain.

Now it is elementary to solve for the radius which minimizes $W_d$

$$R_d^6 = \frac{3\alpha_s}{4\pi B} \sum_{i,j} F_i \cdot F_j \vec{x}_i \cdot \vec{x}_j \quad , \tag{53}$$

and the contribution to the potential energy from the sum of the glue and volume energies becomes

$$V_d(\vec{x}_1, \ldots, \vec{x}_n) = \alpha_s \sum_{i>j} \frac{F_i \cdot F_j}{|\vec{x}_i - \vec{x}_j|} + 2\left(\frac{4\pi B\alpha_s}{3} \sum_{i,j} F_i \cdot F_j \vec{x}_i \cdot \vec{x}_j\right)^{1/2} . \tag{54}$$

The first terms on the right side of (54) represent the <u>two-body color Coulomb potentials</u>; they are directly proportional to the color generators of the quarks. The volume energy and the homogeneous part of the color electrostatic energy contribute equally to the final term in (54), which is an <u>n-body confining potential</u>. This is due to the fact that the size (and shape) of the bag depends upon the positions of all the quarks. To see the nature of this term more clearly, let us look at the $Q\bar{Q}$ and $Q^3$ color singlet systems explicitly. These are especially simple because only one color singlet state can be formed with such a small number of quarks, and hence the color operators become numbers: $F_Q^2 = 4/3$, $F_Q \cdot F_{\bar{Q}} = -4/3$, and $F_Q \cdot F_{Q'} = -2/3$. Putting these values into (53) and (54) yields

$$R_d^{Q\bar{Q}} = \left(\frac{\alpha_s r^2}{\pi B}\right)^{1/6} , \tag{55}$$

$$R_d^{Q^3} = \left(\frac{3\alpha_s \rho^2}{2\pi B}\right)^{1/6} , \tag{56}$$

$$V_d^{Q\bar{Q}}(\vec{x}_1,\vec{x}_2) = -\frac{4}{3}\frac{\alpha_s}{r} + \left(\frac{2}{3}\right)^{1/2} kr , \tag{57}$$

and

$$V_d^{Q^3}(\vec{x}_1,\vec{x}_2,\vec{x}_3) = -\frac{2}{3}\sum_{i>j}\frac{\alpha_s}{|\vec{x}_i - \vec{x}_j|} + k\rho \tag{58}$$

where $r \equiv |\vec{x}_Q - \vec{x}_{\bar{Q}}|$, the hyperspherical radius is defined to be

$$\rho \equiv \left(\sum_{i>j} (\vec{x}_i - \vec{x}_j)^2/3\right)^{1/2} , \tag{59}$$

and k is the slope parameter defined in (45). The simplicity of these formulas in the dipole approximation is very attractive.

As discussed above there is a linear confining potential even for small separations, but the slope (in the $Q\bar{Q}$ case) differs [22] from that at large separations by a factor $(2/3)^{1/2}$.

I want to again emphasize the fact that the $Q^3$ confining term is a genuine three-body potential [23,24], since $\rho$ depends upon the positions of all three quarks simultaneously. Another point to note is that even in the $Q\bar{Q}$ problem the confining term in (57) or (44) is not simply proportional to the product of the color charges of the quarks; only the two-body Coulomb terms have this property. This demonstration that, at least in the MIT bag model, confinement is <u>not</u> represented by a sum of two-body potentials, will be referred to again in Section III.

*Global Approximation*

Having described approximations which are expected to be valid for small and large separations, it is now tempting to write down a global approximation by joining them together [23,27]. For simplicity we match at $R_\infty$, although one could do a better job of fitting any given set of experimental energy levels by choosing a somewhat different value for the matching distance, tailored to the range of distances spanned by the wave functions under consideration. Therefore, Aerts and I proposed [23]

$$V^{Q\bar{Q}}(\vec{x}_1,\vec{x}_2) = -\frac{4}{3}\frac{\alpha_s}{r} + \left(\frac{2}{3}\right)^{1/2} kr, \qquad r \leqq R_\infty \tag{60a}$$

$$= -\frac{4}{3}\frac{\alpha_s}{r} + kr + V_0^{Q\bar{Q}}, \qquad r \geqq R_\infty \tag{60b}$$

and

$$V^{Q^3}(\vec{x}_1,\vec{x}_2,\vec{x}_3) = -\frac{2}{3}\sum_{i>j}\frac{\alpha_s}{|\vec{x}_i - \vec{x}_j|} + k\rho, \qquad \rho \leqq R_\infty \tag{61a}$$

$$= -\frac{2}{3}\sum_{i>j}\frac{\alpha_s}{|\vec{x}_i - \vec{x}_j|} + kr_M + V_0^{Q^3}, \qquad \rho \geqq R_\infty \tag{61b}$$

as global approximations to the potential. $V_0^{Q\bar{Q}}$ is a constant which is chosen to make the potential continuous at $R_\infty$. In the $Q^3$ problem at fixed $\rho$ there are still two angular variables, and $V_0^{Q^3}$ is chosen to make the potential continuous for all geometries.

In the $Q^3$ system it is possible to have two quarks separated by a distance which is small compared to $R_\infty$, with the third quark not close on this scale. In this case the color Coulomb interaction between the two close quarks will be important, and a tube will develop connecting this 'diquark' to the third quark. This is the reason for including the Coulomb terms in (61b). There is no need to do this in the $Q\bar{Q}$ problem, (60b), but we have done so just to maintain the same pattern in the two formulas.

*Numerical Solution for* $Q\bar{Q}$

Equations (60) and (61) represent simple, analytic approximations to the potential energy for the $Q\bar{Q}$ and $Q^3$ systems. For the latter problem we will later make use of the analytic structure of the potential to suggest a way to incorporate a running coupling constant. But first I want to show how accurate (60) is by comparing it with the result of a numerical solution of the bag equations for the heavy $Q\bar{Q}$ problem.

I will describe the procedure followed by Haxton and Heller [28], which includes the cusps in the bag surface that are required by the boundary conditions. The surface is assumed to be axially symmetric about the line joining Q and $\bar{Q}$, which is taken to be the axis of a spherical coordinate system. Instead of using a numerically calculated Neumann Green's function for an arbitrary surface, together with the energy variational principle (which is the procedure that was followed in [24] and [25]) we took the following approach. In (46) use the free space Green's function, but proceed as though the boundary conditions (36) and (37) were satisfied. The first term of the surface integral in (46) vanishes, and $\phi(\vec{x}')$ in the second term is taken to be a fixed constant multiplying the arc length $s(\theta')$ along the surface, in order to give the electric field the constant magnitude required by (36). This leads to [28]

$$4\pi\phi(\vec{x}) = \int dV' \frac{\rho(\vec{x}')}{|\vec{x}' - \vec{x}|} - \left(\frac{2B}{4/3}\right)^{1/2} \int dS' \, s(\theta')\hat{n}'\cdot\nabla' \frac{1}{|\vec{x} - \vec{x}'|} \tag{62}$$

where $\phi^a(\vec{x}) = F_Q^a\phi(\vec{x})$ and $\rho$ is the charge density from (41). For any assumed surface, (62) determines $\phi(\vec{x})$ at an arbitrary interior point. As $\vec{x}$ approaches the surface, it is found that neither (36) nor (37) is satisfied, because the surface is not the correct one. Using a combined measure of the extent to which the boundary conditions are violated, a minimization procedure was used to determine the optimal surface. Four-parameter surface functions--including two which determine the location and sharpness of the cusp--were found to give very accurate fits to the boundary conditions.

A sample bag obtained in this way is shown in Fig. 4, for a $Q\bar{Q}$ separation of 0.3 fm, which is approximately the distance at which the charmonium ground-state wave function peaks. In the neighborhood of the cusp, the surface varies as $R(\theta) - R_{cusp} \sim \theta^p$, with $p = 0.63$ for this value of the separation but varying

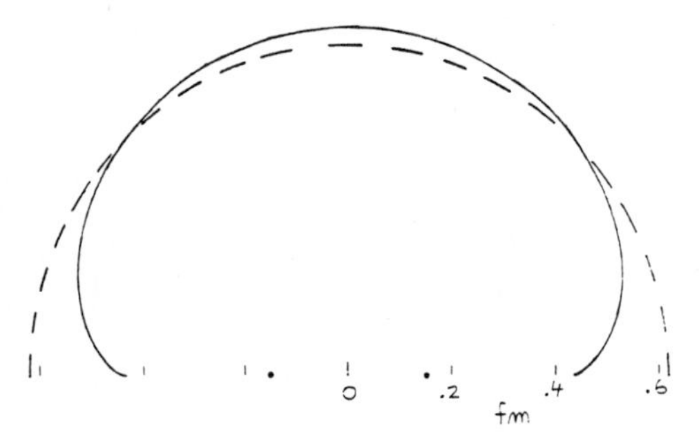


Fig.4 The shape of the bag for a $Q\bar{Q}$ separation of 0.3 fm, taken from Fig. 2 of [28]. The bag is obtained by rotating the figure about the line joining the quarks. Also shown (dashed line) is the dipole approximation to the bag as given by (55). In these calculations $\alpha_s(0.3\text{ fm}) = 0.58$ and $B^{1/4} = 0.145$ GeV.

somewhat for other $Q\bar{Q}$ separations. This value of the exponent is close to that found in a model two-dimensional problem [29] in which the quark and antiquark are taken to be infinite parallel lines of color charge. Using a succession of conformal transformations Giles [29] solved this problem analytically and found the value of the exponent to be 2/3 for all separations.

Also shown in Fig. 4 is the spherical approximation to the shape of the bag, as given by the dipole formula (55). Although the magnitude of the electric field at the surface varies appreciably in going from the pole to the equator [and consequently (36) is satisfied only in an average sense], the value of the energy in the dipole approximation, as given by (57), is in very accurate agreement with that obtained from the more precise bag involving the cusp [28,23]. This confirms the earlier conjecture that when the quarks are far from the surface, the energy is not sensitive to the details of the bag shape.

*Running Coupling Constant*

Before we can examine the numerical calculations of the potential and the spectra, and also the accuracy of the global approximation, it is necessary to discuss the effect of asymptotic freedom. In QCD the potential energy between a static quark and antiquark in a color singlet state at small distances is given by [30]

$$V(r) = -\frac{4}{3}\frac{\alpha_s(r)}{r} \tag{63}$$

where the running coupling constant has the limiting behavior

$$\alpha_s(r) \underset{r\to 0}{=} \frac{4\pi}{11-\frac{2}{3}n_f}\,\frac{1}{\ell n(1/\Lambda^2 r^2)} \quad . \tag{64}$$

$\Lambda$ is a scale constant and $n_f$ is the number of quark flavors which contribute to the vacuum polarization diagram. We chose $n_f = 3$ because the Compton wavelengths of the heavier $Q\bar{Q}$ pairs are appreciably smaller than the sizes of the $c\bar{c}$ and $b\bar{b}$ systems [31]. Since in the $r \to 0$ limit the confining term in (57) vanishes, asymptotic freedom is properly incorporated into the bag model (in this limit) by simply replacing $\alpha_s$ in the Coulomb term of (57) by $\alpha_s(r)$.

But what should be done at finite values of r where confinement does play a role, and also the correct form of $\alpha_s(r)$ is not known? One certainly cannot use (64) beyond $\Lambda r = 1$ because it diverges at that point. Haxton and I used the following phenomenological prescription: we simply replaced the fixed coupling constant which occurs in the charge density, (41), by a variable one

$$\rho(\vec{x}) = g(r)[\delta^{(3)}(\vec{x}-\vec{x}_1) - \delta^{(3)}(\vec{x}-\vec{x}_2)] \tag{65}$$

where $g^2(r)/4\pi = \alpha_s(r)$. The functional form was taken to be

$$\alpha_s(r) = \frac{4\pi}{11 - \frac{2}{3}n_f} \frac{1}{\ell n(\gamma + 1/\Lambda^2 r^2)}, \tag{66}$$

which has the correct asymptotic freedom behavior in the limit $\Lambda r \ll 1$, but is well behaved for all values of r and approaches a constant as $r \to \infty$, provided $\gamma > 1$. From (60b) and (45), this last property means that the potential becomes linear for $\Lambda r \gg 1$. As mentioned earlier, however, such large separations are of no real interest.

For the $Q\bar{Q}$ system in the two limits of small and large separations, the dipole approximation and the tubular approximation are still valid, provided $\alpha_s$ is everywhere replaced by $\alpha_s(r)$. In other words, (60) becomes

$$V^{Q\bar{Q}}(r) = -\frac{4}{3}\frac{\alpha_s(r)}{r} + \left(\frac{2}{3}\right)^{1/2} k(\alpha_s(r))r, \qquad r \leqq R_\infty \tag{67a}$$

$$= -\frac{4}{3}\frac{\alpha_s(r)}{r} + k(\alpha_s(r))r + V_0^{Q\bar{Q}}, \qquad r \geqq R_\infty \tag{67b}$$

where $k(\alpha_s)$ is the function defined in (45). Since, as described earlier, the choice of $R_\infty$ as the matching distance was made for simplicity, there is no additional arbitrariness in taking $R_\infty$ in (67) to be $R_\infty(\alpha_s(\infty))$, as defined in (43).

When we began the study of the $c\bar{c}$ and $b\bar{b}$ mesons, we expected the function $\alpha_s(r)$ to be a link between heavy quark hadrons and light quark hadrons. For the latter systems, in which the quarks are spread out over the entire bag, one can only speak of an average separation between quarks, and this number is approximately 1 fm. Consequently we originally intended to correlate the values of $\gamma$ and $\Lambda$ in (66) in such a way that $\alpha_s(1\ \text{fm}) = 2.2$ (see Fig. 2). But it turned out that we could only obtain a fair representation of the experimental $c\bar{c}$ spectrum with this restriction. Our best results were obtained with $\gamma = 3.36$ and $\Lambda = 0.240$ GeV, and consequently $\alpha(1\ \text{fm}) = 1.0$. This stands as a discrepancy between the work on the light quarks and the work on the heavy quarks. Not seeing any reason to change the value of the bag constant, we maintained $B^{1/4} = 0.145$ GeV.

Fig. 5 shows a comparison of the $Q\bar{Q}$ potential as given by the global approximation (67), with the detailed numerical calculations from reference [28]. For separations less than 1 fm, which are the most important ones for the lowest states of $c\bar{c}$ and $b\bar{b}$, the simple dipole formula (67a) is in error by less than 23 MeV. The discrepancy peaks at $R_\infty = 1.35$ fm, where it is 42 MeV. Since the discrepancy vanishes at $r = 0$ and again at $r \sim 2.5$ fm, the numerical result interpolates smoothly between the two branches of the global approximation. (If a smaller matching distance--such as 1 fm--had been used, the discrepancy would have peaked there and then decreased with increasing separation and gone negative.)

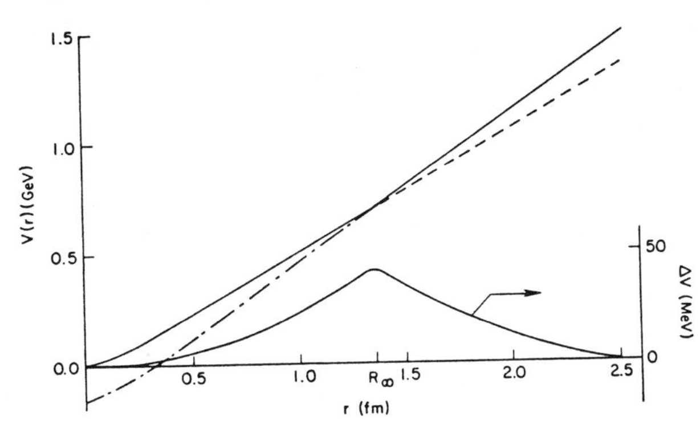


Fig.5 The global approximation to the $Q\bar{Q}$ potential (upper solid curve) taken from Fig. 1 of [27]. For $r < R_\infty$, it is given by the dipole approximation (67a), the continuation of which for $r > R_\infty$ is shown as a dashed curve. For $r > R_\infty$, it is given by the tubular approximation (67b), and its continuation for $r < R_\infty$ is shown as a dash-dot curve. The color Coulomb term is omitted from these curves. The right-hand scale is used for the lower curve, which shows the difference $V_H - V_G$, where $V_H$ is the numerical result from [28], and $V_G$ is the global approximation. All these calculations use the running coupling constant (66) with $\gamma = 3.36$ and $\Lambda = 0.240$ GeV, and consequently $\alpha(1\ \text{fm}) = 1.0$. ($V_H - V_G$ should have a discontinuous derivative, not a rounded top.)

The sequence of bags in Fig. 6 shows the approach to the tubular approximation with increasing $Q\bar{Q}$ separation. Even at the large separation of 2.5 fm there is only a small region near the midplane which is approximately cylindrical. But the potential energy becomes approximately linear long before the cylinder develops.

*Zero-Point Energy*

There is one final ingredient which we added to the potential energy in the calculations of references [28] and [27], and that is the zero-point energy of the confined quark and glue fields. In reference [12] this was parameterized as in (31), and the fitted parameter $Z_0$ acquired the value 1.84. Difficulties with

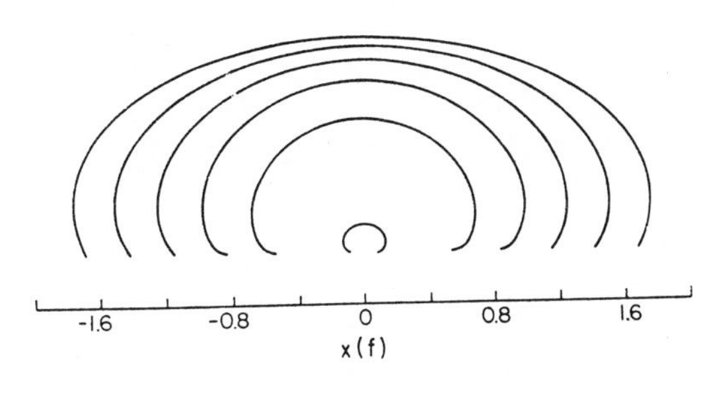


Fig.6 The bag shapes corresponding to $Q\bar{Q}$ separations of 0.01, 0.5, 1.0, 1.5, 2.0, and 2.5 fm, taken from Fig. 3 of [28]. The running coupling constant parameters are the same as in Fig. 5.

Table 1. The spectra of the $c\bar{c}$ and $b\bar{b}$ systems, from Table 1 of [27]. The calculated energies are obtained by solving the Schrödinger equation with the Hamiltonian (70), using the running coupling constant parameters $\Lambda$ = 0.240 GeV and $\gamma$ = 3.36 [thus $\alpha_s$(1 fm) = 1.0]; $m_c$ = 1.872 GeV and $m_b$ = 5.237 GeV. The experimental $c\bar{c}$ energies are from [33], and those for $b\bar{b}$ from [19].

| $c\bar{c}$ State | Energy (GeV) Experiment | Calculated [27] | <r> fm |
|---|---|---|---|
| $1^3S_1$ | 3.095 | 3.095 | .34 |
| $1^3P$ | 3.522 | 3.506 | .57 |
| $2^3S_1$ | 3.685 | 3.669 | .72 |
| $1^3D_1$ | 3.770 | 3.782 | .77 |
| $3^3S_1$ | 4.030 | 4.054 | 1.03 |
| $2^3D_1$ | 4.160 | 4.129 | 1.06 |
| $4^3S_1$ | 4.415 | 4.379 | 1.29 |
| $b\bar{b}$ State | | | |
| $1^3S_1$ | 9.462 | 9.460 | .20 |
| $1^3P$ | - | 9.886 | .36 |
| $2^3S_1$ | 10.015 | 10.005 | .45 |
| $1^3D_1$ | - | 10.133 | .50 |
| $3^3S_1$ | 10.351 | 10.333 | .68 |
| $2^3D_1$ | - | 10.415 | .71 |
| $4^3S_1$ | 10.576 | 10.582 | .88 |

this have already been mentioned [14], and there is a further problem in that an approximate attempt to patch up the violation of translation invariance in the fixed bag calculations of the light quarks leads to another term in the energy which has the same form as (31) [1,32]. Although the bags in the present problem are not spherical, we have seen that the dipole approximation to a spherical bag provides an accurate energy value for small separations, which are the most important ones. So we have simply taken the zero-point energy to be

$$E_0^{Q\bar{Q}}(r) = -\frac{Z_0}{R^{Q\bar{Q}}(r)} \tag{68}$$

with $Z_0$ maintained at the value 1.84, and with $R^{Q\bar{Q}}(r)$ the radius[5] in the dipole approximation as obtained from (55)

$$R^{Q\bar{Q}}(r) = \left(\frac{\alpha_s(r)r^2}{\pi B}\right)^{1/6} . \qquad (69)$$

The final form for the Hamiltonian is[6]

$$H^{Q\bar{Q}} = \sum_{i=1}^{2}\left(m_i + \frac{\vec{p}_i^{\,2}}{2m_i}\right) + V^{Q\bar{Q}} + E_0^{Q\bar{Q}} . \qquad (70)$$

*Results for* $c\bar{c}$ *and* $b\bar{b}$

Using the global approximation (67) to $V^{Q\bar{Q}}(r)$, the resulting spectra for $c\bar{c}$ and $b\bar{b}$ are given in Table 1. The parameters $\gamma$ and $\Lambda$ in (66) were chosen to make the overall fit satisfactory, and the quark masses were fitted to the ground-state energies. The values $\Lambda$ = 0.240 GeV, $\gamma$ = 3.36, $m_c$ = 1.872 GeV, and $m_b$ = 5.237 GeV were determined in this way. The fit is seen to be rather good, especially for $b\bar{b}$, with the largest discrepancy being 36 MeV for the 4S state of charmonium. States above the 2S can undergo decay to charmed mesons, and have significant widths.

Fig. 7 shows the radial wave functions for some of these states, along with the potential energy and the running coupling constant. Note the large variation in the value of the coupling constant in going from the ground state of $b\bar{b}$ to the excited states of $c\bar{c}$.

*The* s *Quark*

It was remarked at the end of Section II.3 that in the fixed bag treatment of the strange quark [12], the product $m_sR$ did not turn out to be small, and therefore the assumption that it should be put into the lowest mode may not be correct. Since (70) represents a flavor-independent potential, why not apply it to the $\phi(1020)$ meson which consists of an $s\bar{s}$ pair? Indeed, why not apply (70) to the $\rho$ meson? This was tried in [28,27], and the quark mass required to fit the $\phi$ meson is $m_s$ = 0.649 GeV, which is considerably larger than the 0.279 GeV found in [12].

---

[5] In reference [28], $R^{Q\bar{Q}}(r)$ was taken as the radius of the spherical bag which best satisfies the boundary conditions, using the same optimizing prescription as for the more complicated bag shapes.

[6] There are several differences between the Hamiltonian (70) and the one employed in [24] and [25]. In that work the coupling constant was held fixed at the value $\alpha_s$ = 0.385; the bag constant was assigned a very different value, $B^{1/4}$ = 0.235 GeV; and no zero-point energy was included.

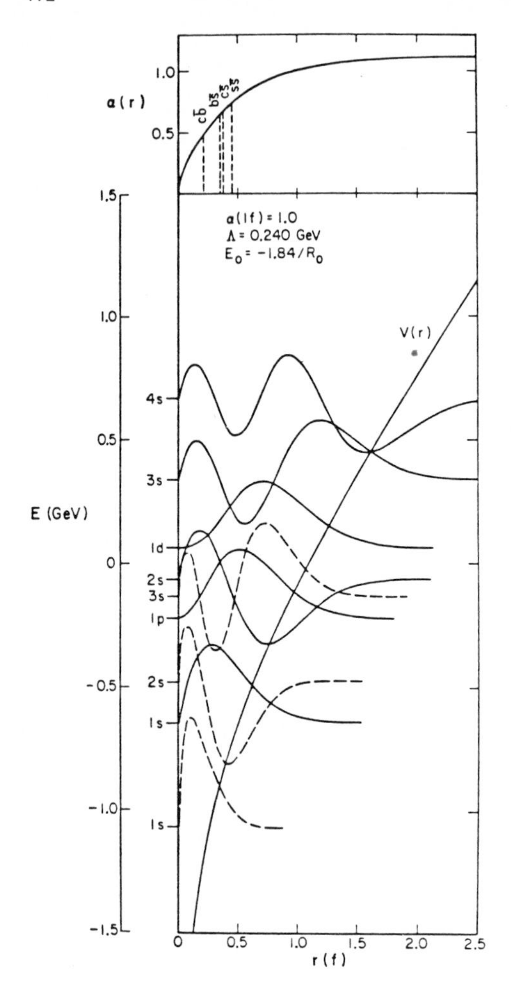


Fig.7 Calculated wave functions and eigenvalues for several states in the $c\bar{c}$ (solid lines) and $b\bar{b}$ (dashed lines) systems, from Fig. 9 of [28]. V(r) includes the zero-point energy. The top part of the figure shows the running coupling constant and the peak of the 1S wave functions for some other systems.

But as the quark mass gets smaller, the wave function extends out farther into the confining region of the potential (see Fig. 7), and the expectation value of $\vec{p}^2/m^2$ increases. See Table 2. While this number is small for the ground states of $b\bar{b}$ and $c\bar{c}$, it is already above 0.5 for the $\phi$ meson; and if extended still further to the $\rho$ meson, $\langle\vec{p}^2/m^2\rangle$ exceeds 0.8 [28]. This is so, even though the mass of the nonstrange quark needed to fit the $\rho$ mass exceeds 0.4 GeV!

When the quark motion becomes quite relativistic, not only the adiabatic approximation which we have employed, but also the entire notion of a static potential energy must be called into question (see Sec. III). I conclude from this that the s quark is a borderline case which is probably not treated accurately by either the fixed bag approximation or the Born-Oppenheimer approximation. For lighter quarks the former approximation is superior, and for heavier quarks the latter should be used.[7]

[7] If one attempts to put the c and $\bar{c}$ quarks into the lowest modes of a spherical bag, as in [34], the $\psi' - \psi$ splitting comes out much too small.

Table 2. $\langle \vec{p}^2/m^2 \rangle$ for a quark in the ground state of heavy $Q\bar{Q}$ and $Q^3$ systems, taken from Table 4 of [27]. $m_s = 0.649$ GeV, $m_c = 1.872$ GeV, and $m_b = 5.237$ GeV.

| | s | c | b |
|---|---|---|---|
| $Q\bar{Q}$ | .565 | .187 | .078 |
| $Q^3$ | .495 | .147 | .053 |

If one forgets about the non-negligible relativistic effects on the motion of the s quark, and simply combines one c quark and one $\bar{s}$ quark using the masses found above from fitting $c\bar{c}$ and $s\bar{s}$, then the solution of (70) provides a prediction for the mass of the $F^*(c\bar{s})$ meson[8]: 2.11 GeV [28]. This is consistent with the experimental value of 2.13 ± 0.05 GeV [19].

### $Q^3$ *and the* $\Omega^-$

It will be recalled that the very same physical arguments which led to (60a) and (60b) as approximations to the heavy $Q\bar{Q}$ potential energy in the limits of small and large separations, also gave (61a) and (61b) as the corresponding expressions for a system of three heavy quarks. Having shown that (60) agrees well with more precise numerical calculations, it is reasonable to expect that (61) is indeed a good approximation to the $Q^3$ potential. But before it can be applied to baryons composed of three heavy quarks, the question of the running coupling constant must be faced again, namely, what is the $Q^3$ analogue of (67)?

The analytic form of the terms in (61) suggests an answer. The two-body color Coulomb terms are due to single gluon exchange, and we expect that the same one-loop correction plus renormalization group argument which applies in the $Q\bar{Q}$ system [30] also applies here. We therefore take $\alpha_s$ in these terms to be the very same function, (66), of the quark-quark separations. For the three-body confining term in (61a) it is less clear what to do, but since it depends only on the hyperspherical radius $\rho$, it seems reasonable that the $\alpha_s$ which appears in the slope k should also be a function of $\rho$. We therefore proposed [27] that (61a) be rewritten as

$$V^{Q^3}(\vec{x}_1,\vec{x}_2,\vec{x}_3) = -\frac{2}{3}\sum_{i>j}\frac{\alpha_s(|\vec{x}_i-\vec{x}_j|)}{|\vec{x}_i-\vec{x}_j|} + k(\alpha_s(\lambda\rho))\rho \quad , \qquad \rho \leqq R_\infty \quad , \tag{71a}$$

[8] Since spin effects have not been included in these calculations, the energies we determine should lie between the triplet- and singlet-state masses. For consistency we choose always to compare to the triplet states, which are displaced less than the singlet states by the noncentral components of the potential.

and in analogy with (67a) the parameter $\lambda$ can be interpreted as a measure of the effective distance at which one quark 'sees' the other two quarks. As such it is expected to lie in the range $\sqrt{(1/3)}$ to $\sqrt{(3/2)}$, with the larger value corresponding to a quark-diquark configuration, and the smaller to an equilateral triangle. Comparing the probabilites of the various configurations, one expects $\lambda$ to be closer to the lower limit. When dealing with states for which the wave function extends beyond $R_\infty$, we make use of the fact that the coupling constant has essentially reached its asymptotic value, and rewrite (61b) as

$$V^{Q^3}(\vec{x}_1,\vec{x}_2,\vec{x}_3) = -\frac{2}{3}\sum_{i>j}\frac{\alpha_s(|\vec{x}_i-\vec{x}_j|)}{|\vec{x}_i-\vec{x}_j|} + k_\infty r_M + V_0^{Q^3} \quad , \quad \rho \geqq R_\infty \tag{71b}$$

where $k_\infty \equiv k(\alpha_s(\infty))$. The very same prescription for k could have been used in (67b) with essentially no change in any of the results in Table 1.

What physical systems can (71) be tested against? Given the success of (70) in describing the $c\bar{c}$ and $b\bar{b}$ systems, it would be very nice to have a baryon composed of three c quarks or three b quarks. Although it seems quite certain that such systems do exist, it may be a long time before they are produced in the laboratory. We must look to the $\Omega^-(s^3)$, therefore, keeping in mind the small mass of the s quark,which leads to sizeable relativistic and spin-dependent effects.

The first question to ask is whether the mass of the $\Omega^-(1672)$ is correctly predicted, using the same value for the s-quark mass which fit the $\phi$ and $F^*$ mesons. Once again the zero-point energy is added to the potential (71) to obtain the Hamiltonian

$$H^{Q^3} = \sum_{i=1}^{3}\left(m_i + \frac{\vec{p}_i^{\,2}}{2m_i}\right) + V^{Q^3} + E_0^{Q^3} \quad . \tag{72}$$

Following the pattern of (68) and (69), $E_0$ is given by

$$E_0^{Q^3} = -\frac{z_0}{R^{Q^3}} \tag{73}$$

and the radius follows from (56)

$$R^{Q^3} = \left(\frac{3\alpha_s(\lambda\rho)\rho^2}{2\pi B}\right)^{1/6} \quad . \tag{74}$$

Since the mass of the $\Omega^-$ is just one number, and we have given ourselves the freedom of an additional parameter $\lambda$, it may appear that (72) is not really being tested. But it is not possible to obtain the experimental value of the $\Omega^-$ mass with any choice of $\lambda$ in the range $\sqrt{(1/3)}$ to $\sqrt{(3/2)}$, with the smaller value coming the closest. If $\lambda$ is fixed at $\sqrt{(1/3)}$, then it is necessary to reduce $m_s$ from 0.649 GeV to 0.571 GeV [27] in order to put the $\Omega^-$ at its physical mass of 1.672 GeV. Examination of Table 2 shows that the quark motion in the $\Omega^-$ is moderately relativistic, only slightly less so than in the $\phi$ meson. (The same quark mass was used in this Table to make the comparison between the $\phi$ and the $\Omega^-$ more meaningful.)

When (72) is applied to the baryons $c^3$ and $b^3$ using the same quark masses as in Table 1, the ground-state masses have the values $M(c^3)$ = 5.04 GeV and $M(b^3)$ = 14.72 GeV. Because the quark motion in these systems is nonrelativistic (see Table 2), we have some confidence in these predictions, especially the energy differences between the ground states and the excited states. Quite a few states in both of these systems are expected to be stable against decays of the type $Q^3 \rightarrow Q^2n + Q\bar{n}$, where n is a u or d quark [27].

## 4.3 Arguments Against a Quark-Quark Confining Potential

To be able to speak about a potential energy--which is a function of the positions (and possibly momenta) of a set of particles--it is necessary first of all that they be localizable. It was argued in Sec. II.2. that if the product of the quark mass and bag radius is small compared to unity, then there is a strong preference, energetically, for the quarks to occupy the lowest mode, in which they are spread out over the entire bag. In the MIT treatment of hadrons composed of u and d quarks, they were treated as either massless or nearly so [12]. But we have also seen that if one tries to assign to these quarks a 'constituent' mass of 0.3 - 0.4 GeV, their motion (in the potential derived for heavy quarks) is quite relativistic. Even the s quark with a mass ~ 0.6 GeV is somewhat relativistic. It appears, therefore, that the concept of a potential energy is not meaningful for the u and d quarks, at least if they are inside the same hadron. The question of quarks in different hadrons will be discussed below.

For heavy quarks, where the potential concept is applicable, there are arguments why the confining part of the interaction cannot be represented as a sum of two-body potentials. In QCD there is a three-gluon vertex, and multiple gluon fission leads to an n-body interaction as in Fig. 8a. In addition, once gluon

a

b

Fig.8 QCD diagrams which lead to many-quark interactions. a) Four quarks interact via two connected three-gluon vertices. b) Three quarks interact via multiple exchange of single gluons.

exchange becomes strong, even the multiple exchange of single gluons leads to many-body forces, as in Fig. 8b. This is the same type of diagram which occurs in nuclear physics, but there the finite mass of the exchanged mesons makes the range of the many-nucleon forces small.

The bag model treatment of heavy quarks in Sec. II.4. gives an explicit confining potential which is an n-body operator. This is clearly seen in (54), (61), and (71), and is due to the fact that the bag boundary depends on the positions of all the quarks simultaneously, in order that the two boundary conditions (36) and (37) be satisfied.

Nevertheless, the simplicity of two-body potentials has led many physicists to try them on quarks, both light and heavy. A great deal of this work is focussed on the single-meson and single-baryon problems, and one branch of this effort is summarized in Isgur's 1978 Erice lectures [35]. There are also attempts to apply quark-quark potentials to the nucleon-nucleon interaction [36]. Quite apart from the question of the theoretical foundation, there is a serious difficulty with this approach because it produces long-range forces between hadrons which strongly conflict with experimental results.

The starting point for these discussions is the assumption that the total potential energy can be written

$$V = -\sum_{i>j} F_i \cdot F_j U(|\vec{r}_i - \vec{r}_j|) \tag{75}$$

where U contains a confining term, usually taken to be of the form $Ar^{\beta}$ with $\beta = 1$ or 2. The first difficulty with (75) appears to be that it makes no objection to the appearance of a single quark; it is simply a free particle. Furthermore, two quarks can and will be bound together in the $\underline{3}^*$ representation where $F_1 \cdot F_2 = -2/3$ (with $A>0$). [But any object with net color will be attracted and bound to one of opposite color (assuming it exists) in an overall color singlet state. Calling the generators for these objects $F_I$ and $F_J$, $(F_I + F_J)^2 = 0$ means that $F_I \cdot F_J < 0$, and consequently the potential (75) confines them.]

Now suppose that two hadrons, each a color singlet, have their centers of mass separated by a distance R which is large compared to their sizes. If the particles were points, (75) would produce no interaction between them because of the color singlet condition, $\sum_i F_i = 0$, where i runs over the quarks in one of the particles. But it is the residual interaction resulting from the finite sizes of the hadrons which must be examined. I will follow the nice discussion in Gavela et al. [37], which does not make use of perturbation theory.

Although (75) is a color singlet operator, it can take the two particles from a state in which both are color singlets, $|11\rangle$, to a state in which both are color octets, $|88\rangle$, but combined into an overall color singlet. Indeed, the terms in which i refers to a quark in one hadron and j to a quark in the other hadron do exactly that because the color generators for each quark are a set of octet operators. If P is a projection operator onto |11> and Q = 1 - P is a projection operator onto |88>, then the effective potential in the |11> state is [38]

$$V_{eff} = PVP + PVQ \frac{1}{E - QHQ} QVP \tag{76}$$

where H is the sum of V and the quark kinetic energies. For the reason just mentioned there is no contribution to PVP from those terms of (75) in which i is a quark in one hadron and j in the other. So all of the dependence of $V_{eff}$ upon R comes from the final term in (76).

One next shows that PVQ ~ $R^{\beta-2}$, that is to say, it is two powers of R lower than the quark-quark potential U. This follows from

$$(\vec{r}_i - \vec{r}_j)^2 = R^2\left(1 + \frac{2\hat{R}\cdot(\vec{\rho}_i - \vec{\rho}_j)}{R} + \frac{(\vec{\rho}_i - \vec{\rho}_j)^2}{R^2}\right)$$

where $\vec{\rho}_i$ is the position of quark i with respect to the center of mass of the hadron it is in. Only terms in the expansion of $(|\vec{r}_i - \vec{r}_j|)^\beta$ which are bilinear in $\rho_i\rho_j$ contribute to PVQ because $\sum_i F_i = 0$ when acting on the |11> state.

The evaluation of $(E - QHQ)^{-1}$ at large separations has a totally different character for $\beta > 0$ and for $\beta < 0$. In the latter case, which is the one encountered in atomic physics, (E-QHQ) becomes independent of R at large separations, and consequently $V_{eff} \sim R^{2\beta-4}$. With $\beta = -1$, one obtains the $R^{-6}$ van der Waals potential. If, on the other hand, $\beta > 0$, then E - QHQ is dominated by -QVQ at large separations, and behaves just like V itself. In this case $V_{eff} \sim R^{\beta-4}$, and a linear confining potential between quarks gives rise to an $R^{-3}$ potential between hadrons. [This discussion makes it clear that second-order perturbation theory cannot be used in the confining case; in place of the energy difference $(E_n - E_0)$ between the ground state and an excited state, one obtains V(R).]

The consequences of a long-range interaction between hadrons can be seen in microscopic and macroscopic experiments. In nucleon-nucleon scattering, for example, the energy dependence of phase shifts at low energy would be $\delta_\ell \sim k^{2-\beta}$ rather than the $k^{2\ell+1}$ which results from short-range forces. With $\beta = 1$, all phase shifts for $\ell \neq 0$ will eventually become larger (as the energy is decreased) than those arising from any finite range potential. To see at what energy this happens, it is necessary to assign a strength to the confining potential $Ar^\beta$. In some of the literature which uses quark-quark potentials, A is taken to be one-half the strength of the confining potential from the heavy $Q\bar{Q}$ problem [39,40], as though one could go from the latter problem to the former by simply multiplying the potential with a factor $F_q \cdot F_{q'}/F_Q \cdot F_{\bar{Q}}$. Taking the value of the slope of the linear confining potential quoted below (45) would make $A = \frac{1}{2}$ k = 0.31 GeV/fm for $\beta = 1$. (This is quite close to the 0.34 GeV/fm which Gavela et al. [37] used in their discussion.)

For a linear confining potential between quarks, the effective nucleon-nucleon potential at large separations is [37]

$$V_{eff}(R) = -\frac{4}{3} A\langle r^2\rangle^2 \frac{1}{R^3} \tag{77}$$

where $\langle r^2\rangle$ is the mean-square radius of a nucleon. Taking $(0.83\ \mathrm{fm})^2$ for this value, and A from above, yields $V_{eff}(R) = 0.20\ \mathrm{GeV\text{-}fm}^3/R^3$. This is ~8 times larger than the potential used in [37], because the model employed there gives too small a value for $\langle r^2\rangle$. With this stronger potential, the inconsistency between the experimental nucleon-nucleon D-wave and F-wave phase shifts, and those calculated from (77), is even greater than that found in [37]. In the D wave, for example, the calculated phase shift is larger than the experimental value for all energies below 50 MeV. For an $r^2$ confining potential the discrepancy is vastly greater.

There is an extensive discussion by Feinberg and Sucher [41] of a number of other experiments which place limits on the allowed strength of a universal nucleon-nucleon potential which varies as $R^{-n}$. For n = 2 or 3 the most stringent limits come from experiments of the Cavendish type, which look for departures from the gravitational force law at distances from 1 cm to 1 m. The discussion in [41] is based upon Long's summary [42] of several such experiments in the interval from 6 to 80 cm. They interpret Long's result to say that

$$\left|\frac{R\frac{d}{dR}(R^2F)}{R^2F}\right|_{R=10\ \mathrm{cm}} < 10^{-2}\ . \tag{78}$$

In more recent literature there is some dispute about how much farther (78) can be pushed, with Long [43] finding a positive effect at the level of $4\times 10^{-3}$, but Spero et al. [44] finding none at $10^{-4}$, at a distance of 2 to 5 cm. This is not critical for the present purpose, since (78) already provides an enormous contradiction with the potential (77).

Introducing a dimensionless strength λ into the non-gravitational part of the interaction

$$V = \frac{\lambda_n}{R}\left(\frac{\mathrm{fm}}{R}\right)^{n-1} , \tag{79}$$

the left-hand side of (78) becomes

$$\frac{n(n-1)\lambda_n}{Gm_p^2}\left(\frac{\mathrm{fm}}{R}\right)^{n-1}$$

where G is the Newtonian gravitational constant and $m_p$ is the proton mass. Eq. (78) now says that

$$\lambda_n < \frac{6.0\cdot 10^{-41}\cdot 10^{14(n-1)}}{n(n-1)} \tag{80}$$

which for n = 3 becomes $\lambda_3 < 10^{-13}$. Examination of $V_{eff}(R)$ below (77) shows that it exceeds this limit by 13 orders of magnitude!

## 4.4 Interaction of Bags

Whatever imperfections the bag model may have, it appears to be the main tool presently available for studying the interaction of hadrons at low and intermediate energies, where confinement is important. One ingredient which is totally absent from the presentation in Sec. II is chiral invariance. According to one point of view, which is summarized in Brown's 1981 Erice lectures [45], incorporating chiral invariance may result in radical departures from the predictions of the bag model with the pion omitted. It is also possible that the pion will make only small changes, a viewpoint which is presented in [1,2]. The first requirement is an understanding of the role of the pion in QCD [46]. This matter will not be considered further in these lectures.

For hadrons composed of heavy quarks, the Born-Oppenheimer approximation discussed in Sec. II. 4 can be applied, in principle, to two or more hadrons at low energy, or even heavy quark matter. With more than three quarks it is possible to have more than one color singlet state, and consequently--depending upon the positions and color state of the quarks--they may all be in one bag or more than one. Although the geometry is complicated, the dynamics is well defined: first calculate the bag energy and then put it into the Schrödinger equation. Of course, one has all the mathematical complexity of solving this N-particle equation. A first look at the possible existence of bound $Q^2\bar{Q}^2$ states has been taken by Ader, Richard and Taxil [47].

For the light quarks, there does not now exist a well-defined procedure for doing bag dynamics, because one does not have a true quantum-mechanical Hamiltonian. (See Sec. II.3.) It is necessary, therefore, to supplement the bag model with some ideas about how bags deform and merge into each other.

In the nucleon-nucleon problem, for example, there are three distinct distance regions to be considered. At large separations where the two 3-quark bags are not overlapping, the highly successful picture of the interaction as resulting from the exchange of virtual mesons must be recovered from the correct theory. The special role of the $\pi$ meson is very prominent here [1,2,45]. At somewhat smaller distances there is probably a region in which the bags begin to overlap, but without completely destroying the identity of the two nucleons. And finally, at very small distances there is just one 6-quark bag.

Starting from the position of complete overlap and working outward, De Tar [48,49] assigned three quarks to a right-handed wave function

$$\psi_R = N(\psi_s + \sqrt{\mu}\,\psi_p) \tag{81a}$$

and three to a left-handed wave function

$$\psi_L = N(\psi_s - \sqrt{\mu}\,\psi_p) \tag{81b}$$

where $\psi_s$ and $\psi_p$ are the lowest s- and p-wave orbitals in the bag. With a proper dynamical theory one could treat $\mu$ as a generator coordinate [50], but De Tar [49] was forced to interpret the bag energy for a fixed value of $\mu$ as a static potential energy between nucleons, even though they are badly deformed. The difficulty with this procedure is that one does not know what inertia to assign to these nucleons. The most striking feature of this 'potential energy' is the large repulsion, ~ 300 MeV at zero separation, which results from the color magnetic interaction.

Jaffe and Low [51] have proposed another way to understand the interaction of hadrons in terms of the bag model, which was inspired by the R-matrix theory of nuclear reactions [52,53]. In that approach, the 3N-dimensional configuration space of N particles is separated into two regions. In the inner region the particles interact strongly; the outer region is taken to be the set of all (2-body) channels into which the system can decay. The boundary separating the two regions is somewhat arbitrary, but is chosen large enough in the relative coordinates of the fragments in each channel so that they are nonoverlapping.

By imposing a condition on the wave function at the boundary, e.g., that it vanish there, a complete set of <u>interior</u> eigenstates $X_n$ of the Hamiltonian, with energies $E_n$, is defined. The exact wave function for the system at any energy E can be expanded in this complete set, in the interior region. If the dynamics is governed by the nonrelativistic Schrödinger equation, for example, then it is straightforward to show that the logarithmic derivative of the wave function on the boundary--which is a matrix in the channel space--can be written [52,53]

$$P_{ij}(E) = P^0_{ij} + E \sum_n \frac{\lambda^n_i \lambda^n_j}{E - E_n} \quad . \tag{82}$$

The residue $\lambda^n_i$ of the nth pole in channel i is determined by the surface integral --<u>on that part of the boundary which corresponds to break-up into channel i</u>--of the overlap of $\hat{n}\cdot\nabla X_n$ with $\phi_i$, where $\phi_i$ is the wave function describing the internal state of the two fragments (as well as their relative orbital motion). Knowledge of the P matrix can be directly translated into the S matrix, provided there is no interaction between the fragments once they separate. But the Coulomb and other potentials can be included there, as discussed below.

(The attempts to make the above discussion relativistic have been based on a single particle interacting with a potential. For a Dirac particle, see Goertzel [54]. For a Klein-Gordon particle interacting with a scalar potential, E in (82) is replaced by $E^2$ [51].)

When Jaffe and Low proposed that bag model eigenstates correspond to poles of the P matrix at a channel radius b which is related to the bag radius R, it was necessary for them to make a number of accomodations. Ultimately, most of them arise from the fact that one is dealing with a set of relativistic particles rather than nonrelativistic ones. But the immediate source of the problems is the fixed bag approximation (Sec. II.3). Rather than being a surface defined by the relative coordinates of the quarks in configuration space, the bag boundary is a fixed barrier in 3-dimensional space. This has two consequences: the residues $\lambda_i^n$ cannot be calculated according to the surface integral rule given above; and it is not clear how the value of the channel radius b (at which the P matrix is evaluated) should be related to the bag radius R. Jaffe and Low presented arguments concerning both of these points, but stressed "the crudity of this model for the residues." The relative strength of the residues in different channels, $\lambda_i^n/\lambda_j^n$, was taken to be determined entirely by the color, spin, and flavor content of the $q^2\bar{q}^2$ wave function $X_n$ when expanded in a $q\bar{q}$ - $q\bar{q}$ basis [55]. For the overall strength of the residue, and the relation between the channel radius b and bag radius R, [51] should be consulted.

For a number of pseudoscalar meson-meson channels there exist sufficient experimental data (phase shifts) to determine the P matrix over a range of energies. The positions of the poles and the residues were then extracted and compared with the predictions of the bag model [51,55]. There is fairly good agreement on the pole positions, but less for the residues. Jaffe and Low stress the fact that a P-matrix pole (and the corresponding bag model state) need not be associated with a nearby S-matrix pole. Only in the case of a narrow resonance does this happen.

Roiesnel [56] has performed a similar analysis for low-energy meson-nucleon scattering, and also finds P-matrix poles at energies which are close to predicted bag model states. Mulders [57] has extended the coupled channel analysis of $\bar{K}N$ - $\pi\Sigma$ to somewhat higher energies and finds some ambiguities in the interpretation.

For the nucleon-nucleon system, there is considerable interaction--due to one- and two-pion exchanges--even when the nucleons are not inside the same bag. In principle, it is straightforward to incorporate such potentials into the R-matrix picture. Starting with the experimental phase shifts at a given energy, which determine the wave function in a given partial wave at very large distances, one uses the Schrödinger equation to integrate the wave function inward to the radius b, and calculates the logarithmic derivative there (the P matrix). Its poles and residues are then compared with the predictions of the bag model. (For $\pi$-$\pi$

scattering, Jaffe and Low [51] estimated the effect of ρ-meson exchange in this way, and found an effect of only a few percent because of the short range of this potential.)

When Mulders [58] tried to study nucleon-nucleon scattering in this way, he concluded that there are too many inelastic channels open, in the energy region where the lowest 6 quark states are expected, to be able to construct the S matrix--and from that the P matrix--from the data. Instead he tried to reverse the procedure by calculating the complete P matrix theoretically, and going from there to the S matrix. But this requires knowledge of all the poles and residues of the bag model P matrix. Mulders assumed that all poles (in a given partial wave) except the lowest one are the same as they would be for free particles; and the lowest one is as given by the bag model. For the details see [58]. In the outer region Mulders includes a theoretical N-N potential, as well as simplified N-Δ and Δ-Δ potentials. Only rough agreement with the experimental NN phase shifts in the intermediate energy region is obtained, but this calculation is a very interesting start on combining the quark structure with the meson exchange aspects of the nucleon-nucleon interaction.

Lomon [59] has emphasized the importance of this approach to the nucleon-nucleon problem, and he is trying to do this by going from the S matrix to the P matrix [60]. He has also raised the possibility that one may be able to dispense with the bag model (that is to say, drop the volume energy term) and treat QCD directly as an R-matrix theory [59].

*Acknowledgements*

My understanding of this material has been greatly aided by many conversations with A. T. Aerts, W. C. Haxton, K. Johnson, and P. J. Mulders. I also want to thank L. Castillejo, T. J. Goldman, E. Lomon, M. M. Nieto, and J.-M. Richard for helpful discussions. A. Martin kindly brought to my attention a preprint by Greenberg and Lipkin [61] which shows that antisymmetrization does not eliminate the long-range forces discussed in Sec. III.

This work was supported by the U. S. Department of Energy.

References

1. R. L. Jaffe, "The Bag," lectures presented at the 1979 Erice Summer School "Ettore Majorana."
2. R. L. Jaffe, "Applications of the Bag Model," Acta Physica Austriaca, Suppl. 22, 269 (1980).
3. C. DeTar, "The MIT Bag Model," in Quantum Flavordynamics, Quantum Chromodynamics, and Unified Theories, K. T. Mahanthappa and J. Randa, Eds., Plenum Press (1979), p. 393.

4. K. Johnson, Phys. Letts. 78B, 259 (1978). In this reference only $\Gamma = 1$ is considered. The generalization to $\Gamma$ of the form in Eq. (6) was discussed by Jaffe in Reference 1 in connection with other formulations of the bag model.
5. J. D. Bjorken and S. D. Drell, Relativistic Quantum Mechanics, McGraw-Hill (1964).
6. E. V. Shuryak, Phys. Letts. 93B, 134 (1980).
7. C. Rebbi, Phys. Rev. D12, 2407 (1975).
8. C. Rebbi, Phys. Rev. D14, 2362 (1976).
9. T. DeGrand and C. Rebbi, Phys. Rev. D17, 2358 (1978).
10. A. Chodos, R. L. Jaffe, K. Johnson, and C. B. Thorn, Phys. Rev. D10, 2599 (1974).
11. A. Chodos, R. L. Jaffe, K. Johnson, C. B. Thorn, and V. F. Weisskopf, Phys. Rev. D9, 3471 (1974).
12. T. De Grand, R. L. Jaffe, K. Johnson, and J. Kiskis, Phys. Rev. D12, 2060 (1975).
13. S. A. Chin, A. K. Kerman, and X. H. Yang, "Many-body theory of confined quarks and gluons--calculation of self and exchange energies," MIT-CTP #919.
14. This problem has been examined by K. A. Milton, Phys. Rev. D22, 1441 and 1444 (1980). His calculated value, $Z_0 \sim -0.26$, has the opposite sign and a smaller magnitude than the fitted value from Reference 12. But there are serious questions about the renormalization procedure and the accuracy of the calculation.
15. R. E. Peierls and J. Yoccoz, Proc. Phys. Soc. (London) A70, 381 (1957).
16. R. E. Peierls and D. J. Thouless, Nucl. Phys. 38, 154 (1962).
17. R. W. Haymaker and T. Goldman, Phys. Rev. 24D, 743 (1981).
18. P. Hasenfratz, J. Kuti, and A. S. Szalay, in Charm, Color, and the J/$\psi$, proceedings of the X Rencontre de Moriond, Méribel-les-Allues, France, 1975, edited by J. Trân Thanh Vân (CNRS, Paris, 1975). This work is also summarized in a review article, P. Hasenfratz and J. Kuti, Phys. Rep. 40C, 75 (1978).
19. K. Berkelman, in High Energy Physics-1980, proceedings of the XX International Conference, Madison, Wisconsin, edited by L. Durand and L. G. Pondrom (AIP, New York, 1981), p. 1499.
20. L. Heller and K. Johnson, Phys. Lett. 84B, 501 (1979).
21. See, for example, D. Husenmoller, Fibre Bundles (Springer, New York, 1966), Chap. 11.
22. K. Johnson, in Current Trends in the Theory of Fields, proceedings of the Symposium, Tallahassee, 1978, edited by J. E. Lanutti and P. K. Williams (AIP, New York, 1978), p. 112.
23. A. T. Aerts and L. Heller, Phys. Rev. D23, 185 (1981).
24. P. Hasenfratz, R. R. Horgan, J. Kuti and J. M. Richard, Phys. Letts. 94B, 401 (1980).

25. P. Hasenfratz, R. R. Horgan, J. Kuti and J. M. Richard, Phys. Letts. 95B, 299 (1980).
26. See, for example, J. D. Jackson, Classical Electrodynamics (Wiley, New York, 1975).
27. A. T. Aerts and L. Heller, "The Spectroscopy of the Three-Heavy-Quark System," Los Alamos preprint LA-UR-81-1775.
28. W. C. Haxton and L. Heller, Phys. Rev. D22, 1198 (1980).
29. R. Giles, Phys. Rev. D18, 513 (1978).
30. See, for example, the discussion in T. Appelquist, R. M. Barnett, and K. Lane, Annu. Rev. Nucl. Part. Sci., 28, 387 (1978).
31. T. Appelquist and J. Carazzone, Phys. Rev. D11, 2856 (1975).
32. J. F. Donoghue and K. Johnson, Phys. Rev. D21, 1975 (1980). In this reference it is estimated that only ~ 1 out of the 1.84 should be assigned to the zero-point energy, the rest arising from the violation of translation invariance.
33. Particle Data Group, Rev. Mod. Phys. 52, No.2, Part II (1980).
34. R. L. Jaffe and J. Kiskis, Phys. Rev. D13, 1355 (1976).
35. N. Isgur, "Soft QCD: Low Energy Hadron Physics with Chromodynamics," in The New Aspects of Subnuclear Physics, edited by A. Zichichi, Plenum Press (1980), p. 107.
36. My colleague D. A. Liberman began this approach in Phys. Rev. D16, 1542 (1977).
37. M. B. Gavela, A. Le Yaouanc, L. Oliver, O. Pène, J. C. Raynal, and S. Sood, Phys. Letts. 82B, 431 (1979).
38. H. Feshbach, Ann. Phys. 19, 287 (1962).
39. R. K. Bharduri, L. E. Cohler, and Y. Nogami, Phys. Rev. Lett. 44, 1369 (1980).
40. D. P. Stanley and D. Robson, Phys. Rev. Lett. 45, 235 (1980).
41. G. Feinberg and J. Sucher, Phys. Rev. D20, 1717 (1979).
42. D. R. Long, Phys. Rev. D9, 850 (1974).
43. D. R. Long, Nature 260, 417 (1976).
44. R. Spero, J. K. Hoskins, R. Newman, J. Pellam, and J. Schultz, Phys. Rev. Lett. 44, 1645 (1980).
45 G. E. Brown, "Nucleon-Nucleon Forces from Bags, Quarks, and Boson-Exchange" unpublished lectures.
46. For a recent discussion of this question, see the talk by K. Johnson presented at the "9th International Conference on High Energy Physics and Nuclear Structure," Versailles, July 6-10, 1981; and private communication.
47. J.-P. Ader, J.-M. Richard, and P. Taxil, "Do Heavy Multiquark States Exist?", TH.3101-CERN.
48. C. De Tar, Phys. Rev. D17, 302 (1978).
49. C. De Tar, Phys. Rev. D17, 323 (1978).
50. J. J. Griffin and J. A. Wheeler, Phys. Rev. 108, 311 (1957).

51. R. L. Jaffe and F. E. Low, Phys. Rev. D19, 2105 (1979).
52. E. P. Wigner and L. Eisenbud, Phys. Rev. 72, 29 (1947).
53. A. M. Lane and R. G. Thomas, Revs. Mod. Phys. 30, 257 (1958).
54. G. Goertzel, Phys. Rev. 73, 1463 (1948).
55. R. L. Jaffe, Phys. Rev. D15, 267, 281 (1977).
56. C. Roiesnel, Phys. Rev. D20, 1646 (1979).
57. P. J. Mulders, "The Quark Structure of the Λ (1405) Resonance," in the Proceedings of the Workshop on Nuclear and Particle Physics at Energies up to 31 GeV: New and Future Aspects, Los Alamos, Jan. 5-8, 1981, edited by J. D. Bowman, L. S. Kisslinger, and R. R. Silbar, p. 388.
58. P. J. Mulders, "Resonances in Baryon-Baryon and Pion-Deuteron Scattering," Los Alamos preprint LA-UR-81-2397.
59 E. Lomon, "Joining the Quark and Hadron Exchange Descriptions of Strong Interactions," in the Proceedings of the workshop cited in [57], p. 400.
60. Private communication from E. Lomon.
61. O. W. Greenberg and H. J. Lipkin, "The Potential Model of Colored Quarks: Success for Single Hadron States, Failure for Hadron-Hadron Interactions," FERMILAB-Pub-81/45 - THY.

# 5. Quark Matter and Nuclei

G. Baym

With 5 Figures

## 5.1 Introduction

In these lectures I would like to focus on the possibility of deconfinement of quarks over extended regions of nuclear matter at high baryon density or excitation energy. While at low densities quarks are confined in individual hadrons, we expect that at high energy density, as may be achievable in ultra-energetic central heavy ion collisions, nuclear matter undergoes a transition to deconfined "quark matter." The detection of this fundamentally new state of matter would be an exciting advance in nuclear physics, and I would like, after describing in detail the basic properties of quark matter, to discuss various ways that one may hope to see quark matter experimentally, by temporary production in nuclear collisions, or in the form of metastable exotic nuclear objects.

## 5.2 Phenomenology of Confinement

As a basis for understanding deconfinement of quarks, let us begin with a brief review of phenomenological pictures of how confinement takes place. The basic idea underlying these pictures is that the vacuum can have two possible states. The first is the normal vacuum outside hadrons, the vacuum in the absence of physical quarks and their color gluon fields; this state by definition excludes quarks and color fields. Quarks and their gluon fields modify the vacuum in their neighborhood, transforming it into the second state, the "perturbative" vacuum, or "bag" state, the form of the vacuum inside hadrons, as described in the MIT bag model (Chodos *et al.*, 1974; Heller's lectures in this volume) and related models of hadrons. This inside state of the vacuum has a higher energy per unit volume than the outside vacuum state; the energy difference, B, is essentially the bag constant of the MIT bag model, although it does not necessarily have the MIT phenomenological value

$$B \approx 56 \text{ MeV/fm}^3 \quad , \tag{1}$$

deduced from fitting hadron masses.

With the concept of the vacuum having two states, one describes hadrons in a bag model by saying that the quarks composing a hadron transform a finite region of space -- the "bag" -- of volume V, from the normal vacuum into the second state. Outside the bag the vacuum remains in the normal state, and the quarks and color fields are confined to the bag. To a first approximation the boundary between the bag and outside can be taken as sharp. Since the energy of the bag, BV, is minimized by making the bag as small as possible, the energy density B acts as an effective pressure of the normal vacuum on the bag.

Essentially two major mechanisms provide support pressure to the bag, preventing it from being squeezed by the normal vacuum pressure to vanishingly small size: the quark kinetic energy, important in light quark systems such as nucleons, and the color electric field energy, important in heavy quark systems such as charmonium. The detailed way in which these mechanisms operate is described in Heller's lectures in this volume.

The confining property of the normal vacuum, the exclusion of color fields, can be characterized in terms of the color dielectric constant, $\varepsilon$, of the two states of the vacuum, by saying that in the normal state the dielectric constant vanishes, and in the bag state it is unity (e.g., Callen *et al.*, 1979; Lee, 1980, 1981). The dielectric constant, as in Maxwell, is defined by the relation

$$\mathbf{D} = \varepsilon\, \mathbf{E} \tag{2}$$

between the color $\mathbf{D}$ field, the field produced by the source charges, and the $\mathbf{E}$ field, the total color electric field including that produced by induced charges.

The vanishing of the dielectric constant of the normal vacuum implies directly that color $\mathbf{D}$ fields vanish in the normal state, and hence are confined to the perturbative vacuum; it also produces confinement of quarks as well. To see how this works, let us first assume, following an argument of Lee (1980, 1981), that in the normal vacuum $\varepsilon$ is finite and $< 1$, while in the vacuum in the vicinity of color charges $\varepsilon = 1$ (the bag state), and calculate the vacuum configuration surrounding a charge of strength g at the origin. Since it costs an energy B to convert the vacuum from the normal to $\varepsilon = 1$ state, the lowest energy state of the vacuum will be a sphere of some radius R centered about the charge, whose interior is in the $\varepsilon = 1$ state, and exterior is in the normal state. The energy of the configuration is

$$E = \frac{4\pi}{3} R^3 B + \frac{1}{2} \int_a^\infty d^3r\, \mathbf{D}\cdot\mathbf{E} \quad , \tag{3}$$

where a is a cutoff at small r. To compute D we temporarily suppress the SU(3) color algebra and write the Maxwell equation

$$\boldsymbol{\nabla}\cdot\mathbf{D} = g\rho(\mathbf{r}) \tag{4}$$

where g is the qcd coupling constant, and $\rho(\underset{\sim}{r})$ is the quark color density, here a delta function at the origin. Thus

$$\underset{\sim}{D} = \frac{g\hat{\underset{\sim}{r}}}{4\pi r^2} \quad ; \tag{5}$$

at the spherical boundary between the two vacuum states the normal component of $\underset{\sim}{D}$ is continuous, so that (5) is the solution throughout space. On the other hand

$$\underset{\sim}{E} = \frac{g\hat{\underset{\sim}{r}}}{4\pi r^2 \varepsilon(r)} \tag{6}$$

is discontinuous across the boundary; only the tangential component of $\underset{\sim}{E}$, which vanishes here, is continuous. With (5) and (6) we find a total energy

$$E = \frac{4\pi}{3} R^3 B + \frac{\alpha_s}{2} \left[\frac{\varepsilon^{-1}-1}{R} + \frac{1}{a}\right] \quad , \tag{7}$$

where

$$\alpha_s = g^2/4\pi \tag{8}$$

is the fine structure constant. If the dielectric constant of the normal vacuum is $> 1$, the energy (7) is minimized by making R as small as possible. However for $\varepsilon < 1$ the minimum occurs at a finite R given by

$$R = \left[ \frac{\alpha_s}{8\pi B} (\varepsilon^{-1}-1) \right]^{1/4} \quad , \tag{9}$$

at which

$$E = \frac{16\pi}{3} BR^3 + \frac{\alpha_s}{2a} \quad ; \tag{10}$$

generally for $\varepsilon < 1$ charges produce structures in the vacuum. The crucial point now is that as $\varepsilon \to 0$, both R and $E \to \infty$, that is, the energy or mass of an isolated charge becomes infinite; thus charges cannot exist in isolation, but rather must form charge neutral combinations with other charges.

In contrast to a single charge, a pair of charges in an overall neutral, or singlet state has a finite energy as $\varepsilon \to 0$. Let us consider a heavy quark and anti-quark, in a color singlet state, separated by a large distance L, along the z axis. The quarks, though separated, remain joined by the color electric field $\underset{\sim}{D}$. If $\varepsilon = 0$, the color D fields are confined to the bag vacuum, and rather than forming a dipolar configuration, as in normal electrostatics, the field, and bag, will be compressed by the normal vacuum pressure into a long, essentially cylindrical configuration, a "flux tube," with axis along $\hat{\underset{\sim}{z}}$ and with the quarks at the ends. The color flux through any slice (between the quarks) of the cylinder in the x-y plane is constant; the energetically most favorable configuration then has, except at the ends, a constant cylinder radius, R, and an approximately constant D. The energy of the cylinder is approximately

$$E = \pi R^2 L\left(B + \frac{1}{2}\underset{\sim}{D}^2\right) \tag{11}$$

where the first term is the bag energy, and the second is the electrostatic field energy.

From (4) we find that the flux across any slice of the cylinder is $\pi R^2 D = g$, and we have $\underset{\sim}{D}^2 = g^2/\pi^2 R^4$. More precisely we should replace $g^2$ by $-(g^2/4)\langle\lambda_1^\alpha\cdot\lambda_2^\alpha\rangle$, where $\lambda_1^\alpha$ and $\lambda_2^\alpha$ are the color generators for the two quarks and $\alpha$, the color index, is summed from 1 to 8. In a color singlet state $\langle(\lambda_1^\alpha + \lambda_2^\alpha)^2\rangle = 0$, so that $\langle\lambda_1^\alpha\cdot\lambda_2^\alpha\rangle = -\langle(\lambda_1^\alpha)^2\rangle = -\langle(\lambda_2^\alpha)^2\rangle = -16/3$, and

$$\underset{\sim}{D}^2 = \frac{4}{3}\frac{g^2}{\pi^2 R^4}\ . \tag{12}$$

The electrostatic contribution to the energy grows with shrinking cylinder radius as $R^{-4}$, and thus exerts an outward pressure on the cylinder. Minimizing (11) with respect to $R^2$ we find an equilibrium energy

$$E = \sigma L \tag{13}$$

where the "string constant" $\sigma$ is given by

$$\sigma = (8g^2B/3)^{1/2}. \tag{14}$$

The equilibrium $\underset{\sim}{D}^2$ is simply 2B.

Inside the flux tube joining the two charges $\varepsilon = 1$, and so $\underset{\sim}{D} = \underset{\sim}{E}$. At the edge of the cylinder, $\underset{\sim}{E}$, which is tangential (except at the ends), is continuous across the boundary, and so is non-vanishing in the normal vacuum. However, since $\varepsilon = 0$ in the normal vacuum, $\underset{\sim}{D} = \varepsilon\underset{\sim}{E}$ must vanish. Although the vanishing of the dielectric constant in the normal vacuum gives rise to confinement of $\underset{\sim}{D}$ fields, it does not lead to confinement of $\underset{\sim}{E}$ fields. [This does not imply though that "information" can be propagated between hadrons through the $\underset{\sim}{E}$ field in the normal vacuum, since the gluon stress tensor, $\sim DE$, must vanish if $\varepsilon = 0$.] Conversely the vanishing of $\underset{\sim}{D}$ in the normal vacuum implies that $\underset{\sim}{D}$ must always be tangential at the inner side of a boundary between normal and $\varepsilon = 1$ vacuum.

The dielectric behavior of the normal vacuum is analogous to the perfect diamagnetism of a superconductor, which when placed in a magnetic field expels the magnetic flux from its interior -- the Meissner effect. This property can be described by a vanishing of the permeability $\mu$, defined by $\underset{\sim}{B} = \mu\underset{\sim}{H}$, inside the superconductor (except within the penetration depth at the surface). If one places a superconducting cylinder in a magnetic field with the cylinder axis parallel to the field, then outside the superconductor $\underset{\sim}{B} = \underset{\sim}{H}$, while inside $\underset{\sim}{B} = 0$; however, since in the absence of applied surface currents the tangential component of $\underset{\sim}{H}$ is continuous across a boundary, $\underset{\sim}{H}$ is non-vanishing inside the superconductor. In qcd the D field is expelled from the vacuum outside bags; the normal vacuum thus exhibits an "electric" Meissner effect.

While, as of yet, there is no microscopic derivation in qcd of the dielectric constant of the normal vacuum, Callen, Dashen and Gross (1979) have given a very appealing argument that indicates why $\varepsilon$ should be $\ll 1$ in the normal vacuum, and that provides good insight into the possible structure of the qcd vacuum. The problem, to determine the dielectric constant $\varepsilon$ of the vacuum in the presence of an applied $\underset{\sim}{D}$ field, is approached as follows. Lorentz invariance of the vacuum implies that the field tensor $F_{\mu\nu}$ formed from $\underset{\sim}{E}$ and $\underset{\sim}{B}$ is proportional to the tensor formed from $\underset{\sim}{D}$ and $\underset{\sim}{H}$, so that $\varepsilon = 1/\mu$, where $\mu$ is the permeability of the vacuum. Thus to determine $\varepsilon$ one can equivalently determine $\mu$. This is done by casting the vacuum problem, through introduction of imaginary time, into Euclidean 4-space (in the same way as one does statistical mechanics at temperature T by employing an imaginary time ranging from 0 to $-i/T$). Qcd in Euclidean 4-space has a class of localized solutions for the gluon field $A^{\alpha}_{\mu}(x)$, the "instantons," which, as Callen, Dashen and Gross point out, behave very much like magnetic dipoles in the 4 dimensional space. In particular, instantons, if present, make the vacuum color paramagnetic, $\mu > 1$, and hence $\varepsilon = \mu^{-1} < 1$. The greater the instanton density, the greater the decrease of $\varepsilon$. The schematic problem, then, to determine the equilibrium density of instantons, n, and hence $\varepsilon(n)$, in the presence of a given $\underset{\sim}{D}$ field, is solved by observing that the electrostatic energy density of the D field

$$E_{e\ell} = \frac{\underset{\sim}{D}^2}{2\varepsilon(n)} \tag{15}$$

increases with n, since $\partial\varepsilon/\partial n < 0$. The derivative $\partial E_{e\ell}/\partial n = -\frac{1}{2}\underset{\sim}{E}^2 \partial\varepsilon/\partial n > 0$ acts as an effective chemical potential for instantons, driving them away from regions of large D through an essentially electrostrictive effect. The net result is that where D is large, $\varepsilon$ will be close to unity, but where D is small, $\varepsilon$ will be decreased from unity, as illustrated in Fig. 1, a plot of the electric field E versus the applied D.

This drawing also shows the possibility, suggested by Callen, Dashen and Gross, of a phase transition in the vacuum as D decreases. If the plot of E versus D loops around, as shown, (and they present some evidence that it might) then points B and C (determined, according to the thermodynamicists, by a Maxwell

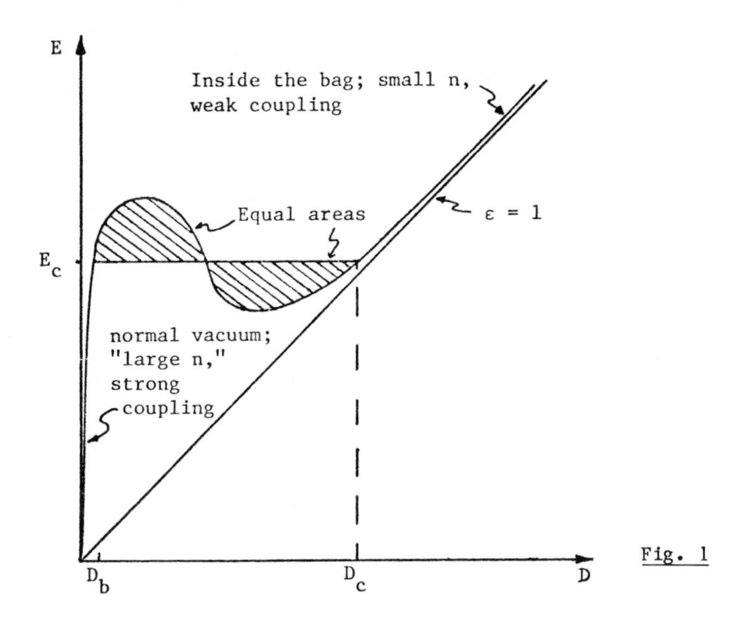


Fig. 1

equal area construction) can be in phase equilibrium, while the region of D from $D_b$ and $D_c$ is physically excluded, in the same way as in a liquid-gas phase transition values of the density between those of the coexisting liquid and gas are excluded. The vacuum in this picture has two distinct phases. For $D > D_c$, as is the case sufficiently near quarks, the field chases away the instantons, and $\varepsilon \approx 1$; this phase corresponds to the vacuum found inside the bag. For $D < D_b$, the vacuum is in the very small $\varepsilon$ phase, corresponding to the normal vacuum. Should $D_b = 0$, then the normal vacuum would have $\varepsilon = 0$ and be confining. The transition from bag vacuum to normal vacuum takes place at the point where the D field just inside the bag has the critical value $D_c$, and for $\varepsilon = 0$ is tangential to the interface. In the simple model earlier in which the color electrostatic pressure balances the pressure of the normal vacuum on the bag, one has

$$D_c = (2B)^{1/2}. \tag{16}$$

Despite its promise, this picture is not yet on a very firm foundation. The large D limit corresponds to a weak coupling regime, and so can be calculated. However, as D decreases, and the instanton density grows, the system becomes strongly coupled; before long the description in terms of an instanton gas breaks down and the problem becomes intractable.

We have so far characterized the two phases of the vacuum in terms of their color dielectric properties, which arise from the underlying structure of the virtual color fields in the vacuum; $\varepsilon$ acts as an order parameter describing the transition, changing from unity in the bag to zero in the normal vacuum. A possibly more fundamental way that the two phases might be characterized is in terms of the way the states of the vacuum realize chiral symmetry. In the normal vacuum chiral symmetry is spontaneously broken, as we know, and realized in the Goldstone mode; the Goldstone boson is the pion. On the other hand, asymptotic freedom strongly suggests that chiral symmetry is fully restored, i.e., realized in the Wigner mode, in the bag vacuum state; the light quarks, u and d, are essentially massless in this state. The two phases would then be characterized by an order parameter $\sim \langle\bar{\psi}\psi\rangle$ (where $\psi$ are the light quark fields) which vanishes inside the bag, but is non-zero in the normal vacuum. The connection, however, between these two characterizations of the states of the vacuum in terms of their dielectric and chiral properties remains an important and difficult problem. [See 't Hooft (1981) for the current status of this question.]

One should note that if the basic criterion to distinguish the two phases of the vacuum is the mode of realization of chiral symmetry, then while bag and normal vacua in equilibrium would, in a static approximation, be separated by a boundary at which chiral symmetry begins to become spontaneously broken, the distance over which the dielectric constant changes from unity to zero could be much longer, and the <u>local</u> dielectric constant not a precise local order parameter to distinguish which phase the system is in. An analogous situation occurs in a superconductor where $\mu$, the local magnetic permeability, and hence a magnetic field, falls to zero only over a distance of a penetration depth in from the surface of the superconductor, although the material within the penetration depth is superconducting.

## 5.3 Decofinement at High Baryon Density of Temperature

Now that we have developed some elementary phenomenology to describe confinement, let us turn to the question of possible deconfinement of nuclear matter with increasing baryon density or temperature. If the density of quarks in a system is low, the quarks are confined in individual hadrons, surrounded by normal vacuum. However as the density is raised, either by increasing the baryon density, or producing hadrons thermally with increasing temperature, the hadrons begin to overlap and one expects the matter eventually to undergo a transition to a state in which the quarks are no longer locally confined. From a bag point of view such a transition would occur when the bags overlap substantially and begin to fill space; then in the transition the individual bags merge into a gigantic bag. Or in terms of the dielectric picture, the transition would occur when the color D fields (in a mean square average sense) exceed a critical value $D_c$ over most of space. The most concrete evidence for the deconfinement transition of the vacuum as it is heated is provided by lattice gauge theories. As Polyakov (1978) and Susskind (1979) have shown analytically, lattice qcd, without quarks, in the

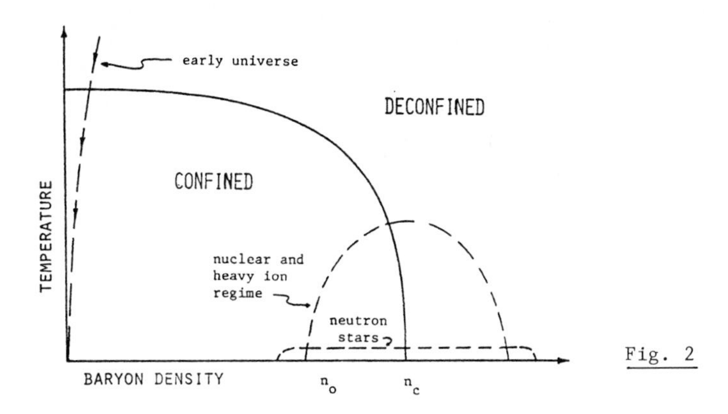


Fig. 2

strong coupling limit exhibits a second order deconfinement transition at a finite (real) temperature $T_c$. Monte Carlo lattice calculations by McLerran and Svetitsky (1981), Kuti, Polónyi and Szlachányi (1981), and Engels et al. (1981), with coupling strength chosen to reproduce a string tension $\sigma \approx 1$ GeV/fm, indicate that the deconfinement transition occurs at $T_c \sim 200$ MeV. However, since these lattice calculations do not include virtual or thermally created fermion pairs in the vacuum, one should not apply this number too rigorously.

We expect the phase diagram of matter to have the structure shown in Fig. 2; at low baryon density $n_b$ and temperature T matter is composed of confined hadrons, but at either large n or T it becomes deconfined quark matter. For example, a proton, dropped into the hot vacuum with $T > T_c$, would evaporate into three quarks. I have also shown on this picture the trajectory followed by the early universe, and the regions of interest in neutron stars, and in nuclear and heavy ion physics (although it is not certain that in either of the latter cases the systems do cross the deconfinement transition line).

We can naively estimate the critical baryon density $n_c$ at $T = 0$ by noting that in nuclear matter, nucleons begin to touch when

$$n_b \approx (4\pi r_n^3/3)^{-1} \equiv n_H , \tag{17}$$

where $r_n$ is the nucleon radius, and $n_H$ is the baryon density inside a nucleon; thus very roughly $n_c \sim 0.24\, r_n^{-3}$. For $r_n = 1$ fm, as in the MIT bag model, this density is only $\sim 1.4 n_o$, where $n_o$ is nuclear matter density ($0.17$ fm$^{-3}$); for $r_n \sim 0.4$ fm, as in the little Stony Brook-Saclay bag (Brown and Rho, 1979; Vento et al., 1980), the touching density is quite a bit higher, $\sim 22\, n_o$.

Similarly, the point $T_c$ at which the hot vacuum will undergo a deconfining transition can be estimated by calculating where thermally created pions begin to fill space. At temperature T, the number density of thermally created (free) pions is given by

$$n_\pi = 3 \int \frac{d^3p}{(2\pi)^3} \frac{1}{e^{(p^2+m_\pi^2)/T}-1} \tag{18}$$

which approaches the value $3\zeta(3)T^3/\pi^2$ at large T. Thus if we say that at $T_c$ the density $n_\pi$ is roughly $(4\pi r_\pi^3/3)^{-1}$, where $r_\pi$ is an effective pion radius, then from (18) we find $T_c \sim 0.87/r_\pi$; for a nominal pion radius $r_\pi \sim 0.6$ fm, $T_c \sim 2m_\pi$, in the same range as that found in the Monte Carlo lattice calculations.

### 5.3.1 Quark Matter

Before discussing in more detail estimates of the transition from hadronic to quark matter, let us consider some of the elementary properties of deconfined quark matter. As a consequence of asymptotic freedom interactions between quarks become arbitrarily weak at sufficiently large momenta or short distances. [See the lectures of Brodsky and Heller in this volume.] The effective "running" coupling constant is given at asymptotically large momentum transfer p by

$$\frac{g^2(p)}{4\pi} \equiv \alpha_s(p) = \frac{6\pi}{33-2N_f} \frac{1}{\ln(p/\Lambda)} , \tag{19}$$

where $N_f$ is the number of flavors of quarks whose (current) mass is less than p. The current masses (as distinguished from the constituent or effective masses, which are larger) of the u and d quarks are $\lesssim$ 10 MeV, while the strange quark mass is $\sim$ 300 MeV. (Charmed and more massive quarks are too heavy to play a role at the temperatures and densities presently attainable in nuclear collisions.) The mass $\Lambda$, which experimentally is in the range 0.1 - 0.5 GeV (see Brodsky's lectures), parametrizes the strength of the asymptotic qcd interactions. As the quark density or temperature becomes large, the characteristic momentum scale grows as $n_b^{1/3}$ or T, and so the mean short range interaction strength falls off logarithmically. Furthermore in quark matter that is locally in a color singlet (or color neutral) state, the interactions between quarks at distances large compared to the interparticle spacing are screened out, analogous to screening of long range Coulomb fields in a plasma in equilibrium. Thus, as Collins and Perry (1975) first pointed out, at high baryon densities or temperatures the net interactions in quark matter in an overall color singlet state should be sufficiently weak that the matter can to a first approximation be taken as a non-interacting relativistic gas.

Imagine then that we start with symmetric nuclear matter, as is approximately the case in a heavy ion collision, and compress it into quark matter, assuming that it remains in its ground state and ignoring for the moment the interactions between the quarks. Since the neutron is composed of one u and two d quarks, and the proton two u and one d, the densities of up and down quarks in the matter obey

$$n_u = n_d = \frac{3}{2} n_b \tag{20}$$

where $n_b$ is the baryon density. In its ground state such "symmetric" quark matter contains twelve Fermi seas, two for u and d, times two for spin, times three for color, e.g., a d-flavor, up-spin, blue sea, etc. Since hadrons are color singlets, all three colors will be present in quark matter equally; thus in the absence of spin polarization, all six u Fermi seas generally have the same Fermi momentum, $p_u$, as do all six d Fermi seas, $p_d$; in terms of $p_u$ and $p_d$, $n_u = p_u^3/\pi^2$, $n_d = p_d^3/\pi^2$. If we assume $p_u$, $p_d \gg m_u$, i.e., fully relativistic quarks [note that at $n_b = n_o = 0.17\ fm^{-3}$, $p_u = p_d = 270$ MeV], then the energy density of the matter is

$$E_o = \frac{3}{4}(n_u p_u + n_d p_d) = (\frac{3}{2})^{7/3} \pi^{2/3} n_b^{4/3}. \tag{21}$$

The energy per baryon is

$$E_o/n_b \approx 1090\ n_b^{1/3}\ \text{MeV}, \tag{22}$$

where $n_b$ is in $fm^{-3}$.

In quark matter at finite temperature, gluons as well as quark- anti-quark pairs are thermally excited. In a non-interacting system in thermal equilibrium, the number density of quark flavor i that is present is

$$n_i = 6 \int \frac{d^3p}{(2\pi)^3} \frac{1}{e^{[(p^2+m_i^2)^{1/2}-\mu/3]/T} + 1}, \tag{23}$$

where $\mu$ is the baryon chemical potential, $m_i$ the quark mass and 6 = 2 (spin) x 3 (color); the number density $n_{\bar{I}}$ of anti-quarks present is given by (23) with $-\mu$. The number of thermal gluons is

$$n_g = 16 \int \frac{d^3p}{(2\pi)^3} \frac{1}{e^{p/T}-1} = \frac{16\zeta(3)}{\pi^2} T^3$$

$$= 245\ (T^3_{GeV})\ fm^{-3}, \tag{24}$$

where 16 = 2 (spin) x 8 (color), and in the latter expression T is measured in GeV, and $n_g$ in $fm^{-3}$.

In a system with zero net baryon number, as might be produced in an energetic $p\bar{p}$ collision, one has $\mu = 0$. If we also assume $T >> m_i$ for u, d and s quarks then

$$n_i = \frac{9\zeta(3)}{2\pi^2} T^3 = 71.3(T^3_{GeV})\ \text{fm}^{-3}, \tag{25}$$

for $i = u, \bar{u}, d, \bar{d}, s$ and $\bar{s}$. The total energy density,

$$E = E_g + \sum_{q_i, \bar{q}_i} E_i \sim T^4 \tag{26}$$

is made up of contributions

$$E_g = \frac{8\pi^2}{15} T^4, \tag{27}$$

from the gluons, and

$$E_i = \frac{7\pi^2}{40} T^4 \tag{28}$$

from each quark flavor i, with an identical contribution from each flavor antiquark.

More generally, for non-zero net quark density $n_i - n_{\bar{i}}$, the free energy density, $F = E - TS$, without interactions, at high densities or temperatures $T >> m_i$ is

$$F_o(T, \{n_i - n_{\bar{i}}\}) = -\frac{8\pi^2}{45} T^4 + \sum_i \left(\frac{3}{4\pi^2} \mu_{oi}^4 + \frac{1}{2} \mu_{oi}^2 T^2 - \frac{7\pi^2}{60} T^4\right), \tag{29}$$

where $\mu_{oi}$ is the zero order chemical potential, given in terms of the net quark density by

$$n_i - n_{\bar{i}} = \mu_{oi}^3/\pi^2 + \mu_{oi} T^2 , \tag{30}$$

and the sum is over flavors. The first term in (29) is the gluon free energy.

### 5.3.2 Effects of Interactions

The above descriptions of both very hot and fully degenerate quark matter neglect any interaction effects. As we see from Eq. (22) however, the energy per baryon in degenerate non-interacting quark matter becomes less than the nucleon rest mass $m_n$ for

$$n_b < (2/3)^7\ m_n^3/\pi^2 = 0.64\ \text{fm}^{-3} . \tag{31}$$

Ordinary matter would thus be unstable against collapse into free quark matter; in order to have stability of ordinary matter compared with quark matter it is necessary to include effects of qcd interactions.

The first effect of interactions is to shift the vacuum energy density of the deconfined phase by an amount B compared with the energy of the vacuum in the normal hadronic vacuum. This energy, just the bag constant, adds onto the kinetic terms (21) or (26). Even with B given by (1), the energy per baryon in cold symmetric quark matter has a minimum $E/n_b \approx 910$ MeV $< m_n$, at a density $n_b = 0.25\ \mathrm{fm}^{-3}$; inclusion of B alone (at least with the MIT value) does not remove the instability of cold nuclear matter. To provide stability it is necessary to include explicit interactions between quarks via gluon exchange. [For the detailed qcd Lagrangian see Brodsky's lectures.] At high densities one can calculate the quark-quark interactions by a perturbation expansion of the ground state energy, or free energy at finite T, in terms of the fine structure constant $\alpha_s = g^2/4\pi$. These calculations are reviewed in detail elsewhere (Shuryak, 1980; Baym, 1977, 1978a) and I will simply summarize certain results. [One should bear in mind that as the density and temperature decrease, the interactions become stronger and stronger, and perturbation theory ceases to be useful. Furthermore, non-perturbative effects can enter at an early stage, for example, possibly introducing a color magnetic field screening length $\sim 1/\alpha_s T$ (Gross, Pisarski and Yaffe, 1981).]

The lowest order correction to the free gas energy is simply the exchange energy, arising from two quarks exchanging a single gluon while exchanging their own states. In the degenerate fully relativistic limit $p_i \gg m_i$ where $p_i$ is the Fermi momentum of flavor i, the exchange contribution from flavor i is

$$E^i_{exch} = \frac{\alpha_s}{2\pi^3} p_i^4 , \tag{32}$$

and in the non-relativistic limit $p_i \ll m_i$,

$$E^i_{exch} = -\frac{\alpha_s}{\pi^3} p_i^4 . \tag{33}$$

The full expression for general $p_i/m_i$ (see, for example, Baym and Chin, 1976b) increases monotonically with $p_i/m_i$, but interestingly changes sign only at the rather large value $p_i = 2.53\ m_i$. This means that the exchange contribution from light quarks, u and d, will be positive in cold quark matter, but strange quarks, if present, can have an attractive exchange energy, which, as we shall see in the next section can lead to metastability of droplets of quark matter with large strangeness.

With (32) the ground state energy density is

$$E = \frac{3}{4\pi^2} \sum_i p_i^4 \left(1 + \frac{2\alpha_s}{3\pi}\right) + B, \tag{34}$$

a result equivalent to applying the MIT model to computing the energy density of quark matter (Baym and Chin, 1976a; Chapline and Nauenberg, 1976; Keister and Kisslinger, 1976). With the MIT parameters, $B = 56\ \mathrm{MeV/fm}^3$ and $\alpha_s = 2.2$ (taken here independent of density), the energy per baryon,

$$E/n_b = 1600\ n_b^{1/3} + 56/n_b, \tag{35}$$

has a minimum at $n_b = 0.18\ \mathrm{fm}^{-3}$, slightly above nuclear matter density, with $E/n_b$ at the minimum equal to 1209 MeV, some 270 MeV above $m_n$. Thus the exchange interaction (32) for light quarks produces the required stability.

To go beyond lowest order in $\alpha_s$ it is useful to apply a simple renormalization group argument (reviewed in Baym, 1977, 1978a) which shows that the ground state energy density can be written in the form

$$E(n_b) = p_F^4\ \varepsilon(\alpha_s(p_f)) \tag{36}$$

where $\varepsilon$ is a dimensionless function of the running coupling constant [see Eq. (19)] evaluated at momentum scale equal to the Fermi momentum $p_f$ (we assume equal numbers of each flavor present). Up to second order in $\alpha_s$, the limit to which perturbation theory has been carried, the energy of the system, for $p_f >> m_i$, is

$$E = \frac{3}{4}\frac{p_f^4}{\pi^2} N_f \left[1 + \frac{2\alpha_s}{3\pi} + \frac{\alpha_s^2}{3\pi^2}\left(N_f \ln \frac{\alpha_s N_f}{\pi} + 0.02\ N_f + 6.75\right) + \mathcal{O}(\alpha_s^3 \ln \alpha_s)\right] + B \tag{37}$$

(Freedman and McLerran 1977, 1978; Baluni 1978a,b). This result is valid only for $\alpha_s/\pi << 1$.

The lowest order interaction corrections to the free energy density are (Chin, 1978; Kapusta 1979)

$$F_1 = \alpha_s \left[\frac{2\pi}{3} T^4 + \frac{1}{2\pi^3} \sum_i \left(\mu_{oi}^4 + 2\pi^2 \mu_{oi}^2 T^2 + \frac{5}{9}\pi^4 T^4\right)\right], \tag{38}$$

in the limit where $m_i$ can be neglected. The first term is the gluon "exchange energy" and the second is the finite temperature generalization of (32); $\mu_{oi}$ is given by (30). To order $\alpha_s$ the free energy density is thus

$$F = F_o + F_1 + B, \tag{39}$$

where we have added in the difference in vacuum energy densities B.

The free energy density of the perturbative vacuum ($\mu_{oi} = 0$) with massless u and d quarks, and entropy density, $S = -\partial F/\partial T$, are to this order

$$F_{vac} = -\frac{37\pi^2}{90} T^4\left(1 - \frac{110}{37\pi}\alpha_s\right) + B, \tag{40}$$

$$S_{vac} = \frac{74\pi^2}{45} T^3\left(1 - \frac{110}{37\pi}\alpha_s\right) . \tag{41}$$

We see that for $\alpha_s > 37\pi/110$ the entropy density becomes negative, indicating a breakdown in the expansion. (The MIT bag model coupling constant $\alpha_s = 2.2$ well exceeds this value.) Whether such a change of sign indicates real physics, or is just an artifact of the perturbation expansion, cannot be answered without a careful analysis of higher order terms. In particular, the coupling strength, from asymptotic freedom arguments, should go to zero at high temperatures as $1/\ell n(T/\Lambda)$, as in Eq. (19), so the high temperature limit is well behaved.

### 5.3.3 Deconfinement Transition

In the absence of a solution to the confinement problem, estimates of the density at which the transition to quark matter might occur, at given temperature, have so far been based on several ad hoc approaches. The simplest, the geometric argument mentioned at the beginning of this section, is to say that the transition from deconfinement to confinement occurs when the density of quarks plus antiquarks is on the order of the density of quarks, $3n_H$, inside a nucleon; an alternative criterion is when the total energy density falls to a critical value. A second approach is based on comparison of the ground state energy per baryon -- and at finite T the free energy per baryon -- of the quark and hadron phases, with the phase of lower (free) energy being the preferred ones. A third, using percolation theory, is also geometric in its point of view.

In the first approach (e.g., Kajantie and Miettinen, 1981), one assumes that the transition occurs when

$$\sum_i (n_i + n_{\bar{i}}) = 3n_H \tag{42}$$

with $n_i$ and $n_{\bar{i}}$ given, in the absence of interactions, by (23) with $\pm.\mu$, and $n_H$ is the baryon density inside the nucleon (or other typical hadron). Since the baryon density, $n_b$, is $\sum_i (n_i - n_{\bar{i}})/3$, this condition becomes

$$2\sum_i n_{\bar{i}} = 3(n_H - n_b). \tag{43}$$

One can approximately evaluate the anti-quark density by letting $\mu = 0$ in (23), since at $n_b = 0$ this is correct, while at the large $n_b$ end of the deconfinement curve, $T \to 0$ and $n_{\bar{i}} = 0$ anyway. Thus using (25) for $n_{\bar{i}}$ we have

$$\begin{aligned} T_c(n_b) &= \left[\pi^2(n_H - n_b)/3\zeta(3)N_f\right]^{1/3} \\ &= 276\ (n_H - n_b)^{1/3} N_f^{-1/3} (\text{MeV}) \end{aligned} \tag{44}$$

where $N_f$ is the number of flavors of light quarks whose anti-particles are present (2 for $T_c \ll m_s$, and 3 for $T_c \gg m_s$), and in the second equation the densities are measured in $fm^{-3}$. At $n_b = 0$, (44) gives

$$T_c(0) = \frac{171 MeV}{N_f^{1/3} r_{n,fm}} , \tag{45}$$

a value in qualitative agreement with the earlier estimate based on the density of thermal pions present, as well as estimates based on Monte Carlo lattice calculations.

The disadvantage of this estimate is that it does not take into account effects of thermal gluons on the transition; indeed at $T_c(0)$, with $n_b = 0$, on the order of half the energy of the system is in thermal gluons. A way of including effects of thermal gluons is to rephrase the transition criterion, as we shall do below, in terms of the energy density, rather than quark density in the system. But before doing this, it is useful to consider estimates of the transition based on comparison of energies per baryon in the confined and deconfined phases.

A typical comparison of the ground state energies of quark matter versus nuclear matter is given in Fig. 3. This drawing shows the energy per baryon versus $n_b^{-1}$, the volume per baryon, for several calculations of quark matter with $n_u = n_d$, $n_s = 0$ -- the "bag model," Eq. (34); Eq. (37) with the parameter $\Lambda$ in the running coupling constant, (19), equal to 180 MeV and 270 MeV; and the $\Lambda = 0$ "free quark" calculation, (22) -- as well as the energy per baryon of normal symmetric matter, calculated in the Chin and Walecka (1974) mean field model, a representative calculation of high density matter in the nucleon phase (shown as the dotted curve). Without an adequate guide to the value of B to use for scale parameter $\Lambda$, I have plotted the three curves $\Lambda = 0$, 180 and 270 MeV with B = 0; these curves are lower bounds to the energy (37).

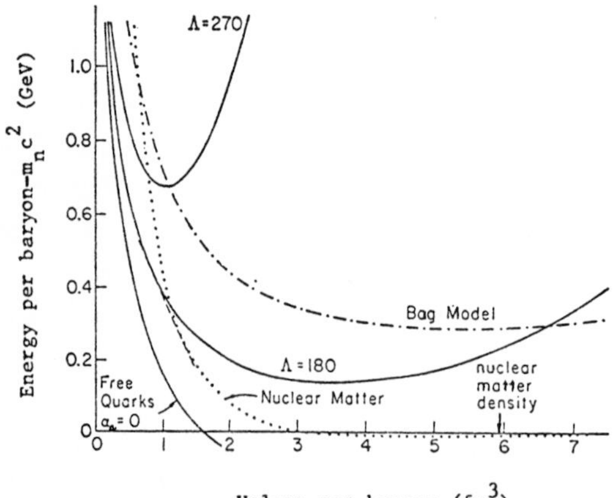


Fig. 3

We see in each case a first order phase transition from the nucleon phase to the quark phase. For example, the "bag model" curve lies above the normal matter curve until a density $\sim 9\ n_o$, above which the quark phase has lower energy. [The density discontinuity at the phase transition is determined by making the standard "double tangent" construction between the two energy curves, illustrated in Fig. 3 as the dashed line bitangent to the normal curve and that labelled $\Lambda = 180$.] Present estimates of $\Lambda$ are in the range $\sim 100 - 500$ MeV; a fit to the bag model value $\alpha_s(p_f) = 2.2$, for density $n_H \sim 0.24\ fm^{-3}$, implies a $\Lambda \sim 220$ MeV. The range of results is large. The transition density in the nuclear phase, $\gtrsim 6\ n_o$ for $\Lambda > 270$ MeV, decreases with decreasing $\Lambda$, falling to about $4\ n_o$ for $\Lambda = 180$ MeV; including a non-zero B in (37) will raise the transition densities. One should note that quark matter becomes unstable to the right of the minima in Fig. 3, where the pressure is negative. [Since $\alpha_s/\pi \lesssim 1$ at the minima, one should not take the perturbative result for the location of the minima too seriously.]

One can, in the same spirit, estimate the location of the transition at finite temperature by comparing the free energies of the quark and hadronic phases. Sample calculations, given by Chin (1978) and Kapusta (1979), indicate a transition curve starting at $T_c \gtrsim 200$ MeV at $n_b = 0$. The details of the transition curve, which one would like in order to estimate whether quark matter can be made in heavy ion collisions, are however quite sensitive to the description of hot nuclear matter employed. In particular, softening of pion degrees of freedom in nuclear matter (Mishustin et al., 1980; Friedman et al., 1981), which can lead toward pion condensation, raises the entropy (and generally the free energy) for given excitation energy; also as Siemens has suggested, if pionic degrees of freedom are more easily excited, the number of quark - anti-quark pairs present at given excitation energy is increased, and the threshold for the transition to quark matter should be correspondingly lowered.

We see from Fig. 3 that the typical energy density of quark matter at the phase transition is $\sim 1.3 - 1.8\ GeV/fm^3$. A simple, reasonable, and experimentally useful phenomenological criterion for the finite temperature phase transition is to say that in quark matter at finite T, confinement takes place when the energy density falls to a certain critical value $E_c$. Such a criterion is likely to be applicable even if a system does not reach complete thermal equilibrium. Then from Eq. (39), taking into account only massless u and d quarks and their antiparticles, we find a transition curve of the general form shown in Fig. 2, with

$$n_c = \frac{(0.94\ fm^{-3})E_{GeV}^{3/4}}{[1+2\alpha_s(p_{fc})/3\pi]^{3/4}} \tag{46}$$

at T = 0, and

$$T_c = \frac{(159\text{MeV})E_{\text{GeV}}^{1/4}}{[1-110\alpha_s(T_c)/37\pi]^{1/4}} \tag{47}$$

at $n_b = 0$, where $E_{GeV}$ is the energy density $E_c$ - B measured in GeV/fm$^3$, and the running coupling constants are evaluated for the appropriate momentum scale. The value (47) for $T_c$ agrees well with other phenomenological estimates.

The comparison of free energies of the hadronic and quark phases generally predicts a first order phase transition. However, in such calculations the descriptions of the two phases start from quite inequivalent bases, so that even if the transition is in fact second order, as the $n_b = 0$ lattice calculations indicate, the free energy calculations would not be expected to show this. Basic to determining the nature of the transition is the identification of an order parameter. As we discussed earlier a likely candidate is the mode of realization of chiral symmetry; the phase transition with increasing density or temperature for this case would be the restoration of non-broken chiral symmetry. [A more thorough discussion of the relations of the chiral and deconfinement transitions, and their order, is given by Pisarski (1981).] One should note though that since small residual masses of the light quarks in the deconfined phase produce a small violation of exact chiral symmetry, the transition from confined to deconfined matter may not be a sharp phase transition but could be slightly smeared out.

Another useful approach to the phase transition is through percolation theory (Baym, 1978, 1981; Çelik et al., 1980). Let us think of a nucleon as a bag (in the sense of the MIT bag model) confining quarks, and ask at what density $n_p$ the nucleon bags overlap sufficiently that they percolate and form an infinite cluster over space, that is, that a quark can move from one side of the system to the other without ever leaving bag vacuum. [We assume, as seems most reasonable from our earlier discussion of the two states of the vacuum, that bags, unlike grapes, have no skin, and merge when in contact into larger bags.] Quarks in this cluster are no longer confined to localized regions of space, but are confined only within the infinite cluster. For a nucleon radius $r_n \approx 1$ fm, we can estimate from percolation theory that $n_p \sim 0.34n_H \approx 0.5n_o$ and hence that a system of nucleons at nuclear matter density $n_o$ is already percolated! As the assumed bag radius decreases the percolation density $n_p$ increases; for $r_n \sim 0.3 - 0.4$ fm, the percolation transition would be at $n_p \gtrsim 3 - 8\ n_o$, which is in the same range as the transition density estimated from comparison of energies of the nucleon and quark phases (Fig. 3).

## 5.4 Production and Detection of Quark Matter

We come now to the difficult question of whether quark matter can be produced, and if so how might it be detected. The most direct possibility is the temporary formation of quark matter in nuclear, and especially heavy ion collisions. A second possibility is the existence of long-lived exotic nuclear objects composed of quark matter, which might be produced either in accelerators or cosmically. Let us first consider collisions, and then discuss some of the suggested exotica.

In an energetic central heavy ion collision nuclear matter will be compressed and heated. The principal questions are: to what extent is the incident energy in the center-of-mass frame deposited as internal excitation energy; to what extent is the deposited energy thermalized; by how much is the matter compressed; does the matter cross the deconfinement curve (Fig. 2), possibly under non-equilibrium conditions; and, finally, what are the signatures of quark matter formation?

To develop some feeling for the possible quark matter attainable in collisions, let us consider a central collision of two identical nuclei each of baryon number A, at total center-of-mass energy $E^*$ per nucleon. The simplest assumption on the dynamics is that the entire incident energy $2AE^*$ is thermalized in the center-of-mass frame. If we also make the simplifying assumption that the nuclei, each Lorentz contracted along the central axis by the factor $\gamma = E^*/m_n$, where $m_n$ is the nucleon mass, completely overlap in the collision without distortion, then the collision volume is

$$V = A/n_o\gamma \tag{48}$$

where $n_o$ is nuclear matter density. The resulting baryon density is

$$n = 2\gamma n_o \tag{49}$$

and the energy density is

$$E = 2\gamma^2 m_n n_o = 0.32\gamma^2 \mathrm{GeV/fm}^3, \tag{50}$$

which, from our earlier estimates, would be adequate to deconfine the quarks for $\gamma > 2.5$. Let us assume that the system is transformed into quark matter with u, d, and s quarks and their anti-particles equally present (we neglect the net excess of quarks over anti-quarks). To compute the temperature of the system we equate (50) with (26), and find

$$T = \left(\frac{18}{19\pi^3 r_o^3 m_n}\right)^{1/4} (E^*)^{1/2} = 0.126 \frac{(E^*_{\mathrm{GeV}})^{1/2}}{(r_{o,\mathrm{fm}})^{3/4}} \mathrm{GeV}, \tag{51}$$

where $4\pi r_o^3/3 = n_o^{-1}$. The total number of quarks plus anti-quarks present per baryon is

$$N_q + N_{\bar{q}} = 6N_i = 13.3\, r_{o,fm}^{3/2}\, T_{GeV} \; . \tag{52}$$

[These results with A = 1 and $r_o$ the nucleon radius have been used to describe the quark-gluon system that can be produced in central pp and p$\bar{p}$ collisions (Mann and Primakoff, 1980).]

The above assumptions substantially overestimate both the energy deposited in internal motion as well as the amount of nuclear compression. Complete thermalization can only occur if the nuclei completely stop each other, whereas in a realistic energetic central collision the two nuclei in a sense pass through each other, each depositing in the other only a fraction of its incident energy; the nuclear fragments emerge from the collision compressed and heated, but with considerable momentum. Anishetty, Koehler and McLerran (1980), using experimental pp inclusive scattering distributions have carried out a more precise estimate of the energy deposition, which points very favorably to the possibility, in very energetic head-on collisions ($E^* \sim 50$ GeV), of production of quark matter in thermal equilibrium. Let us briefly review their arguments.

To estimate the energy deposited by a given projectile nucleon in the target nucleus, Anishetty et al. select out from all final projectile fragments (typically pions) of individual nucleon-nucleon collisions only those that are slow enough, in the target's initial rest frame, to satisfy two conditions. The first is that the fragments move away from the initial projectile nucleon "beam" region before leaving the target nucleus, so that they can no longer be considered part of the incident beam; the second is that the fragments hadronize, and because of the relatively short mean free paths of slow hadrons lose energy, inside the target nucleus.

The net distance that a projectile fragment moves away from the incident beam before leaving the nucleus is $R_\perp \sim R \tan\theta$, where R is the radius of the nucleus, and $\theta$ is the angle the fragment makes with respect to the central axis. The first condition then is that $R_\perp \gtrsim R_o$, the "width" of the projectile beam region, $\sim$ 1 fm; equivalently

$$p_\parallel \lesssim p_\perp R/R_o, \tag{53}$$

where $p_\parallel$ and $p_\perp$ are the longitudinal and transverse components of the fragment's momentum.

To state the second condition, let $t_I$ be the time required, in the fragment's rest frame, for a hadron to form; then in the target's initial rest frame this time is Lorentz dilated to $t_1' \sim (p_\parallel/m)t_1$, where m is the hadron mass. The condition is that $t'$ be less than the time to cross the recoiling target nucleus, $\sim R/c$. {More precisely this time should be extended by the Doppler shift factor $[(1+v/c)/(1-v/c)]^{1/2} = \gamma_f + (\gamma_f^2-1)^{1/2}$ [where v is the recoil velocity of the target, and $\gamma_f = (1-v^2/c^2)^{-1/2}$], and shortened by a similar factor because of the compression of the target nucleus; these two effects tend to cancel.} Thus the second condition becomes

$$p_\parallel \lesssim mR/ct_1. \tag{54}$$

Selecting only those events satisfying both (53) and (54), which clearly underestimates the net energy deposition, they find that in central collisions of uranium on uranium, with incident $E^* \sim 30\text{-}70$ GeV per nucleon, the nuclear fragments are given an excitation energy on the order of 3.5 GeV per nucleon, and the target recoils with $\gamma_f \sim 2$.

The compression of the target can be estimated by noting that during the time $\tau = R/c$ that a projectile nucleon takes to cross the target, the forward edge of the target has recoiled a distance $v\tau$, while the back edge of the target has not begun to recoil. Thus in the initial frame the target is reduced in size in the longitudinal direction from R to $R-v\tau$, i.e., by a factor $1-v/c$. In the rest frame of the recoiling target this distance is a factor $\gamma_f$ greater, so that the longitudinal size becomes $\gamma_f(1 - v/c)$ times the initial size; and the baryon density is increased by the inverse of this factor, i.e.,

$$n = n_o\left[\gamma_f + (\gamma_f^2 - 1)^{1/2}\right] \tag{55}$$

[Compare this result with the crude estimate (49).]

For $\gamma_f \sim 2$, the density is increased to $\sim 3.7n_o$, and the resulting energy density E is $\sim 3.5$ n GeV $\sim 2$ GeV/fm$^3$, increasing very slowly (~20%) with incident energy over the range 30-70 GeV per nucleon. Applying now the criterion for the transition to quark matter that the energy density should exceed a critical value $E_c \sim 1.3 - 1.8$ GeV/fm$^3$, we see that these estimates indicate a strong likelihood that quark matter can, in fact, be made in such collisions.

It is interesting to compare these results for very high energy collisions with those of a Monte Carlo cascade calculation by Gudima and Toneev (1979) done to simulate the time evolution of the central density and "temperature" for central collisions at lower energy. Shown in Fig. 4 are their results for collisions of Ne and U at 2.1 GeV/A energy, and Ar and Ca at 0.5 GeV/A energy. Thermal equilibrium was not established in these simulations, and the temperature shown is rather a measure of the internal excitation energy of the system. The times shown along the trajectories are in units of $10^{-23}$ sec. We see that the

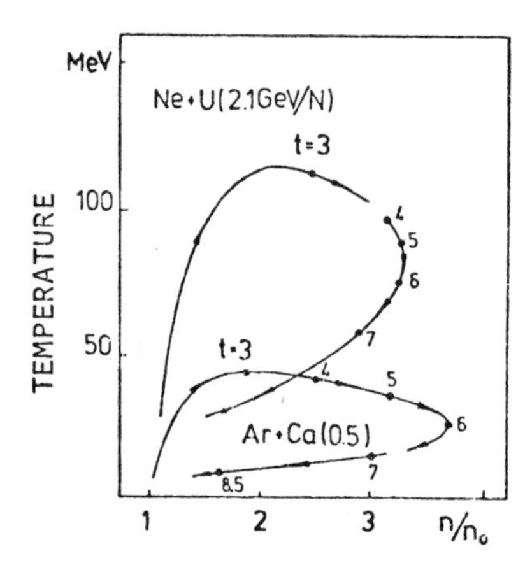


Fig. 4

time spent at high density and "temperature" is only a few times the light travel time across the collision region. While the compression is comparable to that estimated by Anishetty et al. in very high energy collisions, the energy density reached, $\sim m_n n$, is considerably smaller, and consequently so is the likelihood of production of quark matter in collisions in this lower energy range.

Let us now ask what kinds of clues might indicate that quark matter has been formed in a heavy ion collision. Such clues can appear in systematic searches as functions of the incident beam energy and A's of the target and projectile as "phase transitions" in the spectra -- changes in the distributions and multiplicities of particles, change of correlations, and production of unusual final states -- reflecting phase transitions in the colliding matter. [For a general review see Willis (1981).]

One of the most important ways in which quark matter differs from nuclear matter is the difference in the number of degrees of freedom that are excited at various excitation energies. Thus a measurement of the number of degrees of freedom excited can provide a direct clue to the production of quark matter. In condensed matter physics such a measurement is done by determining the specific heat, $C = T\partial S/\partial T$, and hence entropy, S, of a sample as a function of temperature T. The entropy is a direct measure of the number of degrees of freedom excited, while the excitation energy, E*, is related to T and S by $dE^* = TdS$. As Siemens and Kapusta (1979) pointed out an analogous measurement can, in fact, be carried out in heavy ion collisions.

To see their argument in its simplest form, assume that the heavy ion collision is described by a high temperature fireball, consisting in its later stages primarily of a classical gas of nucleons. The entropy per nucleon of the gas is given by

$$S = \frac{5}{2} - \frac{\mu}{T} \tag{56}$$

where $\mu$ is the nucleon chemical potential; the number of protons (and neutrons) is

$$N_p = 2V \int \frac{d^3p}{(2\pi)^3} e^{(\mu - p^2/2m_n)/T} = 2V \left(\frac{m_n T}{2\pi}\right)^{3/2} e^{\mu/T}, \tag{57}$$

where V is the volume of the fireball. In such a gas there will be a small number of deuterons given by

$$N_d = 3V \int \frac{d^3p}{(2\pi)^3} e^{(2\mu - p^2/2m_d)/T} = 3V \left(\frac{m_n T}{\pi}\right)^{3/2} e^{2\mu/T}, \tag{58}$$

where 3 is the spin degeneracy, and we have neglected the binding energy of the deuteron. Taking the ratio of (57) to (58) and solving for $\mu/T$ we find the simple relation

$$S = \frac{5}{2} + \ell n\,(3\sqrt{2}) - \ell n(N_d/N_p) = 3.95 - \ell n\,(N_d/N_p). \tag{59}$$

In other words, the ratio of deuterons to protons in the fireball directly measures its entropy per nucleon. If the expansion of the fireball from its early moments, when possibly exotic states of matter are present, to this later nucleonic stage is adiabatic then the ratio $N_d/N_p$ implies the value of the initial entropy. Furthermore, Siemens and Kapusta show that one can, in fact, use the observed deuteron-proton ratio of the emitted particles in (59) to determine the entropy. The crucial point of this argument is that by a proper analysis of the spectra of emitted particles, one can recover information on the entropy and hence number of degrees of freedom excited in the collision.

We can see the difference in the number of degrees of freedom excited in quark matter, versus nuclear matter, by computing the entropy of the two phases as a function of excitation energy (Mishustin et al., 1980). In the low temperature degenerate regime the entropy per particle of a Fermi liquid is

$$S/\text{particle} = 3m^*T/p_f^2 \tag{60}$$

where for nucleons $m^*$ is the effective mass $m_n^*$ at the Fermi surface; and for quarks

$$m_Q^* = \frac{p_f}{1-2\alpha_s/3\pi}, \tag{61}$$

with only single gluon exchange taken into account (Baym and Chin, 1976b). Similarly the excitation energy per particle is given by

$$E^*/\text{particle} = 3m^*T^2/2p_f = \frac{p_f^2}{6m^*}\,(S/\text{particle})^2. \tag{62}$$

Thus, measuring quantities per baryon, we have

$$S_N = (6m_n^* E^*)^{1/2} / p_f \tag{63}$$

for nucleon matter, and

$$S_Q = [18 m_Q^* (E^* - (E_Q - E_N))]^{1/2} / p_f \tag{64}$$

for quark matter, where here $E^*$ is the experimental excitation energy compared with the nucleon matter ground state energy $E_N$, and $E_Q$ is the ground state energy of quark matter. For symmetric nuclear and quark matter at the same density, $p_f$ is the same in both (63) and (64). At low $E^*$ $(>E_Q-E_N)$ the nucleon phase has the higher entropy, since much of the incident energy must be used to transform nucleon matter to quark matter, but at higher $E^*$:

$$E^* > \frac{E_Q - E_N}{1 - m_n^*/3m_Q^*}, \tag{65}$$

the quark phase has the higher entropy, and is thus the thermodynamically preferred state. Note that the extra color degrees of freedom of quark matter enhance its entropy by the factor $\sqrt{3}$ in (64). While the low temperature thermal equilibrium results (63) and (64) are not particularly applicable to high excitation energy fireballs, and while one needs a detailed dynamical model to determine the density of matter reached for a given excitation energy, these results do illustrate how a measurement of the final entropy can serve to indicate the nature of the matter produced in a heavy ion collision.

More direct information on the formation of quark matter can be obtained by studying leptons and photons emitted in the collision, since, unlike emitted hadrons which undergo substantial cooking in the fireball, these directly reflect conditions such as densities of states and sizes of the emitting region in the collision interior at their time of production. Domokos and Goldman (1981), and Kajantie and Miettinen (1981) show how measurements of massive lepton-pair production can signal the formation of quark matter, and allow one to determine its temperature. In addition two-photon (as well as two-pion) interferometry promises to provide an important handle on the emission region (Willis, 1981).

A further possibility is formation of metastable exotic objects composed of quark matter. Imagine for example that a large fireball of quark matter containing u, d and s quarks and their anti-particles is produced in a collision. Because of the excess of u and d quarks over $\bar{u}$ and $\bar{d}$, such a system would preferentially emit $K^+$ (= $u\bar{s}$) and $K^o$ (= $d\bar{s}$) mesons over $K^-$ and $\bar{K}^o$ (a possible signal of quark matter production), leaving the system with a net strange quark excess. As Chin and Kerman (1979) pointed out, the exchange interaction among the strange quarks, Eq. (33), can, if their density is not too high, make this high strangeness system metastable. [By contrast, the exchange energy of the light quarks is repulsive and has the effect of stabilizing nuclear matter against collapse into quark matter.]

To see this effect, let us examine the energy of a large droplet of quark matter in which the numbers of u and d quarks are equal (to minimize the symmetry energy) and in which a fraction $f_s/3$ of the quarks are strange; $f_s$ is minus the strangeness per baryon. If we neglect the surface and Coulomb energies, the energy per unit volume is

$$E = B + \frac{3}{2\pi^2} p_u^4 \left(1 + \frac{2\alpha_s}{3\pi}\right) + E_o(p_s) - \frac{\alpha_s p_s^4}{\pi^2}, \tag{66}$$

where $p_u$ and $p_s$ are the up and strange quark Fermi momenta, and $E_o(p_s)$ is the free strange particle energy density; in the final term, the strange quark exchange energy, we have assumed $p_s \lesssim m_s$ where $m_s$ is the strange quark mass (~ 280 MeV). The baryon density is

$$n_b = \left(2p_u^3 + p_s^3\right)/3\pi^2, \tag{67}$$

and the quark chemical potentials $\mu_i = \partial E/\partial n_i$ are

$$\mu_u = \mu_d = p_u\left(1 + 2\alpha_s/3\pi\right)$$

$$\mu_s = \left(p_s^2 + m_s^2\right)^{1/2} - \frac{4\alpha_s}{3\pi} p_s. \tag{68}$$

The energy density of non-strange quark matter ($p_s = 0$) can be lowered by converting u and d to strange quarks whenever $\mu_u > m_s$, or $n_b \gtrsim 0.18\ \mathrm{fm}^{-3}$ (using $\alpha_s = 2.2$). Thus as $f_s$ increases from zero, the minimum in the energy per baryon, as shown in Fig. 3, is lowered and shifted to higher $n_b$. A graph of the minimum energy as a function of $f_s$ is shown in Fig. 5, from Chin and Kerman, where the numbers in parentheses along the curve indicate the baryon density. We also see in this plot that the energy per baryon is minimized at $f_s \approx 1.8$; this minimum is determined by the condition that

$$\mu_u = \mu_s = \mu/3 \tag{69}$$

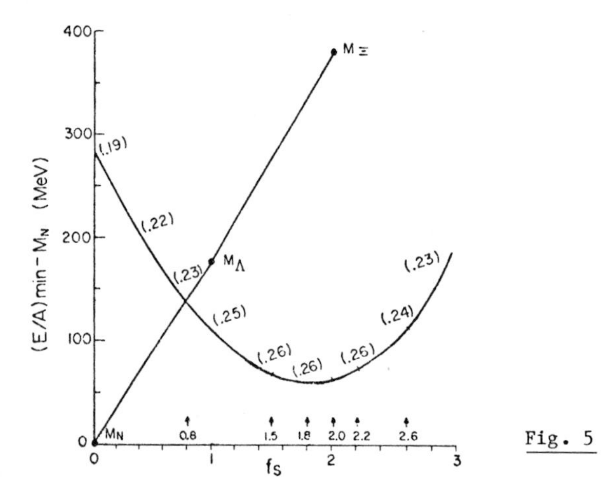


Fig. 5

where $\mu$ is the baryon chemical potential. Furthermore, since the states whose energies are shown in Fig. 5 themselves minimize the energy per baryon at fixed $f_s$, they have zero pressure:

$$P = - n_b^2 \frac{\partial}{\partial n_b} \frac{E}{n_b} = 0 \ . \tag{70}$$

Since $E = - P + n_u\mu_u + n_d\mu_d + n_s\mu_s$, the energy per baryon at the minimum in Fig. 5 is simply $\mu$:

$$E/n_b = \mu = 3p_u(1 + 2\alpha_s/3\pi) = 3(p_s^2 + m_s^2)^{1/2} - \frac{4\alpha_s}{\pi} p_s . \tag{71}$$

Conditions (70) and (71), with $B = 56\ \text{MeV/fm}^3$ and $\alpha_c = 2.2$, fix the parameters of the overall minimum to be:

$$f_s \approx 1.8 \ , \quad n_b \approx 0.26\ \text{fm}^{-3}, \quad \mu \approx 1.00\ \text{GeV}. \tag{72}$$

Also the charge per baryon $Z/A = (1 - f_s)/2$ is $\sim -0.4$. Since as we see in Fig. 5, the energy per baryon at the minimum is less than the rest mass of the $\Lambda$ or $\Xi$, a large glob of quark matter at the minimum (72) would be stable against emission of $\Lambda$'s or $\Xi$'s. [Conversely, a gas of $\Lambda$'s would be unstable under collapse into quark matter with $f_s = 1$.]

Such a "hyperstrange multiquark droplet" is unstable against processes leading to nucleon emission; however these processes increase $f_s$. As Chin and Kerman argue, the primary decay mechanism is strange quark beta decay, $s \to u + e^- + \bar{\nu}$, with an estimated lifetime $\sim 0.7 \times 10^{-4}$ sec. for $\Delta S = 1$.

The strange droplets described above are stabilized by the attractive exchange interactions among the strange quarks. While generally large quark droplets are unstable against breakup into single nucleons, several additional mechanisms, such as finite size effects (Bjorken and McLerran, 1979) could make quark droplets metastable. Smaller quark droplets may very well show quark shell structure tending to make certain configurations more stable; suggested have been quark $^6$Li, which could appear as a resonance in d + $\alpha$ scattering (Chapline and Kerman, 1978), and the $n_u = n_d = n_s = 6$ quark "nucleus" (Mann and Primakoff, 1980).

De Rújula, Giles and Jaffe (1978) have pointed out an interesting mechanism of binding of quark droplets that could occur were confinement of quarks not exact [as the Stanford experiments on niobium spheres suggest (LaRue et al., 1981)]. An unconfined quark would have a large effective radius and mass, given, in a picture in which imperfect confinement is described by a small but non-zero dielectric constant of the normal vacuum $\varepsilon$, by improved versions of Eqs. (9) and (10); it would also have a long ranged attractive dipolar interaction with nucleons, a van der Waals force, due to a nucleon and the quark successively interchanging a pair of gluons, with the nucleon becoming a color octet in the intermediate state. Through this attraction an unconfined quark in nuclear matter would tend to gather around itself a dense ball of nucleons, forming a quark-nucleon system which would be compressed by the dipolar interaction to higher density, perhaps enough to form quark matter in the interior. Such an object would maintain a net color, and thus would have long ranged interactions with nucleons, which should give rise to larger than geometric cross sections for scattering with nuclei. Possibly these objects are candidates for the anomalously short mean free path nuclear fragments reported by Friedlander et al. (1980) to be produced when emulsion stacks are bombarded with $^{16}$O and $^{56}$Fe ion beams of $\sim$ 2GeV/A energy at the Bevalac.

Large metastable quark droplets have been proposed by several authors (Bjorken and McLerran, 1979; Chin and Kerman, 1979; Mann and Primakoff, 1980) as explanations of the famous Centauro cosmic ray events (Lattes et al., 1980) observed in emulsion by the Brazilian-Japanese collaboration on Mt. Chacaltaya in Bolivia during the last decade. These events, 5 of which have been reported so far, have a primary energy in the range 300-370 TeV, multiplicities of $\sim$ 60 to 90 hadrons with a total rest mass $\sim$ 230 GeV, but they have virtually no $\pi^0$ or e at the initial interaction site. In addition 13 "Mini-Centauro" events have been observed, with energies up to $\sim$ 500 TeV, multiplicities of $\sim$ 5 to 20 hadrons with $\sim$ 35 GeV rest mass, and generally a relatively small energy in e and $\gamma$. The large $\langle p_\perp \rangle \sim 1.7$ GeV of the secondaries observed in Centauro I suggest that the initial object is $\sim$ 3-5 times smaller in radius than an ordinary nucleus. As Bjorken and McLerran (1979) show, a metastable highly compressed droplet of quark matter, produced in the upper atmosphere, which undergoes sudden fragmentation into hadrons on colliding with a nucleus near the detector, provides a plausible model for these events.

Let me conclude by remarking that while considerable theroretical work remains to understand the phase transition from hadronic matter to quark matter, and to understand the dynamics of heavy ion collisions and possibilities of exotic objects, the prospects for producing quark matter in very energetic collisions are very promising. Not only would production and detection of such a basic new state of matter be a remarkable achievement in itself, but by enabling one to study qcd in extended deconfined matter over large distance scales it should reveal fundamental aspects of qcd and confinement unattainable from single hadron physics.

This work has been supported in part by U. S. National Science Foundation Grant DMR78-21069.

References

Anishetty, R., P. Koehler, and L. McLerran (1980). Phys. Rev., D22, 2793.
Baluni, V. (1978a). Phys. Lett., 72B, 381.
Baluni, V. (1978b). Phys. Rev., D17, 2092.
Baym, G., and S. Chin (1976a). Phys. Lett., 62B, 241.
Baym, G., and S. Chin (1976b). Nucl. Phys., A262, 527.
Baym, G. (1977). Neutron Stars and the Properties of Matter at High Density, Nordita, Copenhagen.
Baym, G. (1978a). In R. Balian, and G. Ripka (Eds.) Proc. 1977 Les Houches Summer School, North-Holland, Amsterdam.
Baym, G. (1978b). Physica, 96A, 131.
Baym, G. (1981). In H. Satz (Ed.) Statistical Mechanics of Quarks and Hadrons, North-Holland, Amsterdam.
Bjorken, J. D., and L. D. McLerran (1979). Phys. Rev., D20, 2353.
Brown, G.E., and M. Rho (1979). Phys. Lett., 82B, 177.
Callen, C. G., Jr., R. F. Dashen, and D. J. Gross (1979). Phys. Rev., D19, 1826.
Çelik, T., F. Karsch, and H. Satz (1980). Phys. Lett., 97B, 128.
Chapline, G. F., and A. K. Kerman (1978). MIT report CTP #695.
Chapline, G. F., and M. Nauenberg (1976). Nature, 259, 377.
Chin, S. A. and J. D. Walecka (1974). Phys. Lett., 52B, 24.
Chin, S. A. (1978). Phys. Lett., 78B, 552.
Chin, S. A., and A. K. Kerman (1979). Phys. Rev. Lett., 43, 1292.
Chodos, A., R. L. Jaffe, K. Johnson, C. B. Thorn, and V. F. Weisskopf (1974). Phys. Rev., D9, 3471.
Collins, J. C., and M. J. Perry (1975), Phys. Rev. Lett., 34, 1353.
De Rújula, A., R. C. Giles, and R. L. Jaffe (1978). Phys. Rev., D17, 285.
Domokos, G., and J. I. Goldman (1981). Phys. Rev., D23, 203.
Engels, J., F. Karsch, I. Montvay, and H. Satz (1981). Phys. Lett., 101B, 89.
Freedman, B., and L. McLerran (1977). Phys. Rev., D16, 1130, 1147, 1169.
Freedman, B., and L. McLerran (1978). Phys. Rev., D17, 1109.
Friedlander, E. M., R. W. Gimpel, H. H. Heckman, Y. J. Karant, B. Judek, and E. Ganssauge (1980). Phys. Rev. Lett., 45, 1084.
Friedman, B., V. R. Pandharipande, and Q. Usmani (1981). Nucl. Phys. A, (in press).
Gross, D. J., R. D. Pisarski and L. G. Yaffe (1981), Rev. Mod. Phys., 53, 43.
Gudima, K. K., and V. D. Toneev (1979). JINR-Dubna report E2-12644.
Kajantie, K., and H. I. Miettinen (1981). Z. Phys. C9, 341.
Kapusta, J. I. (1979). Nucl. Phys., B148, 461.
Karsch, F. and H. Satz (1980). Phys. Rev., D22, 480.
Keister, B., and L. Kisslinger (1976). Phys. Lett., 64B, 117.
Kuti, J., J. Polónyi , and K. Szlachányi (1981). Phys. Lett., 98B, 199.

LaRue, G. S., J. D. Phillips, and W. M. Fairbank (1981). Phys. Rev. Lett., 46, 967.
Lattes, C. M. G., Y. Fujimoto, and S. Hasegawa (1980). Phys. Reports, 65, 151.
Lee, T. D. (1980). Trans. N.Y. Acad. Sci., 40, 111.
Lee, T. D. (1981). In H. Satz (Ed.) Statistical Mechanics of Quarks and Hadrons, North-Holland, Amsterdam.
Mann, A. K., and H. Primakoff (1980). Phys. Rev., D22, 1115.
McLerran, L. D., and B. Svetitsky (1981). Phys. Lett., 98B, 195.
Mishustin, I. M., F. Myhrer, and P. J. Siemens (1980). Phys. Lett., 95B, 361.
Pisarski, R.D. (1981). Santa Barbara preprint NSF-ITP-81-92.
Polyakov, A. M. (1978). Phys. Lett., 72B, 477.
Shuryak, E. V. (1980). Phys. Reports, 61, 71.
Siemens, P. J., and J. I. Kapusta (1979). Phys. Rev. Lett., 43, 1486.
Susskind, L. (1979). Phys. Rev., D20, 2610.
't Hooft, G. (1981). In Proc. European Phys. Soc. Conf. on High Energy Physics, Lisbon, July 1981.
Vento, V., M. Rho, E.M. Nyman, J.H. Jun, and G.E. Brown (1980). Nucl. Phys., A345, 413.
Willis, W. J. (1981). In R. Bock and R. Stock (Eds.) Proc. Workshop on Future Heavy Ion Experiments, GSI, Darmstadt; and in Proc. European Phys. Soc. Conf. on High Energy Physics, Lisbon, July 1981 (CERN preprint EP/81-120).

# 6. How Should or Will QCD Influence Nuclear Physics?[1]

A. Faessler

With 6 Figures

## 6.1 Introduction

The organizers of this conference have asked me to make some closing remarks and not give a summary talk. This makes it much easier for the speaker, since it allows him to restrict his remarks to points which he understood. As a topic for these final remarks of this summer school session I have chosen:

How should or will QCD influence nuclear physics and why has the impact of QCD on nuclear physics been so small up until now?

There are three possible reasons why the impact of QCD on nuclear physics was until now relatively small.

a) Nuclear physics lives under infrared slavery. At the small momentum transfers $Q^2$ which we have in nuclear physics, the strong coupling constant $\alpha_s(Q^2)$ is large, and perturbative QCD is not allowed. So nuclear physicists have to wait until non-perturbative methods for solving QCD have been developed. Until now not even the real nature of the $\pi$ meson, which plays such a central role in nuclear physics, is understood.

b) The second reason why QCD has not influenced nuclear physics strongly until now might be a psychological one: We always thought that we were handling a strong force in nuclear physics, and if you handle something strong you think you yourself are important also. But we had to learn from QCD as demonstrated in these lectures by Frank Close and Stan Brodsky that the nucleon-nucleon interaction is only a minor remnant of the colour forces. The left part of Fig.1 shows that the lowest-order QCD diagram of one-gluon exchange for the nucleon-nucleon interaction is zero. Both nucleons are colour neutral, and thus they cannot interact by the colour forces. One cannot couple the colour singlet of a nucleon with the colour octet of the gluon to a colour singlet of the second nucleon. This selection rule corresponds in angular momentum to the fact that an electric dipole gamma-ray transition is forbidden between two angular momentum zero states. If one includes the antisymmetrization between the three quarks

1 Closing remarks of the KfK Summer School, "Quarks and Nuclear Forces", in Bad Liebenzell, September 27 - October 3, 1981

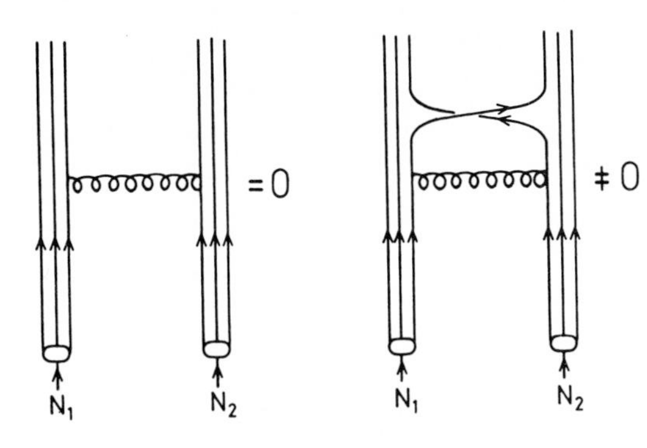


Fig.1. One-gluon exchange diagram between two nucleons. The diagram on the left side without quark exchange is zero due to the colour selectrion rules. The diagram which contributes is on the right. The gluon can also be exchanged between two other quarks in such a way that the colour selection rules are fulfilled

in one nucleon and the three quarks in the other nucleon one gets quark exchange as shown on the right-hand side of Fig.1. This diagram can be different from zero. But if one calculates the nucleon-nucleon interaction using the diagram on the right-hand side of Fig.1 and the other diagrams of the same order with quark exchange, one must keep in mind that such a perturbative expansion of the nucleon-nucleon interaction can only be valid at high momentum transfer. This means that diagrams like the one on the right-hand side of Fig.1 can only describe the extremely short-range part of the nucleon-nucleon interaction where one has high momentum transfer.

c) The third reason why nuclear physicists are so reluctant to take over ideas from QCD may be connected with the emerging new picture of the nucleons: In the fifties and the sixties we described nuclei as a sum of pointlike protons and neutrons which interact with each other by a nucleon-nucleon potential. In the seventies we learned that we also have a large number of mesons inside the nucleus and that a nucleon can be excited into a $\Delta$ resonance. These facts can be seen at higher momentum transfer and better energy resolution with a new generation of accelerators. Will the eighties be the time when we picture the nucleus as indicated in Fig.2? The nucleon bags with three quarks in them fill 60% of the space in the nucleus if we assume the MIT bag radius [1,2] of R = 1 fm. We therefore expect a bag jam in the nucleus. How can the shell model work in such a picture of the nucleus? The nuclear physicists have already started to fight back, and BROWN and RHO [3] invented the little bag with a much smaller radius of the order of 0.4 fm. In addition to the vector current of the quarks $\hat{n}_\mu \bar{q}\gamma^\mu q|_S = 0$ they also conserve the axial vector current

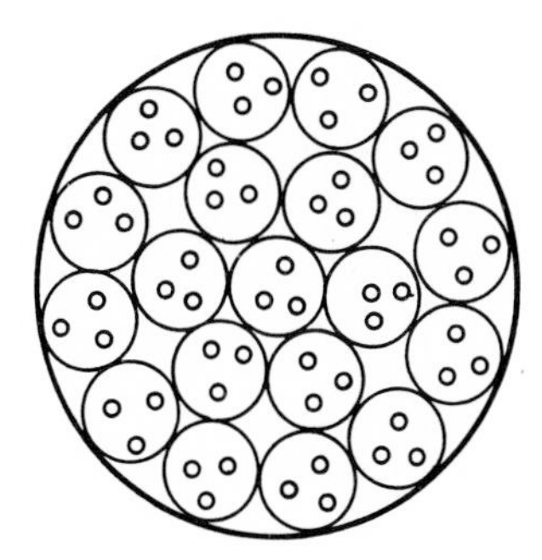

Fig.2. According to the MIT bag model, the nucleons have a radius of 1 fm. They would fill up 60% of the space inside the nucleus. Nuclear physicists are therefore asking: Why does the shell model work?

$\bar{q}\gamma^{\mu}\gamma_5 q|_S = j_{\mu}(\pi)$. Outside of the bag, the axial vector current is carried by the pions. A nuclear physicist finds this picture more sympathetic. On the other hand, we have known since the beginning of the sixties that the root-mean-square radius of the charge distribution of a proton has a value of about 0.8 fm. For the equivalent homogeneous distribution this corresponds to a radius of R = 1 fm. Although we did know that the charge and probably the mass distribution of the nucleon as well are of the extent of the MIT bag radius, we did not worry about the validity of the shell model, since we were able to calculate its validity using many-body methods. We found that a nucleon inside a nucleus polarizes its neighbourhood and moves as a quasi particle with its polarization cloud through the nucleus. This in connection with the Pauli principle guarantees the validity of the shell model. Maybe we will learn in the eighties that a nucleon bag polarizes its neighbourhood and exchanges its identity with neighbouring bags, thus moving as a quasi bag relatively freely through the nucleus.

## 6.2 Quantum Chromodynamics and Nuclear Forces

The lowest-order perturbative diagram for the nucleon-nucleon interaction in quantum chromodynamics is given on the right-hand side of Fig.1. In general, a gluon can be exchanged between any of the quarks and one of the exchanged ones. Since this diagram is in lowest-order perturbative QCD it can only be valid for large momentum transfer. This is only true for the very short-range part of the nucleon-nucleon interaction. Thus we are only able to study the nature of the repulsive core by calculating these diagrams.

QCD-inspired calculations of the short-range part of the nucleon-nucleon interaction have been done in the so-called potential models [4] and in bag models [5]. Since the calculation of the short-range part of the nucleon-nucleon interaction in the bag model was discussed in the lecture by Heller in this school, I will indicate how this is done in the potential models and then turn to the criticism of these calculations, which seems to me quite severe.

In both the potential models [4] and the bag models [5] one starts with the Born-Oppenheimer approximation. The confining potential or the bags of the two nucleons are fixed at two positions with distance r. The six quarks interact by a quark-quark interaction $V_{qq}$. In the bag model, the energy of the gluons is explicitly included, and for the quarks one uses the current quark masses, which are roughly zero for the up and down quarks. In the potential model, the gluon cloud around the quarks is included in an effective so-called constituent quark mass. The expression in the potential model for the nucleon-nucleon interaction is the following:

$$V_{N_1N_2}(r) = \Big\langle A\Big\{ \underbrace{\cup\cdot\cup}_{r} \Big\} \Big| \sum_i t_i + \sum_{i<j} V_{qq}(i,j) \Big| A\Big\{ \underbrace{\cup\cdot\cup}_{r} \Big\} \Big\rangle$$

$$\text{with: } V_{qq}(r_{12}) = \frac{1}{4}\vec{\lambda}_1\cdot\vec{\lambda}_2\left[-ar_{12}^2 + \frac{\alpha_s}{r_{12}} - K\,\vec{\sigma}_1\cdot\vec{\sigma}_2\,\delta(r_{12}) + \ldots\right] \tag{1}$$

The quark-quark interaction contains the colour Fermi-Breit interaction, which is a nonrelativistic approximation to the one-gluon exchange and a confining potential which is added by hand. The confining potential can be linear, logarithmic or quadratic. For the present discussion the detailed form is not important. For simplicity we chose a quadratic potential with a parameter a. LIBERMAN was the first to publish results of such a calculation [4]. He found a repulsion at short distances with a core of heights 350 and 450 MeV for $^3S$ and $^1S$ interactions, respectively. A year later DE TAR obtained a similar result in the MIT bag model [5]. (DE TAR did not remove the relative kinetic energy between the two quark bags from the so calculated interaction energy. CHUNG WA WONG [6] did show that the removal of this spurious kinetic energy reduces the core in the MIT bag model almost to zero.)

Recently NEUDATSCHIN [7] and HARVEY [8] criticized this procedure, as elementary rules of the symmetry group had been violated:

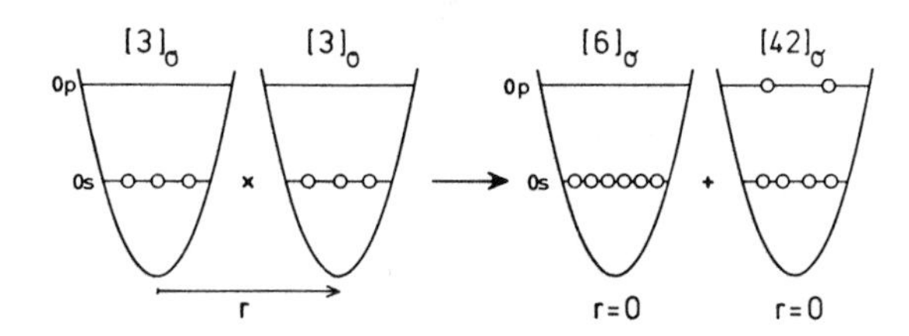


Fig.3. On the left, the two confining potentials for the interacting nucleons are given at a distance r. The spatial wave function for each nucleon is symmetric [3]. At distance zero the spatial symmetry of the six quarks is for an S state either [6] or [42]. Both symmetries have to be taken into account for the calculation of the repulsive core between two nucleons

The left side of Fig.3 shows two nucleons with three quarks confined in an oscillator potential. The spatial symmetry is in each nucleon [3]. If one multiplies two spatial symmetric wave functions with three quarks in each, one obtains the fully symmetric representation [6] and the representation [42] if one restricts oneself to relative S states between the two nucleons. The calculations mentioned above took into account only the fully symmetric spatial representation [6] where all of the six quarks were put into the 0s state. But on the other hand, one obtains automatically for distance $r = 0$ the [42] spatial symmetry also, with two quarks in the 0p state and four quarks in the 0s state. This [42] symmetry is there in the same sense as when one multiplies two spin-1/2 wave functions and obtains a combination of $S = 0$ and $S = 1$ wave functions. One could perhaps argue that the [42] symmetry lies unperturbed at $2\hbar\omega$ and thus plays no important role. But this is not the case. The colour magnetic residual interaction between the quarks lowers the spatial [42] symmetry and pushes the fully symmetric state [6] up, so that both are roughly degenerated. A Born-Oppenheimer calculation in the potential model including this symmetry has been performed by HARVEY [8].

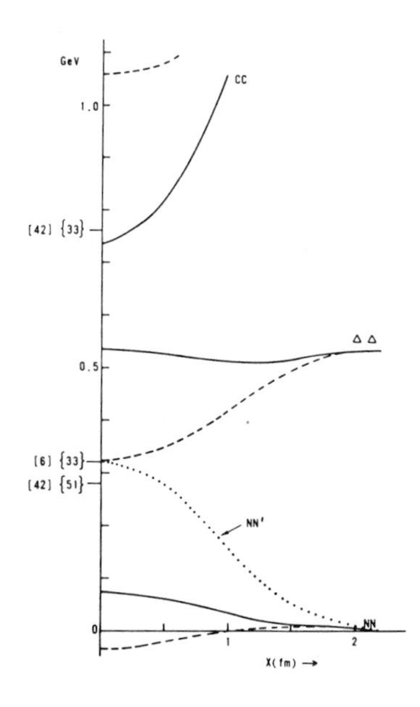


Fig.4. Results of the six-quark cluster calculations by HARVEY [8] for the $^3S$ interaction between two nucleons. The dotted curve for the potential between the two nucleons is calculated without renormalization of the [42] orbital symmetry. That means that at zero distance this symmetry is omitted. The solid line includes this symmetry even at small distances. The dashed line includes the coupling between NN, ΔΔ and the hidden colour (CC) channels

The result is shown in Fig.4. The dotted line is obtained by the usual method in which at distance zero the [42] orbital symmetry drops out. The solid line includes the [42] but does not allow for a polarization of the two nucleons at small distances. The result of the dashed line includes such a polarization by allowing admixtures of two Δ states and two hidden colour states (CC). One sees that in HARVEY's parametrization, the repulsive core, which was originally higher than 300 MeV, turns into an attraction. Details depend naturally on the parametrization. The special choice of the parameters by HARVEY has been criticized [9]. But nevertheless, those groups which criticize him also obtain with the inclusion of the [42] spatial symmetry an appreciable lowering of the repulsive core, so that one can no longer speak of a qualitative explanation of the short-range repulsion between two nucleons.

We thus have to ask once again: What is the nature of the repulsive core? I would like to propose here a mechanism which leads to phase shifts of a repulsive core in the nucleon-nucleon interaction and which is based on the strong admixture of the spatial [42] symmetry. To simplify the argument, let us assume for a moment that we have at zero distance between the two nucleons only the [42] symmetry. The argument works the same way if the fully symmetric state [6] is admixed, but it is easier to understand if we speak for a moment only of the [42] spatial symmetry at zero distance. In such a symmetry two quarks are in the 0p state and four quarks are in the 0s state in the common confining potential for $r = 0$. If the two nucleons are moved apart, the quarks are asymptotic in the two potentials in the 0s-states.
In an oscillator potential energy conservation tells us:
The original states have an excitation energy of $2\hbar\omega$, while the final intrinsic states of the two nucleons have an excitation energy of $0\hbar\omega$. The $2\hbar\omega$ energy therefore has to go into the relative motion. If we consider a relative S-state, then

that means that we have to have a node in the relative wave function. It cannot be an 0s, but has to be a 1s wave function. Outside the interaction range the wave function is naturally oscillating according to the relative kinetic energy of the two nucleons. The phase shifts of these oscillations are determined by the zero in the relative wave function in the interaction range. One could, for example, force such a zero onto a wave function by a hard or a soft core. Thus, we cannot determine asymptotically whether we have a zero of the relative wave function imposed by the [42] spatial symmetry or whether one has a hard or soft core.

To ascertain whether this idea is reasonable, we performed [10] a resonating group calculation [11] for nucleon-nucleon scattering in which we considered the nucleon as a three-quark cluster which is scattered on another three-quark cluster. In such a calculation one does not use the Born-Oppenheimer approximation, but one does take the nonlocality of the nucleon-nucleon interaction fully into account. A technique for solving this problem is well developed in the cluster model for light nuclei [12]. In such a calculation one is not calculating a nucleon-nucleon potential; one directly calculates the phase shifts. This also has the advantage that one does not need to make assumptions about the mass of the two nucleons at small distances as in the two-centre MIT bag model [5] or in the potential model [4]. Results for the $^3S$ phase shifts for a set of parameters are given in Fig.5. Parameters for the oscillator length b = 0.546 fm and the constituent quark mass $m_q$ = 355 MeV, the coupling constant $\alpha_s$ = 1.47 and the confining potential a = 78.5 MeV/fm$^2$ are taken from the work of RIBEIRO [13].

But we have also done calculations for a large number of other parameter sets and obtained qualitatively the same result. This result is that at larger relative energies the phase shift drops like the hard-core phase shift

$$\delta(l=0)_{\text{hard core}} = -r_0 k \propto - r_0\sqrt{E_{rel}} \quad . \tag{2}$$

The calculation has been performed including nucleon-nucleon and $\Delta\Delta$ channels. In this approximation we find a phase shift which corresponds to a hard core of $r_0$ = 0.38 fm. If we include the hidden colour channel also, the phase shift is less repulsive and corresponds to a hard core of $r_0$ = 0.35 fm.

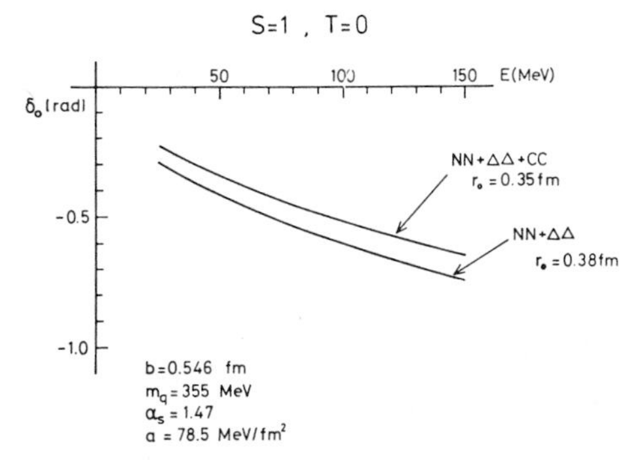


Fig.5. The $^3S$ phase shift is calculated in the potential model using the resonating group method. The curves include the nucleon-nucleon, the $\Delta\Delta$ and the hidden colour channels. The phase shifts which are obtained in this calculation are practically identical to hard-core phase shifts. This identity is due to a zero in the relative wave function which stems from the orbital [42] symmetry and a preference of this symmetry by the colour magnetic interaction. The energy is given in the CM-system

In concluding this chapter we could say that the hard-core phase shifts of the nucleon-nucleon interaction may be due not to a repulsive core but to the importance of the [42] spatial symmetry which stems from the colour magnetic interaction. Although at short distances both symmetries [6] and [42] are present, one still has a zero in the relative wave function between the two nucleons. This zero leads to a phase shift which looks at higher energies like a hard-core phase shift, although such a hard core might not exist. If this explanation of the nature of the hard core is correct, it might also affect the nuclear many-body problem: If the repulsion at short distances is due to a hard core, the relative wave function has to be practically zero and thus contains large admixtures of high relative kinetic energies. If the wave function needs only to have a zero, the relative energy does not need to be so large, since inside the zero the relative wave function does not have to be suppressed. This would lead in the many-body problem to a smaller total kinetic energy and thus to more binding. It could be that the missing binding energy in $^3$H and $^3$He is connected with the nature of the hard core.

## 6.3 Six and Higher Quark Bags in Nuclei

We have already stressed that we need high momentum transfer to apply perturbative QCD to describe the nuclei. Normally in nuclear physics we have quite low momentum transfer at which perturbative QCD is not valid, and thus its application in nuclear physics is restricted. But PIRNER [14] discussed in his seminar a case in which perturbative QCD can be applied successfully in nuclear physics to get information about the percentage of six and higher quark bags in nuclei.

PIRNER examined the inelastic electron scattering data of DAY et al. [15] on $^3$He measured at SLAC. Interesting information is obtained from those data where a very small energy but a large momentum transfer is measured. Such data can only be explained if one has very high momentum components inside the nucleus. They can only be carried by the quarks and not by the nucleons. To describe the data, PIRNER assumed that the nucleons move in a common potential which is represented by an oscillator potential. He then boosted the nucleus into the light-cone reference frame in which each nucleon has practically one-third of the total momentum. The momentum distribution inside the nucleus comes then from the motion of the quarks inside the nucleon. This distribution is taken from the electron-nucleon scattering; the momentum distribution of the quarks

$$f_3(x) = (1 - x)^3 \qquad f_6(x) = (1 - x)^9$$
$$f_n(x) = (1 - x)^{2(n-1)-1} \tag{3}$$

has therefore to be folded with the momentum distribution of the nucleons. Here x is the fraction of the momentum which goes to one quark in a nucleon bag with n quarks. The probability for the distribution is given by $f_n(x)$. If one assumes that only three quark bags exist inside $^3$He, then one cannot explain the relatively high cross section at extremely small energy transfers. But if one assumes that 10%-16% of $^3$He consists of one quark bag with three quarks and another quark bag with six quarks, the data can be explained, since the six-quark bag has two-thirds of the total momentum; thus, by the momentum distribution function of a quark, one quark inside these six-quark bags can with some probability have two-thirds of the total momentum.

In conventional nuclear physics one would have proceeded differently: The nucleons would not have been assumed to move independently of each other in a common potential, but one would have allowed for short-range correlations between them due to the residual nucleon-nucleon interaction. Thus, one would immediately find that one nucleon can have a large part of the total momentum while the other two have only a small part. If we then fold the momentum distribution of the three quark states in a nucleon with the momentum distribution of the nucleons, we can also explain the data, if the momentum distribution of the nucleon contains momentum components which are high enough. But there would still be a difference: In the short-range correlations between the two nucleons, the nucleons still have the quantum numbers of two nucleons. One could explicitly allow them to be changed into two $\Delta$'s. In the six-quark picture all degrees of freedom of six quarks are statistically included. That means 80% of the six-quark bag contains hidden colour states. Thus, the two descriptions are not fully equivalent. Figure 3 suggests an improvement of considerations presented by PIRNER: The hidden colour state which is, in PIRNER's calculation, 80% present in the six-quark bag, is quite highly excited (about 1 GeV) if the residual interaction between the quarks is included. Thus, it seems that a purely statistical consideration overestimates the role of hidden colour.

## 6.4 Static Properties

But QCD has not only dynamic properties which can at the moment only be described in the framework of perturbative QCD. It also contains symmetries which do not need a perturbative treatment to show up. They can be embedded into a simpler quark model. Such symmetry considerations can explain, for example, in rough agreement with the date, the anomalous magnetic moments of the nucleons and the relative coupling strength for different meson-baryon couplings. In the seminars of PIRNER and POVH [17] such properties have been discussed. I would like to stress here a case where perturbative QCD (quark exchange) and nonperturbative QCD (meson exchange) give two totally different results which should be tested in the future.

Figure 6 shows the interaction of a $\Lambda$ or a $\Sigma$ with a nucleus in lowest-order perturbative QCD. Let us consider the spin-orbit part of the $\Lambda$-nucleus and the $\Sigma$-nucleus interaction. If we disregard nucleon-$\Lambda$ and nucleon-$\Sigma$ exchange diagrams, this interaction is due to quark exchange between a nucleon of the nucleus and the hyperon. Since the hyperon contains only an up and a down quark, which can be exchanged, the nucleus-hyperon spin-orbit potential depends on the spin of the up and down quark

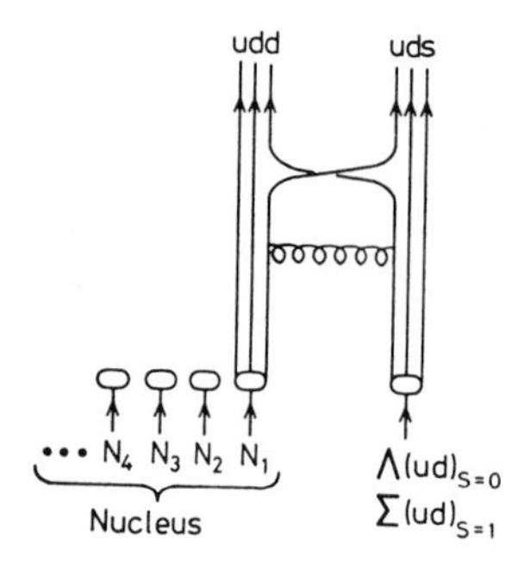


Fig.6. Interaction of a hyperon with a nucleus by one-gluon and quark exchanges. The spin-orbit potential between a Λ and a nucleus should be zero, since the two quarks (up and down) which can be exchanged between a nucleon and a Λ are coupled to angular momentum S = 0. Experimentally, the spin-orbit coupling is very small. This is also in agreement with the meson exchange calculations of BROCKMANN [16]. The spin-orbit potential of a Σ should be four-thirds of the nucleon-nucleus spin-orbit potential, since the up and the down quark coupled to spin S = 1. Thus, we expect the spin-orbit interaction between the up and the down quark and the nucleus to be twice as strong. But since only two quarks can interact, we have to reduce this strength by a factor of two-thirds. Meson exchange calculations by BROCKMANN yield an opposite result: According to these calculations the Σ-nucleus spin-orbit potential should be small. POVH [17] and co-workers are preparing experiments to determine the Σ-nucleus spin-orbit potential experimentally. It will be interesting to see if the meson exchange calculation ("poor man's nonperturbative treatment of quantum chromodynamics") or the quark and gluon exchange calculation is correct

in the hyperon. Since u, d is coupled to S = 0 for the Λ we have no spin-orbit interaction for the Λ-nucleus potential. This is in agreement also with meson exchange calculations by BROCKMANN [16] and the experimental data of POVH [17]. The Σ-nucleus spin-orbit potential is determined in lowest-order perturbative QCD by the fact that the up and down quarks are coupled to S = 1 in the Σ. Thus, we expect that the spin-orbit potential is reduced by a factor of two-thirds, since only two of the three quarks in the Σ can be exchanged. Thus, one expects that the Σ-nucleus spin-orbit potential is about four-thirds of the nucleon-nucleus spin-orbit potential. But on the other hand, BROCKMANN [16] obtains by meson exchange calculations (which is the poor man's nonperturbative QCD treatment) an extremely small spin-orbit potential. Experiments to determine the Σ spin-orbit potential are being prepared by POVH and co-workers. It will be interesting to see if lowest-order perturbative QCD or the meson exchange calculations are correct.

## 6.5 Conclusions

As I already stressed at the beginning of these closing remarks, nuclear physics lives in infrared slavery. Thus, we cannot apply perturbative QCD, since at the small momentum transfers in nuclear physics the coupling constant is large. Since nonperturbative methods for QCD have practically not been developed, possible applications of QCD to nuclear physics are restricted.

QCD can be applied using perturbative methods to high momentum transfer phenomena in nuclear physics. An example of such phenomena is high-energy electron-nucleus scattering, where one can study the probability of there being six or nine quark bags in nuclei. Another application might be relativistic and extremely relativistic heavy-ion collisions, where one could expect to obtain quark matter as discussed in the lectures of GORDON BAYM [18]. In extremely relativistic heavy-ion collisions one also could hope to find large areas of overlapping colour strings. These might lead to new collective phenomena which cannot be seen in the small colour strings obtained in collisions between elementary particles. Another possible application for perturbative QCD is the short-range part of the nucleon-nucleon interaction. The nature of the repulsive core can be inquired into.

A second application of QCD to nuclear physics is connected with the symmetries contained in the theory. Those symmetries are also largely contained in the naive quark model. Facts which can be studied with these symmetries are, for example, the anomalous magnetic moments of the baryons and the relative strength of different meson-baryon coupling and of different baryon nucleus potentials. We saw that perturbative QCD gives a strong spin-orbit potential for the $\Sigma$-nucleus interaction, whil meson exchange yields a very weak one.

Let me come back to the question of the impact which quantum chromodynamics will have on nuclear physics. At the moment and in the near future the impact will definitely be limited, since there are no nonperturbative methods available to solve QCD for the momentum transfers we generally have in nuclear physics. But on the other hand there is a relative new and interesting field of high momentum transfer phenomena in nuclear physics in which QCD will change the theoretical description.

## References


1 A. Chodos, R.L. Jaffe, K. Johnson, C.B. Thorn, V.S. Weisskopf: Phys. Rev. D*9*, 3471 (1974);
A. Chodos, R.L. Jaffe, K. Johnson, C.B. Thorn: Phys. Rev. D*10*, 2599 (1974)
2 T. de Grand, R.L. Jaffe, K. Johnson, J. Kiskis: Phys. Rev. D*12*, 2060 (1975)
3 G.E. Brown, M. Rho: Phys. Lett. B*82*, 177 (1979);
G.E. Brown, M. Rho, V. Vento: Phys. Lett. B*97*, 423 (1980)
4 D. Liberman: Phys. Rev. D*16*, 1542 (1977)
5 C. de Tar: Phys. Rev. D*17*, 323 (1978)
6 C.W. Wong, K.F. Liu: "Nucleon-Nucleon Interactions", in *Topics in Nuclear Physics I*, ed. by T.T.S. Kuo, S.S.M. Wong, Lecture Notes in Physics, Vol.144 (Springer, Berlin, Heidelberg, New York 1981) p.1
7 I.T. Obukhovsky, V.G. Neudatchin, Yu.F. Smirnov, Yu.M. Tchuvil'sky: Phys. Rev. Lett. B*88*, 231 (1979)
8 M. Harvey: Nucl. Phys. A*352*, 301, 326 (1981)
9 I. Bender, H.G. Dosch: Heidelberg reprint 1981
10 A. Faessler, F. Fernandez, G. Lübeck, K. Shimizu: to be published
11 M. Oka, K. Yazaki: Phys. Lett. B*90*, 41 (1980)
12 K. Wildermuth, Y.C. Tang: *A Unified Theory of the Nucleus* (Vieweg, Braunschweig 1977)
13 J.E.F.T. Ribeiro: Z. Phys. C*5*, 27 (1980)
14 Pirner: Seminar at this Conference.
15 D. Day , J.S. McCarthy, I. Sick, R.G. Arnold, B.T. Chertok, S. Rock, Z.M. Zalata, F. Martin, B.A. Mecking, G. Tamas: Phys. Rev. Lett. *43*, 1143 (1979)
16 R. Brockmann, W. Weise: Phys. Lett. B*69*, 167 (1977); Nucl. Phys. A*355*, 365 (1981); private communications
17 B. Povh: Seminar at this Conference
18 G. Baym: Seminar at this Conference

# Classified Index

## Springer Tracts in Modern Physics, Volumes 36–100

This cumulative index is based upon the Physics and Astronomy Classification Scheme (PACS) developed by the American Institute of Physics

## General

### 04 Relativity and Gravitation



### 05 Statistical Physics



### 07 Specific Instrumentation



## The Physics of Elementary Particles and Fields

### 11 General Theory of Fields and Particles



### 11.30 Symmetry and Conservation Laws



### 11.40 Currents and Their Properties

*Verzegnassi, C.:* Low Energy Photo- and Electroproduction, Multipole Analysis by Current Algebra Commutators (Vol. 59)
*Weinstein, M.:* Chiral Symmetry. An Approach to the Study of the Strong Interactions (Vol. 60)

**12 Specific Theories and Interaction Models**

*Amaldi, E., Fubini, S.P., Furlan, G.:* Electroproduction at Low Energy and Hadron Form Factors (Vol. 83)
*Hofman, W.:* Jets of Hadrons (Vol. 90)
*Wiik, B.H., Wolf, G.:* Electron-Positron Interactions (Vol. 86)

**12.20 Quantum Electrodynamics**

*Källén, G.:* Radiative Corrections in Elementary Particle Physics (Vol. 46)
*Olsen, H.A.:* Applications of Quantum Electrodynamics (Vol. 44)

**12.30 Models of Weak Interactions**

*Barut, A.O.:* On the S-Matrix Theory of Weak Interactions (Vol. 53)
*Dosch, H.G.:* The Decays of the $K_0$-$K_0$ System (Vol. 52)
*Gasiorowicz, S.:* A Survey of the Weak Interactions (Vol. 52)
*Gatto, R.:* Cabibbo Angle and $SU_2 \times SU_2$ Breaking (Vol. 53)
*von Gehlen, G.:* Weak Interactions at High Energies (Vol. 53)
*Kabir, P.K.:* Questions Raised by CP-Nonconservation (Vol. 52)
*Kummer, W.:* Relations for Semileptonic Weak Interactions Involving Photons (Vol. 52)
*Müller, V.F.:* Semileptonic Decays (Vol. 52)
*Paul, E.:* Status of Interference Experiments with Neutral Kaons (Vol. 79)
*Pietschmann, H.:* Weak Interactions at Small Distances (Vol. 52)
*Primakoff, H.:* Weak Interactions in Nuclear Physics (Vol. 53)
Renner, B.: Current Algebra and Weak Interactions (Vol. 52)
*Riazuddin:* Radiative Corrections to Weak Decays Involving Leptons (Vol. 52)
*Rothleitner, J.:* Radiative Corrections to Weak Interactions (Vol. 52)
*Segre, G.:* Unconventional Models of Weak Interactions (Vol. 52)
*Stech, B.:* Non Leptonic Decays (Vol. 52)

**12.35 Composite Models of Particles**

*Baym, G.:* Quark Matter and Nuclei (Vol. 100)
*Brodsky, S.J.:* Quantum Chromodynamics at Nuclear Dimensions (Vol. 100)
*Close, F.E.:* The Pauli Principle and QCD for Quarks and Nucleons in Hadrons and Nuclei (Vol. 100)
*Faessler, A.:* How Should or Will QCD Influence Nuclear Physics? (Vol. 100)
*Heller, L.:* Bag Models and Nuclear Forces (Vol. 100)

**12.40 Models of Strong Interactions**

*Ademollo, M.:* Current Amplitudes in Dual Resonance Models (Vol. 59)
*Chung-I Tan:* High Energy Inclusive Processes (Vol. 60)
*Collins, P.D.B.:* How Important Are Regge Cuts? (Vol. 60)
*Collins, P.D.B., Gault, F.D.:* The Eikonal Model for Regge Cuts in Pion-Nucleon Scattering (Vol. 63)
*Collins, P.D.B., Squires, E.J.:* Regge Poles in Particle Physics (Vol. 45)
*Contogouris, A.P.:* Certain Problems of Two-Body Reactions with Spin (Vol. 57)
*Contogouris, A.P.:* Regge Analysis and Dual Absorptive Model (Vol. 63)
*Dietz, K.:* Dual Quark Models (Vol. 60)
*van Hove, L.:* Theory of Strong Interactions of Elementary Particles in the GeV Region (Vol. 39)
*Huang, K.:* Deep Inelastic Hadronic Scattering in Dual-Resonance Model (Vol. 62)
*Landshoff, P.V.:* Duality in Deep Inelastic Electroproduction (Vol. 62)
*Michael, C.:* Regge Residues (Vol. 55)
*Oehme, R.:* Complex Angular Momentum (Vol. 57)
*Oehme, R.:* Duality and Regge Theory (Vol. 57)
*Oehme, R.:* Rising Cross-Sections (Vol. 61)
*Rubinstein, H.R.:* Duality for Real and Virtual Photons (Vol. 62)
*Rubinstein, H.R.:* Physical N-Pion Functions (Vol. 57)
*Satz, H.:* Introduction to Dual Resonance Models in Multiparticle Physics (Vol. 57)
*Schrempp-Otto, B., Schrempp, F.:* Are Regge Cuts Still Worthwhile? (Vol. 61)
*Squires, E.J.:* Regge-Pole Phenomenology (Vol. 57)

**13.40 Electromagnetic Processes and Properties of Hadrons**

*Buchanan, C.D., Collard, H., Crannell, C., Frosch, R., Griffy, T.A., Hofstadter, R., Hughes, E.B., Nöldeke, G.K., Oakes, R.J., Van Oostrum, K.J., Rand, R.E., Suelzle, L., Yearian, M.R., Clark, B., Herman, R., Ravenhall, D.G.:* Recent High Energy Electron Investigations at Stanford University (Vol. 39)
*Gatto, R.:* Theoretical Aspects of Colliding Beam Experiments (Vol. 39)
*Gourdin, M.:* Vector Mesons in Electromagnetic Interactions (Vol. 55)
*Huang, K.:* Duality and the Pion Electromagnetic Form Factor (Vol. 62)
*Wilson, R.:* Review of Nucleon Form Factors (Vol. 39)

**13.60 Photon and Lepton Interactions with Hadrons**

*Brinkmann, P.:* Polarization of Recoil Nucleons from Single Pion Photoproduction. Experimental Methods and Results (Vol. 61)
*Donnachie, A.:* Exotic Electromagnetic Currents (Vol. 63)
*Drees, J.:* Deep Inelastic Electron-Nucleon Scattering (Vol. 60)
*Drell, S.D.:* Special Models and Predictions for Photoproduction above 1 GeV (Vol. 39)
*Fischer, H.:* Experimental Data on Photoproduction of Pseudoscalar Mesons at Intermediate Energies (Vol. 59)
*Flügge, G.:* Experimental Evidence of Quarks and Gluons (Vol. 100)
*Foà, L.:* Meson Photoproduction on Nuclei (Vol. 59)
*Frøyland, J.:* High Energy Photoproduction of Pseudoscalar Mesons (Vol. 63)
*Furlan, G., Paver, N., Verzegnassi, C.:* Low Energy Theorems and Photo- and Electroproduction Near Threshold by Current Algebra (Vol. 62)

*von Gehlen, G.:* Pion Electroproduction in the Low-Energy Region (Vol. 59)
*Heinloth, K.:* Experiments on Electroproduction in High Energy Physics (Vol. 65)
*Höhler, G.:* Special Models and Predictions for Pion Photoproduction (Low Energies) (Vol. 39)
*von Holtey, G.:* Pion Photoproduction on Nucleons in the First Resonance Region (Vol. 59)
*Landshoff, P. V.:* Duality in Deep Inelastic Electroproduction (Vol. 62)
*Llewellyn Smith, C. H.:* Parton Models of Inelastic Lepton Scattering (Vol. 62)
*Lücke, D., Söding, P.:* Multipole Pion Photoproduction in the s-Channel Resonance Region (Vol. 59)
*Osborne, L.S.:* Photoproduction of Mesons in the GeV Range (Vol. 39)
*Pfeil, W., Schwela, D.:* Coupling Parameters of Pseudoscalar Meson Photoproduction on Nucleons (Vol. 55)
*Renard, F. M.:* $\varrho$-$\omega$ Mixing (Vol. 63)
*Rittenberg, V.:* Scaling in Deep Inelastic Scattering with Fixed Final States (Vol. 62)
*Rollnik, H., Stichel, P.:* Compton Scattering (Vol. 79)
*Rubinstein, H.R.:* Duality for Real and Virtual Photons (Vol. 62)
*Rühl, W.:* Application of Harmonic Analysis to Inelastic Electron-Proton Scattering (Vol. 57)
*Schildknecht, D.:* Vector Meson Dominance, Photo- and Electroproduction from Nucleons (Vol. 63)
*Schilling, K.:* Some Aspects of Vector Meson Photoproduction on Protons (Vol. 63)
*Schwela, D.:* Pion Photoproduction in the Region of the $\Delta$ (1230) Resonance (Vol. 59)
*Wolf, G.:* Photoproduction of Vector Mesons (Vol. 57)

**13.75 Hadron-Induced Reactions**

*Atkinson, D.:* Some Consequences of Unitarity and Crossing. Existence and Asymptotic Theorems (Vol. 57)
*Basdevant, J. L.:* $\pi\pi$ Theories (Vol. 61)
*DeSwart, J. J., Nagels, M. M., Rijken, T. A., Verhoeven, P. A.:* Hyperon-Nucleon Interaction (Vol. 60)
*Ebel, G., Julius, D., Kramer, G., Martin, B. R., Müllensiefen, A., Oades, G., Pilkuhn, H., Pišút, J., Roos, M., Schierholz, G., Schmidt, W., Steiner, F., DeSwart, J. J.:* Compilation of Coupling Constants and Low-Energy Parameters (Vol. 55)
*Gustafson, G., Hamilton, J.:* The Dynamics of Some $\pi$-N Resonances (Vol. 57)
*Hamilton, J.:* New Methods in the Analysis of $\pi$N Scattering (Vol. 57)
*Kramer, G.:* Nucleon-Nucleon Interactions below 1 GeV/*c* (Vol. 55)
*Lichtenberg, D. B.:* Meson and Baryon Spectroscopy (Vol. 36)
*Martin, A.D.:* The $\Lambda$ KN Coupling and Extrapolation below the KN Threshold (Vol. 55)
*Martin, B. R.:* Kaon-Nucleon Interactions below 1 GeV/*c* (Vol. 55)
*Morgan, D., Pišút, J.:* Low Energy Pion-Pion Scattering (Vol. 55)
*Oades, G. C.:* Coulomb Corrections in the Analysis of $\pi$N Experimental Scattering Data (Vol. 55)
*Pišút, J.:* Analytic Extrapolations and the Determination of Pion-Pion Shifts (Vol. 55)
*Wanders, G.:* Analyticity, Unitarity and Crossing-Symmetry Constraints for Pion-Pion Partial Wave Amplitudes (Vol. 57)
*Zinn-Justin, J.:* Course on Padé Approximants (Vol. 57)

**14 Properties of Specific Particles and Resonances**

**14.80 Other and Hypothetical Particles**

*Flügge, G.:* Experimental Evidence of Quarks and Gluons (Vol. 100)

## Nuclear Physics

**21 Nuclear Structure**

*Arenhövel, H., Weber, H. J.:* Nuclear Isobar Configurations (Vol. 65)
*Cannata, F., Überall, H.:* Giant Resonance Phenomena in Intermediate-Energy Nuclear Reactions (Vol. 89)
*Faessler, A.:* How Should or Will QCD Influence Nuclear Physics? (Vol. 100)
*Racah, G.:* Group Theory and Spectroscopy (Vol. 37)
*Singer, P.:* Emission of Particles Following Muon Capture in Intermediate and Heavy Nuclei (Vol. 71)
*Überall, H.:* Study of Nuclear Structure by Muon Capture (Vol. 71)
*Wildermuth, K., McClure, W.:* Cluster Representations of Nuclei (Vol. 41)

**21.10 General Properties of Nuclei**

*Donner, W., Süßmann, G.:* Paramagnetische Felder am Kernort (Vol. 37)
*Zu Putlitz, G.:* Determination of Nuclear Moments with Optical Double Resonance (Vol. 37)
*Schmid, D.:* Nuclear Magnetic Double Resonance – Principles and Applications in Solid State Physics (Vol. 68)

**21.30 Nuclear Forces** and

**21.40 Few Nucleon Systems**

*Close, F. E.:* The Pauli Principle and QCD for Quarks and Nucleons in Hadrons and Nuclei (Vol. 100)
*DeSwart, J. J., Nagels, M. M., Rijken, T. A., Verhoeven, P. A.:* Hyperon-Nucleon Interactions (Vol. 60)
*Heller, L.:* Bag Models and Nuclear Forces (Vol. 100)
*Kramer, G.:* Nucleon-Nucleon Interactions below 1 GeV/*c* (Vol. 55)
*Levinger, J. S.:* The Two and Three Body Problem (Vol. 71)

**21.65 Nuclear Matter**

*Baym, G.:* Quark Matter and Nuclei (Vol. 100)

**23 Weak Interactions**

*Gasiorowicz, S.:* A Survey of the Weak Interaction (Vol. 52)
*Primakoff, H.:* Weak Interactions in Nuclear Physics (Vol. 53)

**25.30 Lepton-Induced Reactions and Scattering**

*Theißen, H.:* Spectroscopy of Light Nuclei by Low Energy(70 MeV) Inelastic Electron Scattering (Vol. 65)
*Überall, H.:* Electron Scattering, Photoexcitation and Nuclear Models (Vol. 49)

### 28.20 Neutron Physics

*Koester, L.:* Neutron Scattering Lengths and Fundamental Neutron Interactions (Vol. 80)
*Springer, T.:* Quasi-Elastic Scattering of Neutrons for the Investigation of Diffusive Motions in Solids and Liquids (Vol. 64)
*Steyerl, A.:* Very Low Energy Neutrons (Vol. 80)

### 28.60 Isotope Separation and Enrichment

*Ehrfeld, W.:* Elements of Flow and Diffusion Processes in Separation Nozzles (Vol. 97)

### 29 Experimental Methods

*Panofsky, W.K.H.:* Experimental Techniques (Vol. 39)
*Strauch, K.:* The Use of Bubble Chambers and Spark Chambers at Electron Accelerators (Vol. 39)

### 42 Optics

### 42.50 Quantum Optics

*Agarwal, G.S.:* Quantum Statistical Theories of Spontaneous Emission and their Relation to Other Approaches (Vol. 70)
*Graham, R.:* Statistical Theory of Instabilities in Stationary Nonequilibrium Systems with Applications to Lasers and Nonlinear Optics (Vol. 66)
*Haake, F.:* Statistical Treatment of Open Systems by Generalized Master Equations (Vol. 66)
*Schwabl, F., Thirring, W.:* Quantum Theory of Laser Radiation (Vol. 36)

### 42.72 Optical Sources

*Godwin, R.P.:* Synchrotron Radiation as a Light Source (Vol. 51)

## Atomic and Molecular Physics

### 31 Electronic Structure of Atoms and Molecules, Theory

*Donner, W., Süßmann, G.:* Paramagnetische Felder am Kernort (Vol. 37)

### 32 Atomic Spectra and Interactions with Photons

*Büttgenbach, St.:* Hyperfine Structure in 4d- and 5d-Shell Atoms (Vol. 96)
*Racah, G.:* Group Theory and Spectroscopy (Vol. 37)
*Zu Putlitz, G.:* Determination and Nuclear Moments with Optical Double Resonance (Vol. 37)

### 34 Atomic and Molecular Collision Processes and Interactions

*Dettmann, K.:* High Energy Treatment of Atomic Collisions (Vol. 58)
*Langbein, D.:* Theory of Van der Waals Attraction (Vol. 72)
*Seiwert, T.:* Unelastische Stöße zwischen angeregten und unangeregten Atomen (Vol. 47)

## Classical Fields of Phenomenology

### 41.70 Particles in Electromagnetic Fields

*Olson, C.L.:* Collective Ion Acceleration with Linear Electron Beams (Vol. 84)
*Schumacher, U.:* Collective Ion Acceleration with Electron Rings (Vol. 84)

### 41.80 Particle Optics

*Hawkes, P.W.:* Quadrupole Optics (Vol. 42)

## Fluids, Plasmas

### 47 Fluid Dynamics

*Ehrfeld, W.:* Elements of Flow and Diffusion Processes in Separation Nozzles (Vol. 97)

### 51 Kinetics and Transport Theory of Fluids; Physical Properties of Gases

*Geiger, W., Hornberger, H., Schramm, K.-H.:* Zustand der Materie unter sehr hohen Drücken und Temperaturen (Vol. 46)
*Hess, S.:* Depolarisierte Rayleigh-Streuung und Strömungsdoppelbrechung in Gasen (Vol. 54)

## Condensed Matter, Mechanical and Thermal Properties

### 61 Structure of Liquids and Solids

*Behringer, J.:* Factor Group Analysis Revisited and Unified (Vol. 68)
*Dederichs, P.H., Zeller, R.:* Dynamical Properties of Point Defects in Metals (Vol. 87)
*Lacmann, R.:* Die Gleichgewichtsform von Kristallen und die Keimbildungsarbeit bei der Kristallisation (Vol. 44)
*Langbein, D.:* Theory of Van der Waals Attraction (Vol. 72)
*Leibfried, G., Breuer, N.:* Point Defects in Metals I: Introduction to the Theory (Vol. 81)
*Schroeder, K.:* Theory of Diffusion Controlled Reactions of Point Defects in Metals (Vol. 87)
*Springer, T.:* Quasi-elastic Scattering of Neutrons for the Investigation of Diffusive Motions in Solids and Liquids (Vol. 64)
*Steeb, S.:* Evaluation of Atomic Distribution in Liquid Metals and Alloys by Means of X-Ray. Neutron and Electron Diffraction (Vol. 47)

### 61.12 Neutron Determination of Structures

*Dorner, B.:* Coherent Inelastic Neutron Scattering in Lattice Dynamics (Vol. 93)
*Press, W.:* Single-Particle Rotations in Molecular Crystals (Vol. 92)

### 62 Mechanical and Acoustical Properties and

### 63 Lattice Dynamics

*Dorner, B.:* Coherent Inelastic Neutron Scattering in Lattice Dynamics (Vol. 93)
*Ludwig, W.:* Recent Developments in Lattice Theory (Vol. 43)
*Schramm, K.-H.:* Dynamisches Verhalten von Metallen unter Stoßwellenbelastung (Vol. 53)

### 64 Equation of State, Phase Transitions

*Bussmann-Holder, A., Bilz, H.:* Electronic and Dynamical Properties of IV-VI Compounds (Vol. 99)
*Jantsch, W.:* Dielectric Properties and Soft Modes in Semiconducting (Pb, Sn, Ge)Te (Vol. 99)

### 68 Surfaces and Interfaces

*Engel, T., Rieder, K.-H.:* Structural Studies of Surfaces with Atomic and Molecular Beam Diffraction (Vol. 91)
*Heinz, K., Müller, K.:* LEED Intensities - Experimental Progress and New Possibilities of Surface Structure Determination (Vol. 91)

## Condensed Matter: Electronic Structure, Electrical, Magnetic and Optical Properties

### 71 Electron States

*Bendow, B.:* Polariton Theory of Resonance Raman Scattering in Solids (Vol. 82)
*Bennemann, K.H.:* A New Self-consistent Treatment of Electrons in Crystals (Vol. 38)
*Borstel, G., Falge, H.J., Otto, A.:* Surface and Bulk Phonon-Polarition Observed by attenuated Total Reflection (Vol. 74)
*Claus, R., Merten, L., Brandmüller J.:* Light Scattering by Phonon-Polaritons (Vol. 75)
*Daniels, J., von Festenberg, C., Raether, H., Zeppenfeld, K.:* Optical Constants of Solids by Electron Spectroscopy (Vol. 54)
*Dornhaus, R., Nimtz, G.:* The Properties and Applications of $Hg_{1-x}Cd_xTe$ Alloy Systems (Vol. 78)
*Dornhaus, R., Nimtz, G., Schlicht, B.:* Narrow-Gap Semiconductors (Vol. 98)
Excitons at High Density, *Haken, H., Nikitine, S.,* (Volume Editors). Contributors: *Bagaev, V.S., Biellmann, J., Bivas, A., Goll, J., Grosmann, M., Grun, J.B., Haken, H., Hanamura, E., Levy, R., Mahr, H., Nikitine, S., Novikov, B.V., Rashba, E.I., Rice, T.M., Rogachev, A.A., Schenzle, A., Shaklee, K.L.* (Vol. 73)
*Grosse, P.:* Die Festkörpereigenschaften von Tellur (Vol. 48)
*Kenkre, V.M.:* The Master Equation Approach: Coherence, Energy Transfer, Annihilation, and Relaxation (Vol. 94)
*Pick, H.:* Struktur von Störstellen in Alkalihalogenidkristallen (Vol. 38)
*Raether, H.:* Solid State Excitations by Electrons (Vol. 38)
*Reineker, P.:* Stochastic Liouville Equation Approach: Coupled Coherent and Incoherent Motion, Optical Line Shapes, Magnetic Resonance Phenomena (Vol. 94)
*Schmid, D.:* Nuclear Magnetic Double Resonance - Principles and Applications in Solid State Physics (Vol. 68)

### 72 Electronic Transport in Condensed Matter

*Bauer, G.:* Determination of Electron Temperatures and of Hot-Electron Distribution Functions in Semiconductors (Vol. 74)
*Dornhaus, R., Nimtz, G.:* The Properties and Applications of $Hg_{1-x}Cd_xTe$ Alloy Systems (Vol. 78)
*Dornhaus, R., Nimtz, G., Schlicht, B.:* Narrow-Gap Semiconductors (Vol. 98)
*Grosse, P.:* Die Festkörpereigenschaften von Tellur (Vol. 48)
*Feitknecht, J.:* Silicon Carbide as a Semiconductor (Vol. 58)
*Schnackenberg, J.:* Electron-Phonon Interaction and Boltzmann Equation in Narrow Band Semiconductors (Vol. 51)

### 73 Properties of Surfaces and Thin Films

*Hölzl, J., Schulte, F.K.:* Work Function of Metals (Vol. 85)
*Müller, K.:* How much can Auger Electrons tell us about Solid Surfaces? (Vol. 77)
*Raether, H.:* Excitation of Plasmons and Interband Transitions by Electrons (Vol. 88)
*Wagner, H.:* Physical and Chemical Properties of Stepped Surfaces (Vol. 85)
*Wissmann, P.:* The Electrical Resistivity of Pure and Gas Covered Metal Films (Vol. 77)

### 74 Superconductivity

*Lüders, G., Usadel, K.-D.:* The Method of the Correlation Function in Superconductivity Theory (Vol. 56)
*Ullmaier, H.:* Irreversible Properties of Type II Superconductors (Vol. 76)

### 75 Magnetic Properties

*Dornhaus, R., Nimtz, G., Schlicht, B.:* Narrow-Gap Semiconductors (Vol. 98)
*Fischer, K.:* Magnetic Impurities in Metals: the *s–d* Exchange Model (Vol. 54)
*Schmid, D.:* Nuclear Magnetic Double Resonance - Principles and Applications in Solid State Physics (Vol. 68)
*Stierstadt, K.:* Der Magnetische Barkhauseneffekt (Vol. 40)

### 76 Magnetic Resonances and Relaxation

*Reineker, P.:* Stochastic Liouville Equation Approach: Coupled Coherent and Incoherent Motion, Optical Line Shapes, Magnetic Resonance Phenomena (Vol. 94)
*Schmid, D.:* Nuclear Magnetic Double Resonance - Principles and Applications in Solid State Physics (Vol. 68)

### 77 Dielectric Properties and Materials

*Bussmann-Holder, A., Bilz, H.:* Electronic and Dynamical Properties of IV-VI Compounds (Vol. 99)
*Jantsch, W.:* Dielectric Properties and Soft Modes in Semiconducting (Pb, Sn, Ge) Te (Vol. 99)

### 78 Optical Properties

*Bäuerle, D.:* Vibrational Spectra of Electron and Hydrogen Centers in Ionic Crystals (Vol. 68)
*Borstel, G., Falge, H.J., Otto, A.:* Surface and Bulk Phonon-Polariton Observed by attenuated Total Reflection (Vol. 74)
*Bussmann-Holder, A., Bilz, H.:* Electronic and Dynamical Properties of IV-VI Compounds (Vol. 99)
*Claus, R., Merten, L., Brandmüller, J.:* Light Scattering by Phonon-Polaritons (Vol. 75)
*Daniels, J., von Festenberg, C., Raether, H., Zeppenfeld, K.:* Optical Constants of Solids by Electron Spectroscopy (Vol. 54)
*Dornhaus, R., Nimtz, G., Schlicht B.:* Narrow-Gap Semiconductors (Vol. 98)
*Godwin, R.P.:* Synchrotron Radiation as a Light Source (Vol. 51)
*Jantsch, W.:* Dielectric Properties and Soft Modes in Semiconducting (Pb, Sn, Ge) Te (Vol. 99)
*Lengeler, B.:* de Haas-van-Alphen Studies of the Electronic Structure of the Noble Metals and Their Dilute Alloys (Vol. 82)
*Raether, H.:* Solid State Excitations by Electrons (Vol. 38)
*Reineker, P.:* Stochastic Liouville Equation Approach: Coupled Coherent and Incoherent Motion, Optical Line Shapes, Magnetic Resonance Phenomena (Vol. 94)
*Richter, W.:* Resonant Raman Scattering in Semiconductors (Vol. 78)

## Related Areas of Science and Technology

### 85.70 Magnetic Devices

*Lehner, G.:* Über die Grenzen der Erzeugung sehr hoher Magnetfelder (Vol. 47)

## Geophysics, Astronomy, and Astrophysics

### 95 Theoretical Astrophysics

*Kundt, W.:* Recent Progress in Cosmology (Isotropy of 3 deg Background Radiation and Occurrence of Space-Time Singularities) (Vol. 47)
*Kundt, W.:* Survey of Cosmology (Vol. 58)
*Stewart, J., Walker M.:* Black Holes: the Outside Story (Vol. 69)

### 97 Stars

*Börner, G.:* On the Properties of Matter in Neutron Stars (Vol. 69)